Evolution in the Galapagos Islands

Evolution in the Galapagos Islands

Edited by R. J. Berry

Reprinted from the Biological Journal of the Linnean
Society, Volume 21, Numbers 1 & 2 1984

Published for the Linnean Society of London

ACADEMIC PRESS

(Harcourt Brace Jovanovich, Publishers)

London   Orlando   San Diego   San Francisco   New York
Toronto   Montreal   Sydney   Tokyo   São Paulo

ACADEMIC PRESS INC. (LONDON) LIMITED
24/28 Oval Road
London NW1
(Registered Office)

US edition published by
ACADEMIC PRESS INC.
111 Fifth Avenue
New York
New York 10003

The papers in this volume constitute Contribution Number 363 of the Charles Darwin Foundation for the Galapagos Isles.

Printed in Great Britain by
The Whitefriars Press Ltd., Tonbridge

Contents

Foreword

SIR PETER SCOTT

On my first visit to the Galapagos Islands in 1959—the centenary year of the publication of the *Origin of Species*—I soon became aware that the archipelago represents not only the birthplace of the theory of evolution through natural selection, but also a continuing example of that process in operation. Conservation of the current highly diverse populations of living taxa seemed to me to have an even greater significance there than in any other comparable group of islands. For that reason I was happy to be one of the founders of the Charles Darwin Foundation for the Galapagos Islands.

One of the most striking features of the Galapagos fauna is its innate tameness. The animals evolved without fear of man. Having been back to Galapagos again in 1961, 1974 and 1976 it was encouraging to find on my last visit that the wildlife was for the most part much tamer than it had been in 1959. It was also encouraging to find a greater appreciation of nature conservation among the local human population. Although the presence of man as a resident inhabitant of Galapagos has certainly altered the ecosystems a great deal because of the introduction of rats, cats, dogs, pigs and donkeys, the damage has been less than in more populous parts of the world.

Documentation of the fauna and flora is probably more complete for Galapagos than it is for any comparable island group as a result of the initiative of the California Academy of Sciences and later the work of the C.D.F. and other organizations. This is not to say that there is no need for further research. The results of the meetings of the Linnean Society clearly show that the further study of evolution in Galapagos is likely to be a fruitful source of new insights into the controversial aspects of Darwinian theory.

Biological Journal of the Linnean Society (1984), *21:* 5–27

Islands and evolution: theory and opinion in Darwin's earlier years

A. J. CAIN

Department of Zoology, University of Liverpool, P.O. Box 147, Liverpool L69 3BX

Islands were not of special interest to evolutionists before Darwin. It was he who first appreciated their importance for demonstrating evolution in miniature. They were not of special interest because: (a) their peculiar products seemed no more peculiar than those of continents; (b) there was no special category of oceanic islands, but a continuum from such groups as the Canaries, Madeiras and Galapagos through New Zealand and Madagascar to Australia, Britain, and true continents; and (c) the concept of adaptive radiation, if known at all, was applied only to the higher levels of classification, and then very feebly.

When Darwin was young, classification at the lower levels hardly recognized convergence, and at the higher levels was subject to great changes, while only slowly separating out the major groups. In consequence, many of the facts of geographical distribution were misinterpreted, and numerous theories of the origination of species, groups, and biogeographical provinces were still plausible. It was largely the need for a historical, not ecological, explanation of the distribution of some mammals and plants, plus what he saw for himself in the Galapagos Islands, that convinced Darwin that evolution had occurred. His was a remarkable achievement in recognizing through all this 'noise' the meaning of adaptive radiation.

KEY WORDS:—Darwin – islands – biogeography – classification.

CONTENTS

INTRODUCTION

I suspect the organizer of this conference of a little genial malice in assigning to me 'Islands and evolution', to be followed by 'Darwin and the Galapagos Islands', since before Darwin the subject did not exist. Considering how much

5

was known of animal systematics and geographical distribution, the question
✳ 'Why did it not exist?' must be asked. The answer is an instructive one, since it
clarifies the nature of Darwin's achievement. Both the systematics of all ranks of
living things, and the meaning of their geographical distribution were highly
controversial topics in his younger days and middle age, although for different
reasons at the beginning and the end of that period. Moreover, with unequal
exploration, misidentification, bad taxonomy, and faulty localization, there was
so much 'noise' in the taxonomic system that one might wonder how Darwin
managed to extract anything out of it; without accurate taxonomy,
biogeography seems nearly impossible.

No one who is not well versed both in present systematics and biogeography,
and in the technical literature of the period can appreciate the enormous
differences both in actual knowledge and in the means of unifying it that
separate us from the early nineteenth century. Historians of science may make a
scholarly analysis of the ideas then current; but without the biological facts
(true, or false but believed to be true) on which they were based, the exercise
loses cogency, and may even lead to error. Gale (1982) claims that Charles
Darwin had to 'hypothesize' a previous cold period to explain arctic-alpine
disjunct distributions—but this period was well attested on geological evidence,
and used previously by Forbes (1846) for that very purpose.

For the present topic there are some major differences between then and now
that need to be examined. The impulse was strong in the early nineteenth
century to find some overall plan or scheme of creation; this had a profound
effect on classification, but the faulty taxonomy meant that adaptive radiation,
today so powerful a witness for evolution, was almost unrecognizable then. The
impulse at least led to a closer examination of types of relationship, which in
turn led to Richard Owen's distinction of *homology* from *analogy*. A few years ago,
de Beer (1963) could be scornful of some of the schematists who imposed a non-
existent order on Nature. With Hennigism and transform cladism sweeping
through the museums of the world, it is harder to maintain the note of
superiority; but the present schematic efforts are trivial compared with those of
1820–1860.

Comparative anatomy took a great step forward with Cuvier's splendid works
(*c.* 1812–1828) but so much was still to be done among both invertebrates and
vertebrates, that the most diverse opinions could (and did) exist about the
relationships even of the larger mammals, living and extinct. It was on these
that good biogeography was first founded. It was of course largely the
uncertainties engendered by the variety of classifications that gave many of the
schematists their opportunity for reclassifying according to their own ideas,
without being contradicted by anatomical fact.

Except for some of the larger mammals, and a few birds and groups of plants,
there was extremely little precise information about distribution, and still less
about 'stations' i.e. distribution in relation to local ecological factors. Forbes
(Forbes & Godwin-Austen, 1859) ascribes the first serious investigation of
zonation on the shore and below it, "the first impulse to the scientific
investigation of the distribution of marine animals" to Milne Edwards and
Audouin's *Recherches pour servir à l'histoire naturelle du littoral de la France*, 1832.
Forbes's own investigations, in the Aegean Sea 1841–2 and round the British
Isles, sparked off what became a major industry in biological research in

Britain—dredging. It was through the great works of Humboldt, who was the virtual founder of phytogeography (as well as of so much else) that the influence of temperature and altitude upon vegetation became known (real climatic temperature, not just latitude—Humboldt invented the isothermal map). He also pointed out some curious facts of plant distribution not explicable by temperature, moisture or soil characteristics. Why were cacti confined to the New World but with analogues in the Old? So much of early—and later—biogeography could be explained as straight ecology; it was the element that could not which came gradually to the fore in this period and demanded an explanation.

The biology of the period 1800 to 1860 has been covered summarily by Coleman (1971). Winsor (1976) is useful on some of the issues in classification, but keeps rather closely to the greater people. The great figures of the earlier part of the period are Humboldt (see Botting, 1973, for an introduction), Cuvier (Coleman, 1964) Louis Agassiz (see his own *Essay on Classification* 1859) and Lyell (see Wilson, 1970, 1972). But there is no adequate history of biogeography, and no comprehensive analysis of classification, for the period. There is therefore no substitute for extensive reading in its scientific publications. Yet it was in the middle of this very period that Darwin, a young and impressionable man in his early twenties, was sailing round the world (27 December 1831 to 2 October 1836). It was this fearful mixture of positive knowledge both right and wrong, surmise, dogma, and extensive, but of course unsuspected, ignorance, altered almost daily by fresh descriptions, discoveries and theories, that Darwin encountered. And it was the hard-headed naturalists who, following the example of the geologists, were getting pretty tired of theorists, and the theorists themselves, that he had to convince of the rightness of his own theory.

Charles Kingsley put it well in the first edition of *The Water-Babies* (1863: 289–290). "But stupid old Epimetheus went working and grubbing on. . . always looking behind him to see what had happened, till he really learned to know now and then what would happen next [and he] began to make things which would work, and go on working too. . . And his children are the men of science, who get good lasting work done in the world: but the children of Prometheus are the fanatics, and the theorists, and the bigots, and the bores, and the noisy windy people, who go telling silly folk what will happen, instead of looking to see what has happened already." A glance at the period is sufficient to answer the typically American question (Gale, 1982) "Why did Charles Darwin not publish at once?"

DIFFICULTIES OVER INFORMATION

At all times, systematics and biogeography have been bedevilled by lack of information, false information and misapprehension of specific limits. All these acted powerfully within the period, to an extent now hardly realizable. I give only a few examples, merely to indicate the difficulties caused.

Unequal Information

Large parts of the earth were barely known biologically, large groups of living things hardly investigated. Thompson (1844) pointed out how far the Irish

entozoa (parasitic worms) outnumbered the British, solely because they had
been worked on more intensively. For most groups, Britain, being a small island
with a remarkable density of naturalists, was far better known than most
countries of the continent of Europe, let alone others farther away. Especially in
marine groups, monographers were constantly compelled to admit that the
exact extra-British ranges of British species, particularly of small animals, were
unknown because they had not yet been distinguished abroad (e.g. Forbes &
Hanley, 1853; Hincks, 1868). A committee of the Liverpool Literary and
Philosophical Society, asking seamen of the Mercantile Marine to collect
specimens, said as late as 1862, "Could we but know the range of a single
animal as accurately as Alphonse de Candolle has determined that of many
species of plants, we might begin a new era in Zoology". (Anon. [Moore, T. J.]
1862). Indeed, it was not until 1902 that Robert Lloyd Praeger thought the
distributions of plants in Ireland to be well enough known for analysis into
biogeographical elements.

Correspondingly, many groups of invertebrates were very poorly known, and
it was not always realized how poorly. Whole skeletons of some mammals, birds,
and reptiles had been monographed, and much of the soft anatomy too; but the
majority of molluscs were known only by their shells. The vast genus *Helix*,
comprizing a large proportion of the land-snails of the world, was hardly known
by dissection, except for the closely allied *Helix aspersa* and *Helix pomatia*.
Enormous numbers of convergences went unrecognized, and it was not until
1894 that Henry Augustus Pilsbry, who first brought order into this chaos, said
"We find that the distribution of Helices in space and time is not hap-hazard or
erratic, as has been supposed from the earlier classifications, and from the
erroneous generic and subgeneric references contained in works on the fossil
forms, but that it is orderly and comprehensible", agreeing in short, with the
major zoogeographical regions already mapped out.

False information

Intentional

It was well known later in the period, at least to specialists (e.g. Forbes &
Hanley, 1853: xxiii footnote) that some shell dealers would falsify localities to
enhance the value of their wares. A common Mediterranean shell might fetch
little; but if it was the only known specimen from British waters! Now shells, as
Sir Charles Lyell emphasized (1833, vol II: 111), were of peculiar importance
both because it was largely on fossils that geological strata were
distinguished and identified, and because their living relatives enabled one to
distinguish faunal provinces and argue back, by means of the fossils, to previous
changes in climate.

Unintentional

For much of the earlier part of the period, the importance of precise
geographical localization was simply unrealized. It was the identification and
classification of forms of life that were the overriding considerations. Thomas
Vernon Wollaston (1822–1878), who brought the wingless beetles (and the

endemicity of the fauna) of Madeira to the attention of the scientific world, virtually apologized in the prefatory remarks to his *Insecta Maderensia* (1854) for being over-minute in giving localities for the beetles. When "Habitat in Europa septentrionali" was considered quite sufficient, it was a little odd to be told in exactly which valley, and at what altitude, a species occurs on a small island. The reason he gives is equally remarkable—Madeira was a winter refuge and sanatorium for English invalids, who were there with nothing to do. There was no occupation more therapeutic than the study of Nature; but an invalid would need careful guidance to the right spot in Madeira if he were to succeed in his collecting. This, as far as I know, was the origin of really precise locality labels in zoology.

Such carelessness, or indifference, could have misleading results. A traveller touching at the Cape of Good Hope, Java, and Hong Kong, could well get his collections mixed up, put on labels from memory when he got back, and transfer specimens unwittingly from one zoogeographical region to another. A classic case affected Darwin. Surprised though he was at the variation of the geospizine finches (and the tortoises) from island to island in the Galapagos group, he could not think of the geospizines (nor the tortoises) as a specifically Galapagean group. In his *Naturalist's Voyage around the World* he quotes the excellent ornithologist John Gould (1804–1881), who worked up his birds, as recognizing 13 species, all peculiar to the Galapagos; "and so", he goes on "is the whole group, with the exception of one species of the sub-group Cactornis, lately brought from Bow island in the Low Archipelago". The statement is still found in the editions as late as 1870 (p. 379). So Darwin could not even think of the geospizines as peculiar to the Galapagos islands. (There are no finches in the Low Archipelago.)

False localities, sincerely believed in, were bedevilling even British botany until far later (e.g. Baker & Newbould, 1883; xxviii). The unfortunate Thomas Vernon Wollaston found the greatest difficulty in excluding Madeiran species from the Canaries list (e.g. 1864: 8, footnotes) and in restricting Canarian species to their real distributions (1864: vii); and when it came to the activities of those who described land shells included in imported moss and lichens as native, his language became almost emphatic (1878).

A few misplacings one can only feel sympathetic about; Jeffreys (1859) points out that *Cypraea moneta* (the money cowrie) is not native to Britain, but specimens occasionally come ashore, or are dredged, from wrecks of slave-ships that were carrying them as currency. The total number of exotic species of shells admitted into the British list was given by Clark in 1855 as 138.

Specific limits

Lumping and splitting have always created difficulties for the systematist, which have been only partly overcome by recognizing geographical races (subspecies). Darwin had to see for himself (*Voyage*, 1870: 194) that the *Canis antarcticus* of the Falkland Islands was not the same as the *C. mageklanicus* of Chile, as Molina had believed, but a peculiar species [or subspecies!] confined to the Falklands. Some of the widspread species that caused such difficulty to the theorist biogeographers were in fact complexes of very similar species—the *Helix putris* (i.e. *Succinea putris*) referred to by Lyell (1833) as having a wide range is a case in point. (Others, such as *Helix aspersa,* which he cites from Europe and South America, had, as he thought possible, simply been introduced widely abroad.) The number of species two districts had in common could be raised enormously if a lumper (justified or not) revised the fauna or flora. J. G. Baker, discussing the world's ferns in a paper read to the Linnean Society on 4 April 1867 gives some remarkable figures, e.g. "Where De Vriese has 94 Marattiaceae, we have only been able to define 9; where Van den Bosch has 450 Hymenophyllaceae, we have 149" and so on.

Hooker, in his *Introductory Essay to the Flora of New Zealand* had to devote a whole section to the distribution and variation of species (1853: vii–xxvii) with a special section headed "Species vary in a state of nature more than is usually supposed". Wallace, indeed, in the preface (dated March 1876) to his classic *Geographical Distribution of Animals* gave up species altogether. "Species, as such, are systematically disregarded—firstly, because they are so numerous as to be unmanageable; and, secondly, because they represent the most recent modifications of form... not so clearly connected with geographical changes as are the natural groups of species called genera..." (1876: vii)—and, he could have added, far less certain in extent and range. The exact definition of species was of particular importance in testaceous mollusca, since shells were of the greatest interest palaeontologically. Fleming (1822) based his objection to the theory of a cold spell in Britain (the glacial period) on the specific non-identity of the so-called arctic fossils with living arctic forms. He pointed out, quite rightly, that the mammoth was indeed an elephant, but not specifically identical with living elephants, and specially adapted to a cold climate. The question of how much variation could occur within species was therefore a highly topical one, on which much of the interpretation of past ages could founder. Wollaston's remarkable book *On the Variation of Species with Especial Reference to the Insecta* (1856), dedicated to Charles Darwin, was far from being an isolated 'jeu d'esprit', and has never been given the importance it deserves.

One of the most extensive examinations of biogeographical provinces in the period was by Samuel Pickworth Woodward (1821–1865) in his *Manual of the Mollusca* and he remarks (1851–6: 354) "When the Faunas of the other regions [besides the Mediterranean] have been tested... the result will probably be the establishment of a much greater number of provinces than we have ventured at present to indicate on the map". He goes on to cite examples of great ranges of marine species which are almost certainly erroneous—"when sufficiently investigated, it has usually proved that some of the localities were false, or that more than one species was included".

With these remarks in mind, we can appreciate much more clearly the shock that Darwin felt on finding species limited to particular islands of the Galapagos, and his surprise at finding related species (armadillos, rheas) replacing one another from the tropics to the temperate regions of South America. No wonder that later on in the Voyage his eyes were opened to the remarkable fact that an indigenous (and extinct) land-shell on St. Helena differed from place to place (1870: 488). "It deserves notice that all the many specimens of this shell found by me in one spot, differ, as a marked variety, from another set of specimens procured from a different spot." It really was a new idea for which his reading had given him no preparation whatever.

HIGHER TAXONOMIC RELATIONSHIPS

Relations of the greater groups

Biologists of the early nineteenth century were well aware of what great advances in comparative anatomy were being made by the French school, especially Cuvier, and therefore how unfortunate were parts of Linnaeus's classification—putting *Myxine* among the parasitic worms, for example. Embryology was developing fast under von Baer, Meckel and Serres (see Coleman, 1971). Metamorphosis, hitherto virtually confined to some insects and amphibia, was discovered in cirrhipedes and crustacea, and alternation of generations in tunicates and hydrozoans. Relationships, even of the higher groups, were now continually being called in question. Moreover, palaeontology had brought to light what might well be a succession of different faunas, replacing one another in time, and filling up some of the gaps in the classification of living forms. There seemed to be parallels between different living groups (e.g. the marsupials and the placentals or the flying mammals, birds and fish); living and extinct groups; and stages in development. Some of these relationships were of direct affinity, others seemed to be merely analogy. A wonderful variety of types of classification arose out of their contemplation.

The old eighteenth century idea of a single *scala naturae* was not yet extinct, and as late as 1860 it could be asserted that "an ascending and successive scale or chain of creation is, in the main, correct, when the great classes, and not species or genera, are made the links", and even that "the cartilages of *Sepia* have a true resemblance to those of a Skate, and the Cirrhipede truly connects the Mollusk with the Crustacea...". (Garner, 1860; see also Clark, 1855, also working on testacea.) The first of these statements would of course harmonize well with the idea of a progression of creations or a progressive development of the single creation (see, e.g., Bowler, 1976), but I doubt if the second was received enthusiastically in 1860. At the beginning of the period, the chain of being was mentioned more frequently. In 1790, in Smellie's highly influential *Philosophy of Natural History*, the flying fishes could still be spoken of as directly allied to water birds, but by 1822, in Fleming's *Philosophy of Zoology*, which Darwin took with him on the voyage, it was firmly rejected.

Nevertheless, there was a strong feeling that the living Creation could not be as chaotic as it looked. Newton had worked out the laws governing the heavens, but those governing the living world had not been found. William Sharp Macleay (1792–1865) announced in a paper read before the Linnean Society of London on 5 November 1822 (having said, with a quotation from Linnaeus, that the natural system is the essential aim of the naturalist) "if there be nothing within the whole range of human science more worthy of profound meditation than the plan by which the Deity regulated the creation; so most assuredly no study is more calculated to administer pure and unmixed delight". "That Nature has made use of determinate numbers in the construction of vegetables has long been known empirically" (as indeed in Linnaeus's artificial classification, with definite numbers of stamens and stigmas) and he is therefore delighted to find that determinate numbers have been recognized by a botanist in the distribution (i.e. classification) of the plants themselves. Macleay had discovered that the contents of groups could be arranged by their relations of affinity in circles, that each circle had five constituents, and that their mutual relations were paralleled by the constituents in related circles.

This was the notorious quinary system (there were others based on other numbers) which, although given attention by Winsor (1976), has never been sympathetically examined. Yet Leonard Jenyns (1835) gave a respectful account of it as having stimulated research, and Richard Owen (1859: lxvii) paid a tribute to MacLeay as having promoted the study of affinity (i.e. homology) and analogy. It was a prominent feature of the intellectual landscape from 1820 to 1840, if not later. Several of the best naturalists, e.g. the cryptogamic botanist M. J. Berkeley, the entomologist the Rev. William Kirby, the mammalogist G. R. Waterhouse, and the marine biologist Edward Forbes used it more or less enthusiastically. Many volumes in the Cabinet Cyclopaedia by William Swainson were expositions of it or classifications based on it; and the young Thomas Henry Huxley himself tried it out for a time (Winsor, 1976). Such a scheme had the great advantage of predictability—one could predict what relations of affinity would be found to exist among material not yet classified by it; and one could determine how many forms remained to be discovered in imperfectly known groups. Swainson did indeed take it to extremes; MacLeay himself disagreed with him in part, later on (1842). "I am often afraid of trusting myself to Mr Swainson's method of drawing analogies between things in themselves wide apart. A person may reasonably doubt the legitimacy of any comparison between a fish and an insect, or even between a fish and a bird." However, he admits the legitimacy of comparing fish with fish, and notes the analogy between loaches and lampreys. "Still I am far from denying, that such analogies as he delights in exist in nature... The cause of the greater part of the resemblances which he discovers between objects the most apart from each other in general structure, seems to be a general law of nature, which has ruled that in every group of animals there should be a minor group more essentially carnivorous, another minor group more essentially herbivorous, another more aquatic or natatorial, and so on."

Quinary, septenary and decenary systems were all available in the period, plus numerous others. Oken's fantastic scheme in which groups of animals corresponded to different organs of the human body had quite enough prestige to be translated by Alfred Tulk and published as a Ray Society volume, with a delighted preface by Oken himself, as late as 1847. But Oken was one of the Naturphilosophen, a continental school not only supported in part by Cuvier, but looking back to the revered name of Goethe; and it was "above all things the adoption of the philosophical views of Göthe, together with the recognition of an universal unity of design throughout the vegetable world" that had transformed botany out of recognition since 1800, according to John Lindley (1834).

Other systems were not wanting. Lamarck proposed one composed of two ascending series, one (the parasitic forms) stopping much shorter than the other, which ended in Man. His ardent follower, the anonymous author of *Vestiges of Creation* (Robert Chambers, 1802–1871) proposed a large number of such series, each starting from a separate point by spontaneous generation.

Andrew Murray even suggested (conditionally, it is true) a multiple origin of Mammals (1866: 54). "May they not have sprung from different stocks—the bat from the pterodactyles; the duck-bill, or ornithorhynchus, from birds; the whale from the ichthyosaurus, and the general mass of mammals from terrestrial reptiles?"

Evolution was well known, of course, not only from the *Vestiges of Creation* but from Lyell's exposition of Lamarck in the *Principles of Geology*. Lindley (1834) went so far as to say "The theory of the gradual development of the highest class of organic bodies...has acquired so great a degree of probability among animals, that it has become a question of no small interest whether traces of the same, or a similar law, cannot be found among plants" but this could refer to Cuvier's progression of successive creations catastrophically separated. Derisive references to transformation were not uncommon (e.g. Fleming, 1822; Broderip, 1847) as well as more sober refutations such as Lyell's. The great Agassiz, completely opposed to transformism (evolution), had worked out the different basic ideas in God's mind, and the types of characters which therefore distinguished major groups, lesser groups, and least groups, so that again the whole living Creation became orderly, but the resemblances between forms were due purely to the employment of the same idea in their separate creations (*Essay on classification*, 1859,—Darwin's forthcoming book was linked with it by Sir William Jardine in his address to the British Association in 1859, as another work promising some curious speculations).

When we realize the extent to which the recognition of similarity was carried, it no longer surprises us that some of these systems were eventually abandoned. Swainson, in his *Treatise on Malacology* drew quinarian analogies not only between the groups of the testaceous molluscs (which in his scheme and others of the period included the planarians and liver flukes, giving him great scope for comparison) but also between the testacea and the vertebrata, comparing them in detail in the order given in his table (1840: 49) thus:

Orders of the *Testacea*	Circle of the *Vertebrata*
Gasteropoda	Quadrupeds
Dithyra [bivalves]	Birds
Nudibranchia	Reptiles
Parenchymata [planariae etc.]	Amphibians
Cephalopoda	Fishes

I leave it to the reader to find out Swainson's parallel between bivalves and birds, but must point out that Swainson made some accurate observations, which approximate his scheme to those of others, such as Chambers's, and indeed to present-day arrangements. For instance, he remarked (p. 42) that in all the vertebrated classes (and by implication the rest) Nature "begins, as it were, from a small rudimentary group...which seems to contain such animals as have the very least affinity to those which, standing at the head, exhibit the typical structure; they possess only the rudiments of the perfection to which they gradually, but ultimately, lead...Hence it follows, that such imperfectly formed beings, although found in every large division of animals, must necessarily possess a strong resemblance to each other; although, in reality, they belong to classes widely distinct, when we trace them up to their full development. This theory, although in some respects not new...has never received the attention it deserves. It has consequently resulted, that nearly all those authors who overlook the important differences between *analogy* and *affinity*, have naturally supposed these rudimentary animals, as they may be called, were closely connected to each other *by affinity*...This seems to be the

true cause why we have such a heterogeneous assemblage of animals under
Cuvier's class of *Intestina,* and even in those of MacLeay's *Acrita* and *Vermes*".
They agree, in short, only in primitive and negative characters. Now, this is
good sense, even if its application is faulty, and Swainson hardly deserves de
Beer's condemnation (1963: 13) although Edward Newman may. "There would
be no need to make mention of such abject nonsense were it not that these
notions passed as 'science' and were current."

At lower levels, but still well above the species, there was plenty of doubt and
confusion, derived in part from differing opinions as to which characters were
analogous and which showed affinity, but also, especially with fossils, from
incomplete knowlege of anatomy. The discovery of the dinosaurs was one of the
most spectacular and widely discussed phenomena of the period, but fossil
mammals were even more important for our present themes. Owen (1838),
describing Darwin's South American Fossils, placed *Toxodon* as "A gigantic
extinct mammiferous animal, referrible to the Order Pachydermata, but with
affinities to the Rodentia, Edentata, and Herbivorous Cetacea" (the last being
the Sirenia), and *Macrauchenia* as "referrible to the Order Pachydermata; but
with affinities to the Ruminantia, and especially to the Camelidae". Both, in
fact, are cases of convergence, part of the separate adaptive radiation of South
American mammals. In 1866 Andrew Murray could still include the aard-vark
and pangolins in the same family as the South American ant-eaters. He went so
far as to provide a map of the distribution of all ant-eating mammals, including
the echidna, which did indeed show how they are mostly geographical
representatives of one another. So far from it producing any idea of
convergence, he commented (1886: 294) "The ant-eating structure seems so
special an adaptation that it is difficult to conceive that all the animals in which
it occurs have not sprung from one and the same source. A supposition which
receives support from the horny clothing which many of them possess" which he
then proceeds to describe.

As to detailed support for evolution from the fossil record, none was to be
had; as late as 1859 Huxley had to confirm that *Pteraspis* was a fish as Agassiz
had said, not a mollusc nor a crustacean. And as late as 1851, Woodward still
included *Chiton* and *Dentalium* under the Gastropods, so that living things hardly
helped either. In short, it becomes fairly obvious why local adaptive radiation
was simply not realized as an argument for evolution. It will be shown in the
next section how late even the basis for it was realized.

Types of resemblance

In a paper read to the Linnean Society of London on 17 December 1822, the
Rev. William Kirby (1759–1850) remarked in his introduction "No objects are
more interesting to the scientific naturalist than those which assume the external
appearance of one tribe, while their more essential characters and their habits
indicate that they belong to another". Extraordinary resemblances of unrelated
forms intrigued, indeed fascinated, most biologists throughout the period. But of
course, such resemblances could only be recognized as not being affinities (i.e.
homologies) when systematics and comparative anatomy were good enough to
point them out. As long as brachiopods were included in bivalves, linked in by
the jingle-shells, *Anomia,* (which also have one valve perforated) their shared

characters were those of direct relationship, not convergence. Not until it was realized that bryozoans have a true gut with mouth and anus could one wonder why they should so resemble hydrozoans. As early as 1816, de Blainville had divided the mammals into Monodelphia, Didelphia and Ornithodelphia. But it could still be asked whether the marsupial mode of reproduction was only a characteristic that could turn up independently in different groups, like the different forms of placenta, so that some marsupials belonged in the Rodentia, others in the Carnivora, others in the Insectivora, and so on. According to Owen (1858) who gives a short history of classification of the Mammalia, Cuvier in 1829 noted the parallels between marsupials and placentals; Isidore Geoffroy St Hilaire in 1845 was the first to draw out explicitly the parallels between them considered as two distinct classes.

The extent to which convergence, however superficial, went unrecognized was enormous. Jenyns (1835), presenting a report to the British Association on the recent progress and present state of zoology, could quote MacLeay's *Horae entomologicae* as placing the cephalopods nearest to the chelonian reptiles, adding "He allows, however, that the hiatus occurring between is very considerable". He also referred to D'Orbigny's work of 1825 as *confirming* that the minute polythalamous mollusca (i.e. the Foraminifera) are indeed true cephalopods. Previously, they had been classed together only by analogy on their shells; now it was known that the animal has true arms or tentacula analogous to those of cephalopods. The modern reader is almost overcome with a sense of historic irony as he finds the same author exclaiming vigorously about the necessity for good comparative anatomy as shown by Cuvier (and singing its triumphs) while committing what are now seen to be first-class blunders. Owen himself, in his 1858 paper produces his new classification of the Mammalia on brain characters into Lyencephala (monotremes, marsupials); Lissencephala (rodents, insectivores, bats, edentates); Gyrencephala (cetacea, sirenia, toxodonta, proboscidea, perissodactyla, artiodactyla, carnivora, quadrumana); and Archencephala (Man only), and in support of the low position of the rodents points out that many of them hibernate like reptiles. Sclater, indeed, who first defined the major zoogeographical regions (1858a), backed up Owen with the occurrence of only bats and rodents beside the marsupials and monotremes in Australia, (1858b) "an additional piece of evidence to my mind of the correctness of Professor Owen's recent arrangement of these groups at the base of the Placental Mammalia: for the student of the geographical distribution of animals soon learns to appreciate the value of the old maxim 'noscitur a sociis' [a man is known by the company he keeps.]".

This confusion of convergent and homologous characters, or to use the language of the period (not always precise, as in the quotation from Jenyns above) characters of analogy and affinity was widespread among the major groups, and rampant in the minor ones. Sclater (1858a,b) who defined the primary zoogeographical regions by means of birds, in discussing the zoogeographical position of New Guinea put the brightly coloured, nectar-sucking *Dicaeum* where it appeared to belong, with the sun-birds, Nectariniidae, not as a separate, unrelated group. He included *Pachycephala*, a shrike-like bird, with the shrikes; it is in fact a modified muscicapid, convergent only on the Laniidae. Again, he includes the estrildids of Australia (his genera *Poephila* and *Amadina*) in the family which then included all finches, i.e. all passerine birds with a seed-cracking bill, the Fringillidae, to which all Darwin's finches would also belong.

The idea of representative *species* in different parts of the world therefore became familiar quite early on, but they were regarded as close relatives—for example the puma and the jaguar of the New World were thought of as representatives of the lion and the leopard of the Old. Such representatives had been known from the time of Buffon. Charles Lyell, in his *Principles of Geology* (vol II, 1833) duly mentioned them, but the only local representative of a *group* that he could bring forward (vol II: 118) was from the work of entomologist William Kirby—the confinement of true hive-bees to the Old World, and their replacement by a different group of bees (*Melipona* and *Trigona*) in the New, and by yet another group, at that time undescribed, in New Holland. Humboldt himself had pointed out examples of representative but analogous groups of plants, e.g. the cacti of the New World and the succulent euphorbias of the Old; and the restriction of the true heaths, *Erica*, to the Old World. (I have been shown in the Pine Barrens of New Jersey what at a distance I took to be heather—it was not in flower—but it belongs to. the Cistaceae, not the Ericaceae.) Humboldt was available in more than one English translation (see Freeman, 1980, for a list) but only after 1845. However, he was often quoted by others, or his results were given briefly without acknowledgment, e.g. by Barton (1827); Meyen in Johnston's translation (1846); and Henfrey (1852). These authors also referred extensively to Schouw, who on the basis of differences in species as well as whole groups, had given the first general classification of the botanical regions of the earth (listed in, e.g., Murray, 1866).

Even so, what really impressed botanists (and most zoologists as well) in this period was the extent to which geographical distribution of plants and animals depended on climate. Temperate North America had oaks and maples like, if not specifically the same as, those of Europe and Asia. Tropical America had palms and figs as did tropical Africa and South-East Asia. It was already recognized, however, that Australia was very odd and that different areas had more different floras the farther one went south (e.g. Agassiz, 1850), the northern boreal flora being virtually circumpolar. This could not be put down simply to climate, because plants transplanted to very different geographical regions could flourish happily (e.g. Barton, 1827).

Adaptive convergence and radiation

It was only slowly, therefore, in the 1850s and 1860s that local or general adaptive radiation came to mean anything to naturalists, both because of faulty classification and because so little was known of the actual modes of life and the habits of British animals, let alone ones in far distant parts of the world. Yet parallels there were, and (as can be seen from the quinarians) people felt the need of an explanation of them. (Only a small part of them could be explained, and that only much later on—1862—by Bates's theory of mimicry.) The first serious attempt to give a general theory of them that I have come across was by a convinced Agassizian, Cuthbert Collingwood.

Collingwood was lecturer in botany at the Liverpool Royal Infirmary School of Medicine (*materia medica* being then a necessary study for medical men and largely botanical). He had been present at the meeting of the Linnean Society of

London at which Darwin's and Wallace's contributions were first made public (1 July 1858) and did not accept them; he welcomed Agassiz's *Essay* as counteracting them, and defended it at meetings of the Liverpool Literary and Philosophical Society against the President, the Rev. H. H. Higgins, a well-respected cryptogamic botanist (Higgins, 1860, 1861; Collingwood, 1861a,b). He knew as clearly as did Agassiz and everyone else at that period the difference between homology and analogy, and he too wanted an explanation for analogous similarity—what he called homomorphism and distinguished carefully from homology, and from analogy which implies a *known* agreement in function. His full paper was read to the Liverpool Literary and Philosophical Society on 10 March 1860, and printed in its *Proceedings;* a shorter but substantial version appeared in the *Annals and Magazine of Natural History;* and an abstract of his paper as given to the British Association at its 1860 meeting appeared in the appropriate *Report* (1861). It was therefore brought well before the scientific public.

Collingwood, echoing William Kirby nearly 40 years before, begins his *Annals* version as follows. "No one conversant with zoology can have failed to remark the fact of the recurrence of similar forms in different groups of the animal series. Not only do species of one family resemble species of an allied family, but group with group, order with order, and even class with class, and subkingdom with subkingdom, can produce instances of the most striking homomorphism. The resemblances to which I allude are those of external form, unaccompanied by homologies of internal structure; nevertheless I imagine that this peculiarity, instead of entirely destroying its interest, and rendering it valueless, as some have appeared to consider, only places the subject in a different category of scientific facts, and invests it with a value peculiar to itself."—This was a very just remark, the more so since, as he goes on to say, experienced zoologists differed as to whether such resemblances were to be taken into account in classifying.

Moveover, zoologists had in the past been the source of great errors in classification. "Who can wonder if Pliny spoke of the Bat as "the onely bird that suckleth her little ones", in quaint old Holland's phraseology? What malacologist even can feel surprise that, up to recent times, the Polyzoan Molluscoids [Bryozoa] were mistaken for Zoophytes [Hydrozoa]? or that Lhuyd and at one time the illustrious Ellis, should have regarded them both in the light of "remarkable sea-plants", while his predecessor, Baker, had even looked upon them as the production of "salts incorporated with stony matter"? Who can wonder that, before the time of Savigny, the tunicated *Botrylli* should have been regarded as Polypes? that Linnaeus should have placed *Teredo* [the ship-worm] among the Annelides? that, before the Mémoire of Dujardin in 1835, the Foraminifera should have been classed with the Cephalopodous Mollusca? In all these cases (and others might be brought to swell the list), the animals have been raised, or have sunk, from one *subkingdom* to another.") Collingwood, 1860a.)

Collingwood gives an extensive list, most fully in his Liverpool version, of homomorphies right through the animal kingdom, some of which are highly reminiscent of Swainson, e.g. the great armadillo and the mataco (*Dasypus acar*) being homomorphic with tortoises and turtles. "With fishes, the mammalia are most singularly connected by the cetacea; while a special resemblance appears between the narwhal (Monodon) and the sword-fish (Xiphias)." "We have seen how the loris resembles the sloth; and on the other hand, the edentate genus Bradypus (Ai) bears a singular resemblance to monkeys in general, even in that particular which is so characteristic of them, viz., their physiognomy,—while it has a carnivorous homomorph in the sloth bear (Ursus labiatus), called by Pennant the ursiform sloth, and by Shaw, Bradypus ursinus." But he is no monomaniac. He notes that, with all the wonderful variety of fishes "the forms of fishes are *sui generis,* and in no class is it less easy to find homomorphic shapes". Besides general homomorphism, he points to its occurrence in particular organs and appendages, for example the central nasal horns of the rhinoceros, the rough-billed pelican, some hornbills, the extinct *Iguanodon* (this was actually a mistake for the thumb-spike misrestored as a nasal horn), and some fishes. "And what closer resemblance can be looked for than that between the head of the vulture, and the avicularia, or bird's head appendages so common in the polyzoa, especially Bugula, and which are considered by Huxley to be truly a part of these molluscoids, and not of a parasitic nature."

What is the reason for these resemblances? "On no principle of gradation of *form* can these resemblances, unaccompanied as they are by homologous relations, be accounted for. Some are *advances,* others *degradations* of form; and we must look for some deeper and more subtle cause which shall connect animals so widely separated as are the members of distinct subkingdoms. There is one circumstance, however, which cannot fail to strike the thoughtful inquirer... in not a few cases, striking deviations from typical *form* are accompanied by no less striking modifications of typical *habits;* and further, that these *modified* habits have a strong tendency to assimilate with the habits naturally exhibited by those animals whose form they assume." He gives as examples the habits of the Ursine Opossum (*Dasyurus ursinus*) as bear-like, the cat-like habits of the cat-like douroucouli, the rabbit-like habits of the Hyrax which is nevertheless a pachyderm, and so on. "In all these cases—and the list might be greatly swelled—the agreement between form and habit, independent of homological relations, is so striking that one is almost led to the conclusion that a certain external configuration necessitated certain habitual movements. I do not mean to say that this is the case; but I am inclined to think that a more careful review will lead us to the conviction that the converse of this proposition is the secret, not only of these, but of the other striking cases of *homomorphism,* as it has been called, to which reference has already been made" namely "*That agreement of habit in widely-separated groups is accompanied by similarity of form.*" In the working-out of his proofs, he comes close to validating MacLeay's explanation of Swainson's more distant analogies quoted above.

He also points out (1860a) the difficulties—"... how little do we know of the habits of the Invertebrate classes generally? The majority of them are marine; and it is only quite recently that they have even been *seen,* except through the medium of pictures, by the majority of persons... we know not why a *Chiton* resembles an *Aphrodite,* because we are equally ignorant of the habits of either"—and it takes him a long paragraph to set out his reasons for believing that the homomorphy of bryozoa and hydrozoa relates primarily to the tentacles and therefore to "a very great similarity, nay, almost *identity,* in one of the most important of habits, namely the mode of procuring food".

It was, of course, a commonplace that different groups of animals lived on different things and in different ways over the face of the earth—wolves fed on sheep which fed on grass, etc.—and even that there were land carnivores and sea carnivores, land herbivores and sea herbivores (the sirenia). The idea that a fauna was made up by adaptive radiation, whether from a single stock or several, was not explicit, even though there were plenty of advances towards it, as for example in the recognition of the parallels between the marsupials in Australia and the placentals elsewhere. Far too little was known, however, about the different habits of closely related species in the same area, for it to be seen that there was a general principle involved at all classificatory levels. Gilbert White's excellent observations of (for example) different warblers had appeared in 1789, and the *Natural History of Selborne* was rapidly acclaimed as a classic, but there was no conceptual framework to give such observations scientific weight until far later, in fact until Darwin. No wonder Darwin was so astonished that in the Galapagos Islands, it almost looked as though "from an original paucity of birds in this archipelago, one species had been taken and modified for different ends. In a like manner it might be fancied that a bird originally a buzzard, had been induced here to undertake the office of the carrion-feeding Polybori of the American continent" (*Naturalist's Voyage*, chapter 17). The dramatic significance of this discovery for Charles Darwin can only be appreciated against the literature of the period.

GEOGRAPHICAL DISTRIBUTION

Centres of creation

As already mentioned, a great deal of distribution (but not all) could be explained by reference to climate and soil. Richard Owen even explained the prevalence of the marsupial mode of reproduction in Australia as related to the long droughts and vast areas without water in Australia, an explanation rather remarkably accepted by Bennett (1860), who had much direct experience of them, and more remarkably rejected by the egregious Andrew Murray (1866: 285) who was right for once, since Owen did not explain why all desert mammals were not marsupials.

Nevertheless, the recognition of distinct provinces, biogeographically characterized, (see Woodward's *Treatise*, 1851–6; Forbes & Hanley, 1853; Forbes & Godwin-Austen, 1859; and the summaries in Murray, 1866) immediately raised the question of how they had originated, if the living world was saved only by the Ark. Linnaeus could, with some ingenuity, accept the dispersal of all living things from a single point somewhere in the mountainous Near East, and Fleming (1822, vol II: 104) could accept the Flood as one of the revolutions of the earth's surface. Louis Agassiz, like most others in our period, knew far too much to be able to agree. In a remarkable paper in the Edinburgh New Philosophical Journal (1850) he pointed out valid reasons why the present fauna and flora could not have achieved its present distributions from any single centre—if it was in the cold regions, tropical forms would die before they could reach their destinations, and *mutatis mutandis* for hot ones.

Agassiz also believed, like many others, that the further one went back in time, the closer uniformity in biota do we find all over the world, the earliest fossils being more or less the same anywhere, whereas recent fossil biotas agree with present day ones, as in the edentates of south America. But a few separate centres of creation—e.g. for the New World, or for South America—would be open to exactly the same objections as a single one. "Now these facts in themselves leave not the shadow of a doubt in our mind, that animals were primitively created over all the world, within those districts which they were naturally to inhabit for a certain time." He also finds it easy to show that they could not have been in primal pairs, but in much the same densities over the same areas as we find them in, varying of course with variations in climate and with the depredations of man—lions no longer occur in Europe.

Now quite early on in the period, disjunct distributions, especially of northern or arctic plants reappearing on more southern mountain ranges in Europe, had been reported and confirmed (e.g. Watson, 1832; Henfrey, 1852; de Candolle, 1855). More spectacular ones included the tapirs in tropical America and South-East Asia, the marsupials in the New World and Australasia, the Camelidae in the centre of the Old World and the Andes, the great apes in Africa and the Sunda Islands. Agassiz himself, as a world expert on fishes, knew of similar distributions in Europe, for example in the basins of the Rhine, Rhone and Danube. Some species (which he lists) are peculiar to each basin; others, such as the pike, are found in all; but some occur in only two out of the three, the common perch, *Perca fluviatilis*, for example, which is represented in the Danube by a different species. Now, he says, some have set these peculiar disjunct distributions down to migration; but if one species can migrate, why not all? Yet they have not. The only possible explanation, therefore, is the creation of some species of fish in two different basins at once.

Although this paper appeared comparatively late in the period, such ideas had been in circulation for a long time. Watson (1832: 2) refers to the divergent opinions of botanists, "one party imagining all plants to have originated in some central point from which they have been gradually spread over the earth's surface; others conceiving that several of such centres must have existed; and a third party believing species for the most part to have originated where they now appear as the natural and untransported products of the soil and climate". Watson, and others before him, knew of the transport of West Indian seeds by the Gulf Stream to the western shores of Britain—he knew also that they need the right conditions to germinate (1832: 4–5). The transport by natural means of plants and animals was a major subject from very early on, and a whole series of experiments was made on behalf of the British Association, ending in 1857 (Baxter, 1858).

There was a considerable divergence of opinion, in part inevitable because of the uncertain limits of species, as to how many were really widespread (see Hooker, 1853, in contrast to de Candolle, 1855). The question was far from academic—it bore upon the varieties (or species?) of Man and the nature of his origins. The anthropologist James Cowles Prichard, to whose writings on geographical distribution Sir Charles Lyell was greatly indebted, reported to the British Association on philological and anthropological research (1833) on this very topic, remarking that "it has been confidently assumed that these tribes of men [in the South Seas], like the bread-fruit and coconut trees by which they are fed, are the indigenous produce of the coralline or volcanic soil on which they exist. This notion might have been strenuously maintained, if researches into the structure and affinity of languages had not furnished its refutation and displayed, in the idioms of these insular tribes, sufficient evidence of their mutual relationship and of the derivation of the whole stock of people from a common centre".

Prichard was no doubt right, but a little premature in the triumph of his conclusion. Seventeen years later, Agassiz (1850), having settled the issue in favour of diffuse creation, pointed out that what held for the zoogeographical provinces held also for the races of Man, since they coincided. Sclater, a disciple in this of Agassiz, in setting up the modern zoogeographical regions (1858a) added that Agassiz's solution, which he supposed "few philosophical zoologists, who have paid attention to the general laws of the distribution of organic life, would now-a-days deny," got rid of the "awkward necessity" of introducing the red man "into America by Behring's Straits, and of colonizing Polynesia by stray pairs of Malays floating over the water like cocoa-nuts". (Such a mixture of right and wrong in a classic paper occurs, of course, over and over again in the history of science.)

Andrew Murray went one better (1866: 6). Fully alive to the facts of geographical variation within species, and, I am afraid, to the popular newpapers of the time, he saw "a faint glimmering of light, because we have seen a race of man formed under our own eyes, the Anglo- or rather the Europeo-American nation, as distinct and well marked a race as any other; and yet the change has been effected over the whole of the United States without any transition men having been observed; and what is still more extraordinary, it has been effected over the whole of the region where it occurs at the same time . . . there he is, a nation *per se;* known to "Punch",—known to passport officers,—known to ourselves,—easily identified, easily figured, and easily caricatured". (Did he really think of that diversity of individuals as all close copies of Uncle Sam?) This was published in a standard textbook of mammalian geographical distribution, referred to with respect (if with disagreement) by Alfred Russel Wallace, by a man who achieved a serious entry in the Dictionary of National Biography. If the intellectual landscape could be so different as to allow such a statement as late as 1866, how much more do we need to study to understand that of Beagle days!

Restricted species

That some species appear to be peculiar to very restricted areas was known very early on, and steadily confirmed as exploration increased and they were still not found—the peculiar species of trees in St Helena are noted as an example by Barton (1827). But a species on a small island cannot have a large distribution. Vast numbers of endemics were indeed discovered on many islands, notably the Canaries, Madeira group and Azores by Thomas Vernon Wollaston and others; but then, continental species often had highly restricted distributions as well, for example in the Alps. De Candolle, in his monumental *Géographie botanique raisonnée* (1855) devoted a special article to species of greatly restricted distribution and came to the conclusion that they were very frequent both on continents and on islands, much more so than genuinely and natively widespread ones. There was nothing special about islands, then, in this respect. Lyell had already pointed out in 1833 that such confined ranges were more easily explicable by single than by multiple centres of creation, but otherwise all he could say about island floras was that they had large numbers of endemic species, and, with the exception of St Helena, resembled in their affinities the plants of the nearest continent. If single centres are the rule, then remarkable disjunct distributions need to be explained, and we have seen how Agassiz did it. A more scientifically convincing argument for the European arctic-alpines was put forward in a remarkable memoir in 1846 by Edward Forbes. He explained their present disjunction by warming up after a previous cold spell, the glacial period, for which he was able to quote convincing geological evidence. Charles Darwin had come to the same conclusion, and regretted, (*More Letters*, I: 408–409) that this was one of the results of his theory on which he had been forestalled by prior publication.

It is true that Forbes's explanation was not quite the modern one—he envisaged a cold sea over the whole of northern Europe except for the higher mountains, but in that he was no different from his contemporaries and successors. He was able to quote a great deal of geological and palaeontological evidence, including the Arctic fossils found in British deposits by James Smith of Jordanhill; the recognition that the earlier 'Crag' deposits of East Anglia contained shells, as described by Searles Wood (e.g. 1842), which on Forbes's own research were related to the present Portuguese-Mediterranean marine fauna; and a classic paper by Sir Charles Lyell showing that the land in Sweden on the edge of the Baltic was even now slowly rising.

Now it was a commonplace that so much present dry land in so many parts of the world was shown by its fossils to have been laid down under the sea, that no one needed to doubt the previous occurrence of massive differences in the distribution of land and sea, which might well explain many biogeographical anomalies. Many organisms in Madeira were believed to be identical with European ones—how could this be if there had not been a landbridge? To the end of his life, Forbes (see Forbes & Godwin-Austen, 1859: 110–120) was convinced of a former Lusitanian land to explain the biotic affinities of Madeira and the Azores. Woodward (1856) accepted the necessity of a now foundered land connection, and gave references to geologists, including Lyell, for corroborating evidence (1856 Part III: 387). Alfred Russel Wallace (1860) in his classic paper on the Malay Archipelago was able to give excellent reasons why two land masses in that region must have had far greater extensions formerly (and he was right). He also refers specifically to Madeira—"Yet of the comparatively rich insect-fauna of Madeira, 40 per cent are continental species; and of the flowering plants more than 60 per cent. Nothing but a former connexion to the continent will explain such an amount of specific identity" and he points to the directions of the Atlas Mountains and the Sierra Nevada, towards the Canaries and the Madeira group respectively, as possible supporting evidence. Not until his *Geographical Distribution* (1876) and *Island Life* (1880) did he realize the powers of natural distribution, the unbalanced nature of these island faunas (effectively precluding a land bridge), and the depths of the seas around the islands. Thomas Vernon Wollaston, one of the principal researchers in the Macaronesian Islands and an admirer of Edward Forbes, also believed in a land bridge; but I hope I have said enough to indicate how reasonable such a belief could be in those days. Hull's derogatory comments (1973) on Wollaston's review of the *Origin,* in relation to this particular point, seem to me to show insufficient comprehension of the period.

Types of islands

Moreover, until Darwin, there was no clear-cut classification of islands, biologically or geologically. Certainly, some remote islands were known to have (or have had) some very strange animals—the dodo is the outstanding example. But the moas and *Apteryx* of New Zealand, and the bizarre monotremes and the multiplicity of marsupials in Australia were equally strange. And so were the giraffe, the tapir, the sloth, the armadillo and a host of others.

Moreover, what was to be counted as an island? Was Australia an island or a continent? If it was an island, how could one separate it into a different category from, say, the mainlands of the New World taken together? If it was not in this sense an island, then islands had no monopoly of strange creatures. Bory de St. Vincent (1778–1846) had pointed out just after the beginning of the century that on remote volcanic islands such as the Mascarenes, which he had visited, one had not only strange flightless birds but an absence of amphibia which would find it difficult to cross salt water. But he also believed in a former Atlantean continent even including St Helena, and in the earlier part of our period, his reputation as a scientist was not high. He had published a very elaborate classification of the infusoria based only on external shape, which was quickly superseded by the classic work of Ehrenberg; and Jenyns's reference to him in his survey of the progress of zoology (1835) was not complimentary.

Darwin, of course, does refer to Bory's remarks on amphibia, which may well have set him thinking; but a much more substantial reference, used by many authors, was to the findings of the U.S. Exploring Expedition under Lieutenant Wilkes. Some of the results were retailed in the Edinburgh New Philosophical Journal for January 1844, particularly the observation than all the small islands of the South Seas and New Zealand were without native mammals except for bats. William Thompson, for example, noticed this in his report to the British Association on the fauna of Ireland (1844). Now only part of New Zealand is volcanic and there is an indigenous frog. The issue was still not clear-cut.

Another anomalous island, rather in the same way as New Zealand geologically, but with a wealth of indigenous mammals almost recalling Australia, was Madagascar, of which Andrew Murray (1866: 82–83) says "That country is distinguished not only by the peculiar endemic types which it possesses, but perhaps, even more so, by the absence of other forms, which we might naturally think that it ought to contain. Its vicinity to Africa, and the fact that there is a comparatively shallow submarine neck of land, which would on a small rise connect it with that continent...mark it *prima facie* as an African dependency...If it were formerly connected with Africa, why are so many of the special types of the neighbouring land wholly wanting?—Where are its Antelopes and its Pachyderms?". Perhaps they had not yet come into existence when Madagascar separated, or perhaps it would depend "on the character of the portion of Madagascar not submerged, what animals would be present and what could survive. If all but wooded peaks were under water it is plain that there could be none of the Antelopes which feed on grassy plains: they would not be there at all".

Equally anomalous in a different way were the Azores; volcanic and remote, they should be characteristic oceanic islands, yet the degree of endemism was strikingly low compared with the Canaries and Madeira, or indeed with St

Helena. Not until 1870 did Frederick Du Cane Godman explain it by the high degree of immigration likely into these windy islands, an explanation accepted by Wallace (1876, vol. I: 209).

It is perhaps fortunate that St Paul's Rocks, visited by Darwin, and mid-Atlantic but not volcanic, have so little terrestrial biota. By distinguishing carefully between remote volcanic islands (including with them remote coral islands) on the one hand, and continental islands on the other, he made a valid grouping of so many islands that the few anomalous ones could be left as anomalous (and largely explained in recent years by plate tectonics).

CONCLUDING REVIEW

If the intellectual landscape inhabited by Charles Darwin as a young man was so wildly different from that of today, what did he have to go on when thinking about the possibility of evolution? Adaptive radiation, now one of the strongest arguments for evolution, was not available until he saw evidence for it himself in the Galapagos. Classification, in the sense of actual taxonomic relations, was much too uncertain, and unstable for the most part, to be useful. But no-one had ever disputed the classification of vertebrates into mammals, birds, reptiles (including amphibians or not) and fish, and those groups certainly gave independent evidence for biogeographical provinces, as did plants. The evidence that something other than climate and soil, powerful though they were, was needed to explain plant and animal distribution was becoming really compelling. No one who had actually seen the pampas and savannahs of South America with no existing indigenous grazing mammals (except a few peculiar rodents), those of Australia with the kangaroos and wallabies, or those of Africa with antelopes, zebra, giraffes and pachyderms, could doubt that some historic cause was also necessary to explain the present distribution of animals (and some plants). There was indisputable evidence for great changes in the relative distribution of sea and land, and in the climate of the northern hemisphere, so that historic processes were a real possibility. The dodo had become extinct, the moas were shown to have become so since Man inhabited New Zealand, the kiwis were retreating before civilization, and the great auk was decreasing rapidly in numbers if not already extinct. That extinction was a natural process not due only to Man, was thoroughly illustrated by the fossil record. One or two survivors such as stalked sea-lilies, or the Nautilus, or the trigoniid clams of Australia might turn up, but there was nothing on the earth like the great faunas of the Palaeozoic or Mesozoic periods, or even the early Tertiary.

Widely disjunct distributions, such as that of the arctic-alpine plants of Europe, discussed above, were susceptible to various explanations, but only one was acceptable if supernatural interference was ruled out. Some islands, as Bory de St. Vincent had suggested, had arisen *de novo* from the sea bottom, and must have acquired their biota across the seas. At the beginning of the period, the problem was to explain highly localized and endemic species, since so much could be seen of natural dispersal of both plants and animals. At the end, endemics could be understood without difficulty—it was the widespread forms (if not distributed by Man) that needed explanation. And many, of course, were explicable from the new information given by the fossil record, showing ranges far more widespread formerly.

It has often been said that the *Origin of Species* convinced people that evolution occurred because it provided a purely natural mechanism (natural selection) to account for it. This is a half-truth. Those people who were already well aware of the facts of geographical distribution (past and present) and inclined to evolution may have been converted by natural selection. Thomas Henry Huxley and Hewett Cottrell Watson seem to be cases in point. But as the numerous papers and articles published in 1909, 50 years after the *Origin*, show clearly, it was primarily the facts of geographical distribution that had been found convincing; natural selection was accepted by only a tiny minority.

Progress towards a real knowledge of biogeography was of course not uniform. We can distinguish, rather artificially, three major stages in it. The first is marked by Buffon's celebrated essay on the differences, largely specific, between the mammals of the Old and New Worlds, but including the presence of the opossum and the absence of the horse in the Americas. The next begins with the work of Humboldt and Schouw, and is followed by the writings of Cuvier and Prichard, and Lyell's essays in early editions of the *Principles of Geology*. (For these, see Lyell, 1833, vol II.) The third is the period of consolidation, of geological or palaeontological support, and of inspiration of both Darwin and, years later, Wallace. It is specially marked by the work of Forbes, Watson, Wollaston and J. D. Hooker, culminating in the geographical chapters in the *Origin*, Wallace's paper on the Malay Archipelago, and Wallace's great books (1876 and 1880). Throughout all this, and interacting with it, we have the complex history of classification, the natural system, and modes of origination which I have sketched above. A good history of these interacting themes would be useful, not only in itself but in order to show philosophers and historians the complexity of the controversies they deal with so inadequately in science.

ACKNOWLEDGEMENTS

I am grateful to Professor R. J. Berry, Dr R. G. Pearson and Dr G. M. Davis for comments on this paper.

REFERENCES

AGASSIZ, L. 1850. Geographical distribution of animals. *Edinburgh New Philosophical Journal, new series*, **49:** 1–33.

AGASSIZ, L., 1859. *An Essay on Classification*. London: Longman, Brown, Green, Longman & Roberts, & Trübner.

ANON. (CHAMBERS, R.), 1844. *Vestiges of the Natural History of Creation*. London: Churchill.

ANON. (MOORE, T. J.), 1862. *Suggestions offered on the part of the literary and philosophical society of Liverpool, to members of the mercantile marine, who may be desirous of using the advantages they enjoy for the promotion of science, in furtherance of zoology*. Appendix II to *Proceedings of the Literary and Philosophical Society of Liverpool, 51st session, 1861–2*. 49 pp.

BAKER, J. G., 1868–70. On the geographical distribution of ferns. *Transactions of the Linnean Society of London*, **26:** 305–352 (Read 4 April 1867).

BAKER, J. G. & NEWBOULD, W. W. (Eds) 1883. *Topographical Botany: Being Local and Personal Records Towards Shewing the Distribution of British Plants Traced Through the 112 Counties and Vice-Counties of England, Wales, and Scotland. By Hewett Cottrell Watson* (2nd edition). London: Quaritch.

BARTON, J., 1827. *A Lecture on the Geography of Plants*. London: Harvey & Darton.

BAXTER, W. H., 1858. Sixteenth and final report of a committee, consisting of Professor Daubeny, Professor Henslow, and Professor Lindley, appointed to continue their experiments on the growth and vitality of seeds. *Report of the 27th meeting of the British Association . . . 1857:*, 43–56.

DE BEER, G. R., 1963. *Charles Darwin. Evolution by Natural Selection.* London: Nelson.

BENNETT, G., 1860. *Gatherings of a Naturalist in Australasia: Being Observations Principally on the Animal and Vegetable Productions of New South Wales, New Zealand, and Some of the Austral Islands.* London: van Voorst.

BOTTING, D., 1973. *Humboldt and the Cosmos.* London: Sphere Books.

BOWLER, P. J., 1976. *Fossils and Progress. Paleontology and the Idea of Progressive Development in the Nineteenth Century.* New York: Science History Publications.

BRODERIP, W. J., 1847. *Zoological Recreations:* London: Colburn.

DE CANDOLLE, A., 1855. *Géographie Botanique Raisonnée ou Exposition des Faits Principaux et des Lois Concernant la Distribution Géographique des Plantes de L'Epoque Actuelle.* 2 vols. Paris: Masson; Genève: Kessmann.

CLARK, W., 1855. *Mollusca Testacea Marium Britannicorum. A History of the British Marine Testaceous Mollusca, Distributed in their Natural Order, on the Basis of the Organization of the Animals; With References and Notes on Every British Species.* London: van Voorst.

COLEMAN, W., 1964. *Georges Cuvier, Zoologist. A Study in the History of Evolution Theory.* Cambridge, Mass.: Harvard University Press.

COLEMAN, W., 1971. *Biology in the Nineteenth Century: Problems of Form, Function, and Transformation.* New York: Wiley.

COLLINGWOOD, C., 1860a. On homomorphism; or, organic representative form. *Proceedings of the Literary & Philosophical Society of Liverpool, 49th session,* 181–216.

COLLINGWOOD, C., 1860b. On recurrent animal form, and its significance in systematic zoology. *Annals and Magazine of Natural History, 3rd series, 6:* 81–91.

COLLINGWOOD, C., 1861a. On recurrent animal form and its significance in systematic zoology. *Report . . . British Association 1860, Transactions of the sections:* 114–115.

COLLINGWOOD, C., 1861b. On Agassiz' views of Darwin's theory of species. *Proceedings of the Literary and Philosophical Society of Liverpool, 50th session:* 81–99.

DARWIN, C., 1870. *Naturalist's Voyage Round the World* (spine title). *Journal of researches into the natural history and geology of the countries visited during the voyage of H.M.S. Beagle round the world, under the command of Capt. Fitz Roy, R.N.* (New edition), London: Murray.

DARWIN, F. (Ed.), 1903. *More Letters of Charles Darwin. A Record of his Work in a Series of Hitherto Unpublished Letters.* 2 vols. London: Murray.

FLEMING, J., 1822. *The Philosophy of Zoology; or a General View of the Structure, Functions, and Classification of Animals.* Edinburgh: Constable; London: Hurst, Robinson.

FORBES, E., 1846, On the connexion between the distribution of the existing fauna and flora of the British Isles, and the geological changes which have affected their area, especially during the epoch of the Northern Drift. *Memoirs of the Geological Survey of Great Britain, 1:* 336–432.

FORBES, E†, & GODWIN-AUSTEN, R., 1859. *The Natural History of the European Seas.* London: van Voorst.

FORBES, E., & HANLEY, S., 1853. *A History of British Mollusca and Their Shells.* Vol. 1. London: van Voorst.

FREEMAN, R. B., 1980. *British Natural History Books 1495–1900. A handlist.* London: Dawson, Archon Books.

GALE, B. G., 1982. *Evolution Without Evidence. Charles Darwin and the Origin of Species.* Albuquergue, N. M.: University of New Mexico Press.

GARNER, R., 1860. On the shell-bearing Mollusca, particularly with regard to structure and form. *Journal of the Proceedings of the Linnean Society of London, 4:* 35–43. (No date of reading).

GODMAN, F. Du C., 1870. *Natural History of the Azores, or Western Islands.* London: van Voorst.

HENFREY, A., 1852. *The Vegetation of Europe, its Conditions and Causes.* London: van Voorst.

HIGGINS, H. H., 1860. On Darwin's theory of the origin of species. *Proceedings of the Literary and Philosophical Society of Liverpool, 50th Session,* 42–49.

HIGGINS, H. H., 1861. On Darwin's theory of the origin of species (continued). *Proceedings of the Literary and Philosophical Society of Liverpool, 50th session.* 134–140.

HINCKS, T., 1868. *A History of the British Hydroid Zoophytes.* 2 vols. London: van Voorst.

HOOKER, J. D., 1853. *Introductory Essay to the Flora of New Zealand.* London: Lovell Reeve.

HULL, D., 1973. *Darwin and his Critics.* Cambridge, Mass.: Harvard University Press.

JARDINE, Sir W., 1860. [Presidential address]. *Report . . . British Association 1859, Transactions of the sections,* 126–129.

JEFFREYS, J. G., 1859. Further gleanings in British conchology. *Annals & Magazine of Natural History, 3rd series, 3:* 30–43.

JENYNS, L., 1835. Report on the recent progress and present state of zoology. *Report . . . British Association 1834,* 143–251.

JENYNS, L., 1857. On the variation of species. *Report . . . British Association 1856, Transactions of the sections,* 101–105.

KINGSLEY, C., 1863. *The Water-Babies: a Fairy Tale for a Land-baby.* London & Cambridge: Macmillan.

KIRBY, W., 1825. A description of some insects which appear to exemplify Mr William S. MacLeay's doctrine of affinity and analogy. *Transactions of the Linnean Society of London, 14:* 93–110. (Read 17 Dec. 1822.)

LINDLEY, J., 1834. On the principal questions at present debated in the philosophy of botany. *Report . . . British Association 1833,* 27–57.

LYELL, Sir C., 1832-3. *Principles of Geology, Being an Attempt to Explain the Former Changes of the Earth's Surface, by*

Reference to Causes Now in Operation. Vol. 1, (2nd edition), 1832. Vol. II, (2nd edition), 1833. Vol. III, (1st edition), 1833. London: Murray.

MACLEAY, W. S., 1825. Remarks on the identity of certain general laws which have been lately observed to regulate the natural distribution of insects and fungi. *Transactions of the Linnean Society of London, 14:* 46–68. (Read 5 Nov. 1822).

MACLEAY, W. S., 1842. On the natural arrangement of fishes. *Annals and Magazine of Natural History, 9:* 197–207.

MEYEN, F. J. F., translated by JOHNSTON, M., 1846. *Outlines of the Geography of Plants: With Particular Enquiries Concerning the Native Country, the Culture, and the Uses of the Principal Cultivated Plants on Which the Prosperity of Nations is Based.* London: Ray Society.

MURRAY, A., 1866. *The Geographical Distribution of Mammals.* London: Day.

OWEN, R., 1838. Fossil Mammalia. In C. Darwin, (Ed.) 1838. *The Zoology of the Voyage of H.M.S. Beagle Under the Command of Captain Fitzroy, During the years 1832 to 1836.* Part 1, no. 1, 13–40. London: Smith, Elder.

OWEN, R., 1858. On the characters, principles of division, and primary groups of the Class Mammalia. *Journal of the Proceedings of the Linnean Society of London, Zoology, 2:* 1–37. (Read 17 February and 21 April 1857).

OWEN, R., 1859. [Presidential address]. *Report . . . British Association 1858,* xlix–cx.

PILSBRY, H. A., 1894. *Manual of Conchology: Structural and Systematic . . . Second series: Pulmonata. vol. IX (Helicidae, vol. 7). Guide to the study of the Helices.* Philadelphia: Academy of Natural Sciences.

PRAEGER, R. Ll., 1902. On types of distribution in the Irish Flora. *Proceedings of the Royal Irish Academy, 24B:* 1–60.

PRICHARD, J. C., 1833. Abstract of a comparative review of philological and physical researches as applied to the history of the human species. *Report of the 1st and 2nd meetings British Association,* 529–544.

SCLATER, P. L., 1858a. On the general geographical distribution of the members of the class Aves. *Journal of the Proceedings of the Linnean Society of London, Zoology, 2:* 130–145.

SCLATER, P. L., 1858b. On the zoology of New Guinea. *Journal of the Proceedings of the Linnean Society of London, Zoology, 2:* 149–170.

SMELLIE, W., 1790. *The Philosophy of Natural History.* Edinburgh: Charles Elliot heirs etc.

SWAINSON, W., 1840. *A Treatise on Malacology; or the Natural Classification of Shells and Shell-Fish.* London: Longman, Orme, Brown, Green and Longmans.

THOMPSON, W., 1844. Report on the fauna of Ireland: *Division Invertebrata. Report . . . British Association 1843,* 245–291.

TULK, A. (translator), 1847. *Elements of Physiophilosophy, by Lorenz Oken . . .* London: Ray Society.

WALLACE, A. R., 1860. On the zoological geography of the Malay Archipelago. *Journal of the Proceedings of the Linnean Society of London, Zoology 4:* 172–184.

WALLACE, A. R., 1876. *The Geographical Distribution of Animals With a Study of the Relations of Living and Extinct Faunas as Elucidating the Past Changes of the Earth's Surface.* 2 vols. London: Macmillan.

WALLACE, A. R., 1880. *Island life: or, The Phenomena and Causes of Insular Faunas and Floras, Including a Revision and Attempted Solution of the Problem of Geological Climates.* London: Macmillan.

WATSON, H. C., [1832]. *Outlines of the Geographical Distribution of British Plants; Belonging to the Division of Vasculares or Cotyledones.* Edinburgh: privately printed.

WILSON, L. G., 1970. *Sir Charles Lyell's Scientific Journals on the Species Question.* New Haven & London: Yale University Press.

WILSON, L. G., 1972. *Charles Lyell. The Years to 1841: The Revolution in Geology.* New Haven and London: Yale University Press.

WINSOR, M. P., 1976. *Starfish, Jellyfish, and the Order of Life. Issues in Nineteenth-Century Science.* New Haven & London: Yale University Press.

WOLLASTON, T. V., 1854. *Insects Maderensia.* London: van Voorst.

WOLLASTON, T. V., 1856. *On the Variation of Species with Especial Reference to the Insecta; Followed by an Inquiry Into the Nature of Genera.* London: van Voorst.

WOLLASTON, T. V., 1864. *Catalogue of the Coleopterous Insects of the Canaries in the Collection of the British Museum.* London: British Museum.

WOLLASTON, T. V., 1878. *Testacea Atlantica or the Land and Freshwater Shells of the Azores, Madeiras, Salvages, Canaries, Cape Verdes, and Saint Helena.* London: Reeve.

WOOD, S. 1842. A catalogue of shells from the Crag. *Annals & Magazine of Natural History, 9:* 455–461, 527–544.

WOODWARD, S. P., 1851-6. *A Manual of the Mollusca; or, Rudimentary Treatise of Recent and Fossil Shells.* Pt. I (1–58), 1851; Pt II (159–330) 1854; Pt III (331–486) 1856. London: Weale.

Biological Journal of the Linnean Society (1984), *21:* 29–59. With 11 figures

Darwin and the Galapagos

FRANK J. SULLOWAY

*Department of Psychology and Social Relations,
Harvard University, Cambridge, Massachusetts, U.S.A.**

Accepted for publication 15 August 1983

Charles Darwin's historic visit to the Galapagos Islands in 1835 represents a landmark in the annals of science. But contrary to the legend long surrounding Darwin's famous Galapagos visit, he continued to believe that species were immutable for nearly a year and a half after leaving these islands. This delay in Darwin's evolutionary appreciation of the Galapagos evidence is largely owing to numerous misconceptions that he entertained about the islands, and their unique organic inhabitants, during the *Beagle* voyage. For example, Darwin mistakenly thought that the Galapagos tortoise—adult specimens of which he did not collect for scientific purposes—was not native to these islands. Hence he apparently interpreted reports of island-to-island differences among the tortoises as analogous to changes that are commonly undergone by species removed from their natural habitats. As for Darwin's finches, Darwin initially failed to recognize the closely related nature of the group, mistaking certain species for the forms that they appear, through adaptive radiation, to mimic. Moreover, what locality information he later published for his Galapagos finch specimens was derived almost entirely from the collections of three other *Beagle* shipmates, following his return to England. Even after he became an evolutionist, in March of 1837 (when he discussed his Galapagos birds with the eminent ornithologist John Gould), Darwin's theoretical understanding of evolution in the Galapagos continued to undergo significant developments for almost as many years as it took him to publish the *Origin of Species* (1859). The Darwin–Galapagos legend, with its romantic portrait of Darwin's 'eureka-like' insight into the Galapagos as a microcosmic 'laboratory of evolution', masks the complex nature of scientific discovery, and, thereby, the real nature of Darwin's genius.

KEY WORDS:—Charles Darwin – Galapagos Islands – evolutionary theory – biography.

CONTENTS

*Present address: Department of Psychology, University College London, Gower Street, London WC1E 6BT

0024–4066/84/010029+31 $03.00/0

FRANK J. SULLOWAY

THE MYSTIQUE OF THE GALAPAGOS ISLANDS

All remote oceanic islands possess a certain romantic mystique. But Charles
Darwin has raised the mystique of the Galapagos Islands to a level that finds
virtually no rivals in the history of scientific thought. Both for the biology
textbooks and for the history of science, these 'enchanted islands' have become
the highly acclaimed symbol of one of the greatest revolutions in Western
intellectual thought. Indeed, that this momentous scientific revolution sprang
from insights that Charles Darwin garnered during a brief five-week visit to the
Galapagos in 1835 has made them into the symbolic equivalent of Newton's
famous apple among the great stories of scientific discovery. But unlike Newton's
apple, with its fleeting but historic fall, the Galapagos Islands have remained a
permanent showcase of the fundamental scientific insights that Charles Darwin
first divined. For those who have visited the Galapagos in Darwin's historic
wake, there is inevitably a feeling of being on hallowed scientific ground and
also of standing intellectually in Darwin's awesome and ever-present shadow. As
one Galapagos researcher has recently commented, with a curious mixture of
reverence and irritation: "It's like studying art at the Louvre. It's hammered
into you—Darwin, Galapagos—Galapagos, Darwin" (cited in Stoppard,
1981:7).

Nevertheless, much of the undeniable mystique that Darwin has given to the
Galapagos Islands, while certainly well deserved, is based on a considerable
historical misconception. Indeed, the story of Darwin and the Galapagos is
shrouded in a series of interrelated myths—myths that are part of a subtle but
pervasive politics of science that surrounds all great scientific discoveries. In this
essay I will begin by examining the myth of Darwin's Galapagos conversion to
the theory of evolution; for it is this myth that forms the very core of the
Darwin–Galapagos legend—a legend that tends to obscure both the true nature
of Darwin's genius and the important role that these famous islands have indeed
played, and continue to play, in the development of evolutionary theory.

The sources of the widespread myth that Darwin was converted 'eureka-like'
to the theory of evolution during his Galapagos visit in 1835 are manifold,
stemming from certain of Darwin's own autobiographical accounts of his
conversion experience, as well as from a century of subsequent biographical
writings.[1] Even the scientific evidence—Darwin's *Beagle* collections—also tend
to support this view (Lack, 1947:9, 23). But during the last decade or so,
Darwin scholars have generally come to the conclusion that Darwin's conversion
did not occur during his visit to the Galapagos Islands in 1835; or even, for that
matter, during the voyage of the *Beagle* itself.[2] Exactly when the conversion *did*
occur has nevertheless remained the subject of doubt. Above all, the real story
behind Darwin's conversion—even more important than the timing of that
conversion—has not been sufficiently understood; and it is this story,

---

[1]See, for example, Darwin, 1859 : 1; 1868, *1* : 9–11; 1958[1876]: 118–19; 1887, *2* : 23, 34; *3* : 159; 1903, *1* : 118–19, 367; F. Darwin, 1888 : 74, 76; 1903, *1* : 37–38, 1909 : xiv; Barlow, 1933 : xiii; 1945 : 262–64; 1963: 204–5, 277; 1967 : 12; Irvine, 1955 : 50; de Beer, 1958 : 5; 1962 : 323; Wichler, 1961 : 85–87; and Huxley, 1966 : 3.

[2]See Himmelfarb, 1959 : 107–23; Smith, 1960 : 392; Gruber & Gruber, 1962 : 200; Sulloway, 1969 : 99–102; 1979 : 26–27; 1982a : 19–20, 22–23; 1982b : 57–58; 1982c; Ghiselin, 1969 : 32–36; Limoges, 1970 : 7–20; and Herbert, 1974 : 249; 1980 : 7–12. Hodge (1983 : 61–63, 104, n.102), however, still holds to the view that Darwin was converted to the theory of evolution during the last year of the *Beagle* voyage. See further note 4.

disentangled from legend, that helps to distinguish Darwin's genius from the intellectual talent of his contemporaries. In what follows I will reconstruct the story of Darwin's conversion from the time of his visit to the Galapagos Islands in the fall of 1835 (Fig. 1) to his decision in July of 1837 to begin the first of a series of notebooks on the transmutation of species. I will also argue that much of Darwin's theoretical understanding of the Galapagos occurred only gradually, during the twenty-two years separating his conversion to the theory of evolution in 1837, from his eventual publication of the *Origin of Species* in 1859.

THE *ORNITHOLOGICAL NOTES*

Approximately nine months after leaving the Galapagos Islands, while preparing a separate catalogue for his ornithological specimens, Darwin made the following famous entry as he recopied a portion of his voyage zoology notes dealing with the Galapagos mockingbirds:

> I have specimens from four of the larger Islands....The specimens from Chatham and Albermale [*sic*] Isd appear to be the same; but the other two are different. In each Isld. each kind is *exclusively* found: habits of all are indistinguishable. When I recollect, the fact that from the form of the body, shape of scales & general size, the Spaniards can at once pronounce, from which Island any Tortoise may have been brought. When I see these Islands in sight of each other, & possessed of but a scanty stock of animals, tenanted by these birds, but slightly differing in structure & filling the same place in Nature, I must suspect they are only varieties. The only fact of a similar kind of which I am aware, is the constant asserted difference—between the wolf-like Fox of East & West Falkland Islds.—If there is the slightest foundation for these remarks the zoology of Archipelagoes—will be well worth examining; for such facts [would *inserted*] undermine the stability of Species.[3]
>
> (1963[1836]:262)

This passage, which contains the first explicit hint of the evolutionary views that Darwin was later to expound with such revolutionary consequences, should not be taken, as it sometimes has, as the statement of a confirmed believer in the theory of evolution. Rather, Darwin apparently drew face to face with this possibility, some nine months after leaving the Galapagos, only to reject it on the grounds that the mockingbirds were probably "only varieties".[4] To

---

[3] The composition of Darwin's *Ornithological Notes*, which has long been the subject of conjecture and divergent opinion, can now be assigned to the thirty-one day period between 18 June and 19 July 1836, while the *Beagle* was sailing from the Cape of Good Hope to St. Helena and Ascension islands (Sulloway, 1982c : 327–37).

[4] Hodge (1983 : 62–63), in contrast to certain other Darwin scholars cited in note 2, believes that Darwin's expression "I must suspect they [the mockingbirds] are only varieties" is a statement *favoring* transmutation, since the formation of local varieties is the first step in the process of transmutation. Even within the creationist system, however, species were widely held to vary at the subspecific level according to local conditions. Hence the only kind of evidence that could ever convincingly shatter the creationist theory was solid proof that local varieties, under certain circumstances, were capable of breaking the immutable 'species barrier'. Darwin's reluctance to accept this conclusion when writing about the Galapagos mockingbirds in mid-1836 is reinforced by the fact that the relevant *Ornithological Notes* passage, drafted nine months after Darwin's zoology diary entry for these birds, has actually backed away from his initial judgment that "This bird...is singular from existing as varieties or distinct species in the different Is^ds. . ." (DAR 31.2 : MS p. 341; all DAR numbers refer to the Darwin manuscript collection at Cambridge University Library). In short, without the backing of an expert taxonomic opinion, Darwin, in mid-1836, was clearly giving the benefit of the doubt to the prevailing dogma of the fixity of species. Hodge's (1983) differing opinion about this passage is based on his erroneous assumption that Darwin, in 1836, already knew about the highly endemic nature of the Galapagos flora and

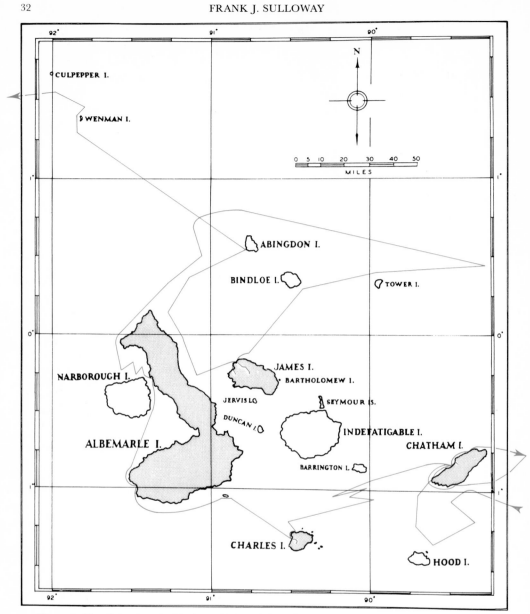

Figure 1. The Galapagos Archipelago. (Adapted from Lack, 1945: Frontispiece.) Of the sixteen principal islands, Darwin visited the four shaded ones, proceeding via the route indicated on the map. Although Darwin glimpsed Hood, a low island, from a fairly close distance (16 September), he did not disembark there. At Chatham, the first island on which Darwin landed, he spent five days collecting and geologizing, mostly on the western side of the island (16–22 September). There he visited four different localities, including a brief landing near Wreck Point on the southwestern tip of the island (16 September); Stephens Bay (17 September); Terrapin Road, near the northern tip of the island (18 September); a possible brief landing at Freshwater Bay, on the southeastern side of the island (20 September); and, after returning once again to Stephens Bay (21–22 September), an area several miles to the northeast that is studded with small volcanic cones (Fig. 2). En route to Charles Island (23 September), which Darwin visited next, he would have had a good glimpse of the southern side of the low but steep-cliffed Barrington Island. On Charles Island, Darwin collected for one day in the area around Post Office Bay (24 September). On 25 and 26 September he visited the highlands, going first by longboat to Black Beach, where he

understand *why* Darwin was hesitant to see his four Galapagos mockingbird specimens in an evolutionary light, it is necessary to review the impression that various other Galapagos organisms had already made upon him during his five-week visit to these islands.

THE TORTOISES

The reported differences among the various island populations of the Galapagos tortoise are particularly relevant in this connection. Darwin was first informed of the possibility that the numerous islands of the Galapagos group

---

(note 4, *contd.*) fauna (see page 45). Darwin, Hodge therefore argues, assumed all these endemic Galapagos forms—including the debatable island "varieties"—had arisen by colonization and subsequent transmutation. Years later, Darwin *did* consider the presence of varieties or "doubtful species" on the different islands of archipelagoes as supporting the theory of evolution (1909[1844]: 160, 187, 197; 1845 : 397; 1859 : 404). But in contrast to his thinking in 1836, when he still did not know whether the Galapagos organisms were endemic or whether the Galapagos mockingbirds, tortoises, or Falkland foxes would be considered by expert systematists as more than local varieties, Darwin had long possessed the decisive "facts" (or rather the sworn judgments of reliable systematists) that were ultimately necessary to "undermine the stability of Species".

What Darwin had in mind when writing the crucial *Ornithological Notes* phrase "*such facts* would undermine the stability of Species" was the common nineteenth-century view that all *constant* forms not united to one another by intermediate morphological links are presumably 'good' species. Since Darwin's day this strictly 'morphological' species concept has been rejected in favor of a more flexible, 'biological' species concept. Darwin, who himself later adopted the biological species concept after he became an evolutionist (see note 13), was clearly aware in mid-1836 that a purely morphological species concept was not apparently suitable to assessing the Galapagos mockingbirds and tortoises unless one was also willing to give up the whole foundations of pre-Darwinian biology.

Once again, Darwin was more inclined to stretch his species concept a bit rather than let four mockingbird specimens, collected from a small and perhaps unrepresentative series of islands within the Galapagos group, overthrow the doctrine of the immutability of species. At present some ornithologists still consider the Galapagos mockingbirds to be a single species (*Nesomimus trifasciatus*) possessing nine insular races (see Davis & Miller, 1960 : 447–48).

---

followed a trail to the settlement. On 27 September, after briefly reboarding the *Beagle* at Black Beach, Darwin ascended Saddle Mountain, the highest point on the island. The *Beagle*, leaving Charles Island on the morning of 28 September, subsequently anchored for part of a day in Iguana Cove on the southwestern tip of Albemarle Island (29 September). Darwin, however, disembarked only at Tagus Cove (1 October), mainly exploring the geological terrain a few miles to the south and collecting plants. The *Beagle* then proceeded toward Abingdon Island, where strong currents caused the ship to drift twenty miles to the northwest before favorable winds allowed the vessel to sail within close sight of Tower and Bindloe islands (4–7 October). After passing Bindloe Island, Darwin and three other men were landed on James Island, while the *Beagle* returned to Chatham, Hood, and Charles islands. On James Island Darwin and his companions set up camp for nine days (8–17 October) at Buccaneer Cove, just a few miles north of James Bay. From this base camp Darwin made two trips to the highlands, where he spent three days collecting (9 and 12–13 October). Darwin also took a trip by longboat to James Bay, where he examined the lava flows and the salt mine (11 October). The *Beagle*, after picking up Darwin and his three companions at Buccaneer Cove, surveyed the northeastern side of Albemarle Island (17–18 October). The following day the *Beagle* met up with its surveying yawl at Abingdon Island; and finally, on 20 October, the ship completed its work in the Galapagos Archipelago with a survey of the two northernmost islands (Wenman and Culpepper). Altogether, Darwin spent nineteen days—some only in part—on land in the Galapagos Islands, and another eighteen days on board the *Beagle* during the ship's surveying activities and movements from island to island. Although Darwin landed on only four islands, he had relatively good views of eight of the other twelve main islands of the Galapagos group, including all six of the largest volcanic craters. (I have reconstructed the *Beagle*'s movements from Admiralty survey charts L 945–950, 953–954, 958, and map 1375 [Hydrographic Department, Taunton]. In doing so, I have not included certain minor perturbations in the *Beagle*'s general route.)

Figure 2. The heavily cratered region northeast of Stephens Bay that Darwin visited on 21 September 1835. In his *Journal of Researches* Darwin described this area as "part of the island where some black cones—the former chimneys of the subterranean heated fluids—were extraordinarily numerous. From one small eminence, I counted sixty of these truncated hillocks, which were all surmounted by a more or less perfect crater. . . . From their regular form, they gave the country a *workshop* appearance, which strongly reminded me of those parts of Staffordshire where the great iron-foundaries are most numerous" (1839:455). Photographed by the author on Chatham Island.

might be tenanted by slightly different forms when Nicholas O. Lawson, the vice-governor of the islands, told Darwin that "the tortoises differed from the different islands, and that he could with certainty tell from which islands any one was brought" (1845:394). "I did not for some time," Darwin later asserted, "pay sufficient attention to this statement, and I had already partially mingled together the collections from two of the islands. I never dreamed that islands, about fifty or sixty miles apart, and most of them in sight of each other, formed of precisely the same rocks, placed under a quite similar climate, rising to a nearly equal height, would have been differently tenanted. . . . [B]ut I ought, perhaps, to be thankful that I obtained sufficient materials to establish this most remarkable fact in the distribution of organic beings" (1845:394). What Darwin did not go on to relate in his *Journal* account of this episode, however, are the various reasons that caused him initially to disregard the vice-governor's comments.

The key to Darwin's oversight lies in the specific name—*Testudo indicus*—by which the Galapagos tortoise was known at the time. In the 1830s two different species of giant land tortoise, one now known to have come from the Aldabra Islands in the Indian Ocean and the other from the Galapagos, had been confused under this name. This error in systematics had in turn encouraged the mistaken belief that the Galapagos form of giant tortoise was not actually native to these islands but had been transported there by buccaneers and, earlier, by the oceangoing peoples of the Pacific islands. Captain Robert FitzRoy reiterated this view in his own published account of the *Beagle* voyage. There he emphasized that virtually no animal was more suited for extensive oceanic transport, since the giant land tortoise was easily caught and good to eat, and required little food or water for long periods (1839:505). FitzRoy also cited the views of the buccaneer William Dampier, who claimed to have seen another variety of this species in Madagascar and elsewhere in the Indian Ocean (1729:102). Like Dampier, FitzRoy had no doubt that the Galapagos form of tortoise was a mere variety of this other race, slightly altered by removal to its new environment.

It was this widespread confusion regarding the original habitat of the Galapagos tortoise that apparently caused Darwin, like FitzRoy, to dismiss the reported island differences as a phenomenon readily explained by the changes in form that customarily accompany an animal's introduction into a new and dissimilar country. Since *Testudo indicus* was already known to be a single species, Darwin seems to have concluded that the differences found on the various islands of the Galapagos were merely varietal peculiarities somehow dependent upon the harsh and by no means identical conditions of each island.[5]

Darwin's attitude toward the reported differences among the tortoises was

---

[5]Although Darwin later stressed in his *Journal of Researches* (1845 : 394, 397) how similar all the islands of the Galapagos were, he was fully aware at the time of his visit that these islands varied considerably in size, height, terrain, availability of fresh water, and degree of vegetation. For example, in a voyage specimen catalogue he referred to Chatham Island as "a dry Isld" compared with some of the other islands (Porter, 1980 : 87). Moreover, Darwin knew that such geographic and climatic differences were occasionally associated with local variations of an apparently subspecific nature among the organic inhabitants of the islands. In his voyage *Diary* he recorded that "those [plants] of the same species" attained a much greater size on James Island than elsewhere in the archipelago (1933 : 340). Similarly, he noted that marine iguanas grew to a larger size on Albemarle Island than on other islands he had visited (DAR 31.2 : MS p. 333). At the time, none of these observations prompted any particular surprise or evolutionary speculations on Darwin's part.

reinforced by one other circumstance that has gone unrecognized in connection with his visit to the Galapagos. The first island that Darwin visited, and the place where he saw his first tortoise, was Chatham Island (Fig. 3). From there the *Beagle* proceeded to Charles Island, where the vice-governor was residing. The Charles Island tortoise, like the nearby Hood Island variety, has its shell turned up in front like a Spanish saddle. This is an adaptation found on the smaller and drier islands, allowing the tortoise to stretch its neck much higher in search of food (Fig. 4). At the time of the *Beagle*'s visit to Charles Island, this distinctive saddleback tortoise was nearly extinct, and apparently no live ones were seen by Darwin or FitzRoy. Nevertheless, Darwin did have at least one opportunity to observe the unusual Charles Island form of tortoise shell. Carapaces were readily visible at the settlement, where they were being used as flower pots (FitzRoy, 1839:492). Unfortunately, neither Darwin nor FitzRoy thought it important to procure a specimen for scientific purposes, or even to record the form of the shell as compared with the Chatham Island form. Within about ten years of the *Beagle*'s visit, the Charles Island tortoise became extinct, and herpetologists had to wait nearly a century to find remains of this subspecies in a lava cave (Broom, 1929).

From Charles Island Darwin proceeded to Tagus Cove on Albemarle Island, where he spent only part of a day on shore (1 October) and did not see any tortoises. Then on 8 October, Darwin, his servant, and three other men were left on James Island for nine days in order to collect specimens from this large and central location. In the highlands on James Island Darwin saw many tortoises. As luck would have it, the James Island tortoise is fairly similar to the Chatham Island race, the only other form that Darwin had personally seen (Fig. 5). Both have a carapace that is relatively dome-shaped, the other of the two morphological extremes found in the archipelago. Darwin, noticing no real difference based on his memories of the Chatham Island tortoise, probably concluded that whatever distinguishing features there were could not be all that pronounced. In fact, herpetologists can by no means tell at a glance what island any Galapagos tortoise is from, and the vice-governor's claim was something of an exaggeration. For example, of the more than fifty tortoises that zoos have recently repatriated to the Galapagos in order to assist in the tortoise-rearing and conservation program, only one has thus far been identified by island.

Darwin had one last opportunity, after his departure from James Island, to follow up the vice-governor's claims about the tortoises. FitzRoy, returning to Chatham Island for fresh water while Darwin was collecting specimens on James, had taken on thirty large tortoises to be stored in the ship's hold as a supply of fresh meat during the *Beagle*'s cruise across the Pacific (1839:498). But Darwin and his *Beagle* shipmates gradually ate their way through the evidence that eventually, in the form of hearsay, was to revolutionize the biological sciences. Regrettably, not one of the thirty Chatham Island carapaces reached England, having all been thrown overboard with the other inedible remains.

Two small tortoises, evidently kept as pets by Darwin and his servant, did survive the *Beagle* voyage. When Darwin, back in England, finally realized the necessity for having an expert compare the various forms of tortoise, these, together with two specimens that FitzRoy had procured for the British Museum, were his only remaining sources of evidence. Although the four tortoises were from three different islands (Hood, Charles, and James), those brought home by

Figure 3. The Chatham Island tortoise (*Geochelone elephantopus chathamensis*), a relatively dome-shaped form. Photographed by the author in the interior of northwestern Chatham Island, near where Darwin did most of his collecting and geological observations. Three days after a hunting party had brought eighteen live tortoises on board the *Beagle*, Darwin went for an extensive geological walk (21 September), during which he first saw Galapagos tortoises in the wild. He later described this episode in his *Journal of Researches:* "The day was glowing hot, and scrambling over the rough surface and through intricate thickets, was very fatiguing; but I was well repaid by the strange Cyclopean scene. As I was walking along I met two large tortoises....These huge reptiles, surrounded by the black lava, the leafless shrubs, and large cacti, seemed to my fancy like some antediluvian animals" (1845:374–75).

Figure 4. The Hood Island tortoise (*Geochelone elephantopus hoodensis*), an extreme saddleback form of tortoise similar to the now-extinct Charles Island race (*G. elephantopus galapagoensis*). Photographed by the author at the Charles Darwin Research Station, Isla Santa Cruz.

Darwin and his servant were too young to allow any meaningful scientific comparison (1839:465).

It was also upon his return to England that Darwin was told by Thomas Bell and other herpetologists that, in their opinion, the Galapagos tortoise was almost certainly native to that archipelago and, furthermore, that at least two species were endemic there. Once informed of this opinion, Darwin soon realized the evolutionary significance of the Galapagos tortoise evidence. But it was not

Figure 5. The James Island tortoise (*Geochelone elephantopus darwini*), a dome-shaped form. Photographed by the author in the intermediate zone, three miles southeast of the salt mine on James Island.

until the second edition of his *Journal of Researches* (1845:394) that he was finally able to describe the different dome-shaped and saddleback carapaces based on Captain David Porter's informative report thirty years earlier (1815, *1*:215). So it took Darwin nearly a decade to resupply the scientific evidence that he had initially allowed to slip through his fingers when he visited the Galapagos Islands in 1835.

DARWIN'S FINCHES

Analogous to the legendary roles assigned to the Galapagos tortoises and mockingbirds, it has widely been claimed that Darwin's finches played a key role in converting Darwin to the theory of evolution.[6] According to David Lack (1947:23), Darwin began to separate his finches by island shortly after hearing the vice-governor's testimony on Charles Island that the tortoises from the different islands could be differentiated. Lack based his assertion on certain of Darwin's own statements about his collecting procedures, as well as on the fact that many of Darwin's type specimens at the British Museum (Natural History) are labeled as coming from James Island, the last of the four islands Darwin visited. Still, Lack himself had doubts about the accuracy of a number of

---

[6]See, for example, Swarth, 1931 : 10; Huxley, 1954 : 6; 1960 : 9; Eibl-Eibesfeldt, 1961 : 18; Eiseley, 1961 : 172–73; de Beer, 1963 : 132; Peterson, 1963 : 11–12; Moody, 1970 : 303; Leigh, 1971 : 136; Thornton, 1971 : 12, 161–62; Grzimek, 1973 : 359; Gruber & Barrett, 1974 : 130; Dorst, 1974, *2*: 252; Silverstein, 1974 : 505; Dobzhansky *et al.*, 1977 : 12; Kimball, 1978 : 587; and Ruse, 1979: 164; 1982 : 24, 115.

Darwin's localities—including specimens from the one island (James) where Lack believed Darwin had carefully segregated all of his finch specimens. Contrary to the impression Darwin may have given Lack and others, he derived virtually all of his locality information for the Galapagos finches by borrowing, after his return to England, the carefully labeled collections of three other *Beagle* shipmates (Sulloway, 1982a; 1982b). This circumstance has nevertheless served, in a highly ironic manner, to reinforce the myth that Darwin's conversion to the theory of evolution occurred during his visit to the Galapagos. Curators at the British Museum naturally assumed that Darwin's published locality information had come from his *own* finch specimens. Hence, whenever Darwin, in the *Zoology of the Voyage of H.M.S. Beagle* (1841), indicated that a species had come from one island only, the curators subsequently entered that island on the labels of Darwin's type specimens. As time went on, it increasingly appeared that Darwin must have appreciated the evolutionary significance of his Galapagos finches while he was still in the archipelago, if he had gone to the trouble of segregating most of his specimens by island.

On the other hand, this circular derivation of the localities of Darwin's types largely accounts for the taxonomic nightmare these specimens have caused later ornithologists (e.g., Swarth, 1931; and Lack, 1945; 1947). The dubious localities entered on many of Darwin's type specimens even gave rise to vigorous debates about whether certain subspecies of Darwin's finches had evolved since his visit to these islands! Fortunately, clarification of the borrowed nature of Darwin's localities has resolved these debates, together with certain related problems concerning the systematics and geographic distribution of two now-extinct subspecies of Darwin's finches.

Elsewhere I have identified and described in detail the four separate *Beagle* collections of Galapagos specimens and, through caliper measurements or manuscript evidence, have been able to supply an accurate island locality for almost every *Beagle* specimen of Darwin's finches (Sulloway, 1982b). From these measurements and manuscript records—especially Captain FitzRoy's meticulously kept catalogue of his own official voyage collection—it is now evident that *Geospiza magnirostris magnirostris*, the extinct and unusually large-billed subspecies of the large ground finch, was once endemic to two islands—Chatham and Charles.[7] This extinct form was never present, as some ornithologists have believed, on James Island or any of the other northern islands of the Galapagos group. Of additional ornithological interest, John Gould's (1837a) "*G. nebulosa*" is indeed, as David Lack (1947:23) strongly suspected, an extinct and unusually large-billed form of the sharp-beaked ground finch (*G. difficilis*) once endemic to Charles Island. Besides the two surviving *Beagle* specimens, I have recently located a third and virtually identical specimen, collected on Charles Island in 1852 during the voyage of the Swedish frigate *Eugenie*. Steadman (1982) has also found six fossil specimens of

---

[7]At least 90 percent of the *Beagle* specimens of *Geospiza magnirostris magnirostris* are distinguishable from *G. magnirostris strenua*, thus more than satisfying David Lack's criterion of subspecific status (1947 : 17). David Steadman (1981) has recently found fossilized remains of *G. magnirostris magnirostris* on Charles Island, where, prior to its extinction, it was by far the most common species of Darwin's finches in the arid zone. If suitable paleontological sites can be found on Chatham Island, fossilized remains of this subspecies should also be encountered there.

this large-billed subspecies on Charles Island. In the light of these facts, the name *nebulosa* (Gould, 1837a) has valid priority over the name *difficilis* (Sharpe, 1888) and accordingly should be restored as the official name of this taxon (Sulloway, 1982b:69–70).[8]

The imprecise nature of many of Darwin's Galapagos finch localities is a large part of the reason why Darwin, contrary to the legend, never mentioned these birds in the *Origin of Species* (1859).[9] For without more accurate and extensive locality information, it was simply not possible to put these finches forward as a convincing example of speciation through geographic isolation. The two cases of geographic replacement among the finches that Darwin tentatively argued for in his *Journal of Researches* (1845:395) are, in fact, spurious, being based on inaccurate locality records and insufficient collecting. It is no wonder, then, that Darwin was so excited and relieved in 1845 when he finally learned the results of Joseph Hooker's rigorous demonstration of geographic representation in Darwin's several hundred species of Galapagos plants. To Hooker he responded in July of that year, "I cannot tell you how delighted and astonished I am at the results of your examination; how wonderfully they support my assertion on the differences in the animals of the different islands, *about which I have always been fearful*" (1887, 2:22; italics added).[10] Darwin's subsequent decision not to use the Galapagos finches to bolster his evolutionary argument in the *Origin* proved well advised. After Drs. Kinberg and Habel brought back extensive collections of birds from the Galapagos in the 1850s and 1860s, Osbert Salvin was forced to comment that "Mr. Darwin's views as to the exceedingly restricted range of many of the species must be considerably modified" (1876:461). By then, however, Darwin's evolutionary views, and his generalized claims about the Galapagos, had won the day, owing largely to Joseph Hooker's indisputable botanical evidence.

As for the claim that Darwin was immediately impressed by the morphology of the finches as a classic case of adaptive evolutionary radiation, nothing could

---

[8]The international code of zoological nomenclature states that the name of the first described subspecies shall be the specific name of all subsequently described subspecies of the same species-group. The convention of disallowing otherwise valid names that have not been used for more than fifty years is not applicable in this instance, since *G. nebulosa* is by no means a forgotten or overlooked name. The name has been used by Lack (1945, 1947, 1969) and other ornithologists, throughout the last three decades, when referring to the Charles Island subspecies of the sharp-beaked ground finch; and it has also been proclaimed by Paynter & Storer (1970 : 162) to be the obligatory name of this taxon should Lack's views be confirmed and accepted. Similarly, clarification of the long-disputed localities of the *Beagle* specimens of *G. magnirostris magnirostris* and *G. magnirostris strenua* has now legitimated beyond any doubt the tentative trinomials proposed by Lack (1947, 1969) and Paynter & Storer (1970 : 161n.). See further Sulloway (1982b: 68, n.53).
[9]The following authors, for example, have mistakenly assumed that Darwin honored the Galapagos finches with a prominent place in the *Origin*: Lack, 1945: 4; Gillsäter, 1968 : 85; and Ruse, 1982 : 115. See also Williamson (1983 : 566), who perpetrates a slightly modified version of this myth (Sulloway, 1983). The various island forms of the Galapagos tortoise are also not mentioned in the *Origin*, presumably because Darwin remained rightfully doubtful about whether the different populations could really be considered separate species. Today the eleven living and four extinct forms of the Galapagos tortoise are considered, just as Darwin himself suspected, to be subspecies of a single species (*Geochelone elephantopus*). See Thornton (1971 : 115).
[10]Darwin's comment that he had "always been fearful" about his previous claims with regard to the differences between the animals on the various islands of the Galapagos group clearly shows that he believed *specific*, not *varietal*, distinctions were necessary to support the doctrine of the mutability of species. Darwin's 1845 remark to Hooker therefore reinforces the interpretation I have given earlier concerning Darwin's *Ornithological Notes* (1963[1836] : 262) passage about the "varieties" of the Galapagos mockingbirds. See note 4.

be further from the truth.[11] While in the Galapagos Islands Darwin was more impressed by the apparent differences than by the similarities among these unusual finch species. At the time, he actually believed he was dealing with a highly diverse group having at least three or four separate subfamilies. For example, Darwin identified the cactus finch in his voyage notes as an "Icterus", a genus in the family of the orioles and blackbirds; and he mistook the warbler finch for a "Wren" or warbler. In fact, Darwin correctly identified *as finches* only six of the thirteen species—less than half the present total—and he placed these six species in two separate groups of large-beaked and small-beaked Fringillidae. Thus, it was only after his return to England, when the eminent ornithologist John Gould astutely recognized the close affinities of this bird group, that Darwin's finches became Darwin's *finches* in the sense that we now comprehend.

Furthermore, with the exception of the cactus and warbler finches, Darwin had failed to observe any differences in diet among the various species, mistakenly believing that their food consumption was largely identical (1841:99–100). For this reason he could never argue in the *Origin* that the different beaks were necessarily adaptive and hence produced by natural selection. Ornithologists continued to discount the adaptive nature of the beak differences in Darwin's finches for nearly a century.

It was in the wake of Darwin's failure to appreciate the evolutionary significance of the tortoises and the finches that he was also inclined to reject such an interpretation for the mockingbirds and hence to suspect, in the *Ornithological Notes* (1963[1836]) passage I have already cited, that these birds must be "only varieties". This brings me to another more general point. What was crucial to Darwin's eventual decision to accept the mutability of species was not really the mockingbirds or any other single group of Galapagos organisms. Rather, his appreciation of the mockingbirds required his concomitant appreciation of the Galapagos evidence as a whole. This was not possible, however, until he had been freed of the many misconceptions he had entertained about these unusual organisms during his brief visit to these islands.

DARWIN'S CONVERSION

Darwin's conversion to the theory of evolution can, I believe, be dated with considerable certainty to the second week of March 1837, nearly a year and a half after Darwin had departed from the Galapagos Islands, and five months following his return to England (Sulloway, 1982c). What finally catalyzed Darwin's conversion was a meeting he had at the Zoological Society of London with John Gould, the society's eminent ornithologist and taxidermist.

At Cambridge University Library there exists a piece of paper on which Darwin, in a somewhat hurried hand, recorded the various taxonomic conclusions that Gould had already reached by early March regarding Darwin's Galapagos birds (Figs 6, 7). On this same sheet Darwin later added various memoranda that clearly indicate the revolutionary new direction his thoughts

---

[11] Most of the authorities cited in note 6 make this claim, sometimes emphasizing, in addition, how impressed Darwin must have been by the habits of the tool-using 'woodpecker' finch (*Camarhynchus pallidus*). In actual fact, Darwin never saw this species, which was first collected in 1868, nine years after Darwin published the *Origin of Species*. The earliest report of the woodpecker finch's remarkable tool-using behavior had to wait another fifty years (Gifford, 1919:256).

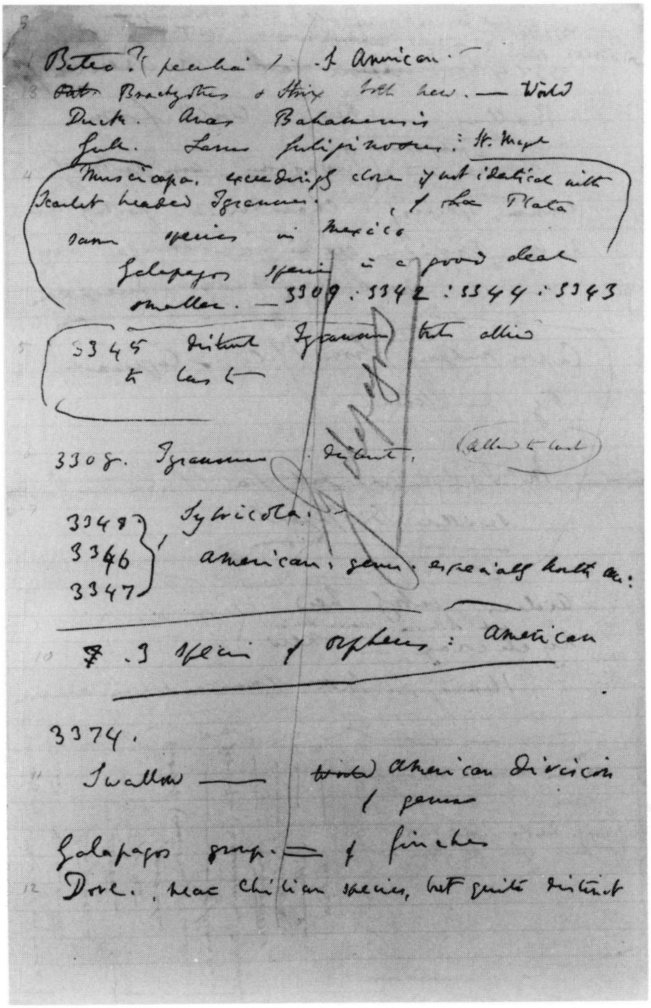

Buteo ? peculiar / of American
Brachyotus & Strix both here. — World
Duck. Anas Bahamensis
Gull. Larus fuliginosus: St. Magd
Muscicapa. exceedingly close & not identical with Scarlet headed Tyrannus of La Plata
same species in Mexico
Galapagos species is a good deal smaller. — 3309 : 3342 : 3344 : 3343
3345 distinct Tyrannus) but allied to last
3308. Tyrannus distinct. (allied to last)
3348, 3346, 3347 Sylvicola. american genus. especially North Am:
7. 3 species of Orpheus : american
3374. Swallow — North American division / genus
Galapagos group. — of finches
Dove. near Chilian species, but quite distinct

Figure 6. Darwin's manuscript record of John Gould's Galapagos species designations (*front*). (Courtesy of the Syndics of Cambridge University Library.)

began to take following his meeting with Gould. The meeting, and the document recording what was discussed, can be dated to within a six-day period, namely, 7 to 12 March, immediately after Darwin had moved from Cambridge to London in order to be near the various systematists who were working on his *Beagle* collections.[12] The document (Fig. 6) begins with a listing

[12]As early as 4 January 1837, when Darwin delivered his collection of birds and mammals to the Zoological Society, he had evidently discussed his Galapagos birds in general terms with Gould. In a letter of 9 January 1837 to William Jardine, John Stevens Henslow reported that "Darwin has just returned from a visit to London & tells me that Gould pronounces all his Galapagos animals to be so entirely novel & curious that he will undertake an express work to illustrate them" (Jardine letters, Royal Scottish Museum, Edinburgh). It was not until March 1837, however, that Darwin learned the full details of Gould's taxonomic judgments, as recorded on Darwin's handwritten "Galapagos" sheet (Figs 6, 7). I have discussed the dating of this document (DAR 29.3 : 27), as well as the timing of Darwin's conversion, more fully in Sulloway (1982c).

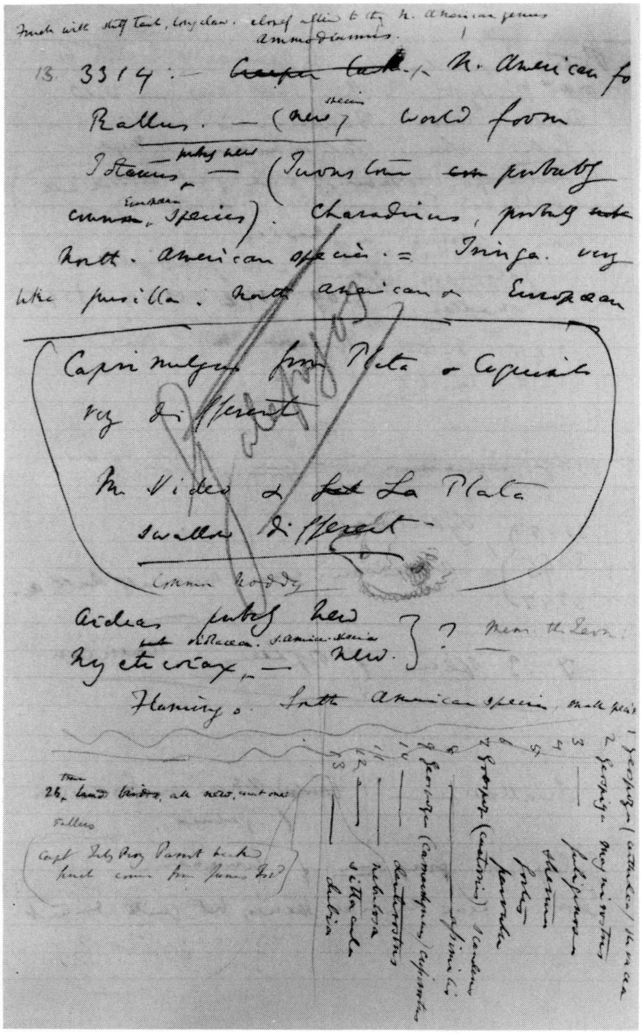

Figure 7. Darwin's manuscript record of John Gould's Galapagos species designations (*back*). (Courtesy of the Syndics of Cambridge University Library.)

of the Galapagos land birds in almost precisely the same order, and with the same commentaries about their American continental alliances, as in Darwin's *Journal* (1839:461–62). The oral nature of Darwin's discussion with Gould is indicated by Darwin's omission, within the list of finches, of the silent '*p*' in his spelling of the name *psittacula* (Fig. 7). As Gould began to tell Darwin about the Galapagos birds, I believe that Darwin, whose interests were undoubtedly aroused by Gould's remarks, asked for a sheet of paper on which to record the details of their discussion. Since someone, perhaps Gould, had already made a pencil sketch of a small animal on one side of the sheet, Darwin began writing on the unused side, proceeding later to the used side of the sheet, eventually writing over the drawing.

From the very first entry on the list—"Buteo?"—it is clear that Darwin was in for a surprise during his meeting with Gould. In his voyage ornithological notes, Darwin had previously classified the Galapagos hawk among the caracaras or Polybori owing to the bird's unusual tameness and carrion-feeding habits (1963[1836]:238, 262). During a 24 January meeting of the London Zoological Society, John Gould—not knowing of Darwin's voyage observations about the bird's behavior—had rightly assigned this new species to the Buteos; and Gould's assignment is why Darwin, who was initially unpersuaded, entered a question mark after Gould's judgment. Based, however, on Darwin's testimony about this hawk's habits, Gould subsequently convinced himself that the species shared various morphological characters with both the Buteos *and* the Polybori (1837b). Not long afterwards, in his first notebook on the transmutation of species, Darwin correctly recognized this Galapagos species as a case of convergent behavioral evolution. Thus he wrote of the "principle of animal having come to island where it could increase, but there were causes to induce great change, like the Buzzard [*Buteo*] which has changed into Caracara at the Galapagos" (de Beer *et al.*, 1967: B 55e). Later, in the second edition of the *Journal of Researches*, Darwin hinted publicly at his evolutionary interpretation of this unusual species when he remarked that "it might be fancied that a bird originally a buzzard, had been induced here to undertake the office of the carrion-feeding Polybori of the American continent" (1845:380).

As John Gould continued to discuss the various land birds recorded on Darwin's March 1837 "Galapagos" sheet, he emphasized that virtually all of the land birds were endemic to the Galapagos, although clearly of American character. This was a fact that Darwin, who had not visited the coast of South America north of Lima, could not have fully appreciated without access, like Gould, to large museum collections. Darwin, who was evidently very impressed by this information, made a special note of it at the bottom of the verso side of the sheet ("26 true land birds, all new except one"). Even several of the waders and waterbirds were new, Gould also informed Darwin—something that likewise surprised Darwin, and to which he later called special attention in his *Journal of Researches* (1839:461). Equally startling was the conclusion, backed by Gould's expert authority, that some of the Galapagos species (namely, the mockingbirds) indeed replaced one another geographically on the different islands of the Galapagos group. It was facts such as these, Darwin had already acknowledged eight months earlier, that, if ever confirmed, "would undermine the stability of Species". Finally, Gould must surely have piqued Darwin's interests when he informed him that his Galapagos "Wren" and "Icterus" were actually finches, and that the whole group constituted a unique and unusual genus of thirteen species (*Geospiza*), comprising three separate but closely related subgenera (*Cactornis, Camarhynchus,* and *Certhidea*).

Darwin was frankly stunned by Gould's various taxonomic conclusions. On the one hand, Gould's judgments convinced him beyond a doubt that transmutation must be responsible for the presence of similar but distinct species on the different islands of the Galapagos group. The supposedly immutable 'species barrier' had finally been broken, at least in Darwin's own mind. At the same time Darwin was prompted by Gould and other systematists who examined and discussed Darwin's Galapagos species at the London Zoological Society, to realize just how little agreement there was, even among experts, as to what constitutes a 'species'. It was this experience that Darwin apparently had in mind when he later wrote in the *Origin of Species:* "Many years ago, when comparing, and seeing others compare, the birds from the separate islands of the Galapagos Archipelago, both one with another, and with those from the American mainland, I was much struck how entirely vague and arbitrary is the distinction between species and varieties" (1859:48).[13]

It was also in reference to the crucial changes in his thinking that occurred in the spring of 1837 that Darwin subsequently wrote in a pocket "Journal": "In July [1837] opened first notebook on 'Transmutation of Species'—Had been greatly struck from about Month of previous March on character of S. American fossils—and species on Galapagos Archipelago. These facts origin (especially latter) of all my views" (de Beer, 1959:7). Darwin's reference in this passage to the "character of S. American fossils" is clearly an allusion to Richard Owen's recent paleontological findings about Darwin's collections, communicated to Darwin in mid-February 1837. By that date, Owen had assertained that Darwin's fossil Mammalia were virtually all extinct prototypes of the smaller, present-day forms that now characterize this continent (llamas, sloths, armadillos, and so forth). But this information about the close bond between the living and the extinct fauna of the South American continent had apparently not, on its own, been sufficient to convince Darwin of the mutability of species—although he may have had growing suspicions in this direction; hence Darwin's later insistence in his pocket "Journal" that it was "especially" the Galapagos facts, fully clarified by Gould three weeks later, that had provided the "origin . . . of all my views".

Consistent with this historical interpretation, Darwin's first evolutionary entry in another pocket notebook—the *Red Notebook*, which precedes the transmutation series—can be dated to about 15 March. This date is within a few days of Darwin's meeting with John Gould and is also the day after he and Gould had delivered papers on the two species of South American rheas that replace each other geographically in Patagonia (Darwin, 1837; Gould, 1837c). Ironically, Darwin had actually been eating (and had almost finished consuming) his only specimen of the southern form of rhea, when he had suddenly recalled, in 1834, what the gauchos and indians had previously told him about the existence of a *second* species of rhea in South America. Darwin managed to salvage most of the bones and some feathers for Gould's later examination; and this half-eaten specimen became the type of the species, which Gould named *Rhea darwini* in honor of Darwin's energetic labors in natural history. In his *Red Notebook* Darwin now pondered whether the model of speciation by geographic isolation would suffice for such cases of continental speciation, thus beginning a long chain of evolutionary deliberations that were never to cease (Darwin, 1980[1836–37]:127–28).

---

[13]Soon after accepting the transmutation of species, Darwin adopted a biological species concept (based on the criterion of actually or potentially interbreeding populations) as a means of reconciling the 'reality' of species with their sometimes 'arbitrary' classification by different naturalists. In the late 1840s and early 1850s Darwin gave up this biological species concept in favor of a predominantly morphological concept—one that defined species as "only strongly marked and permanent varieties" (1859:469). This view was in many ways a 'tactical' species concept, aimed at convincing numerous frustrated systematists that species distinctions *are* arbitrary and hence that systematic work could never succeed in discovering the true 'essence' of each and every species (1859:48). See further Kottler (1978) and Sulloway (1979).

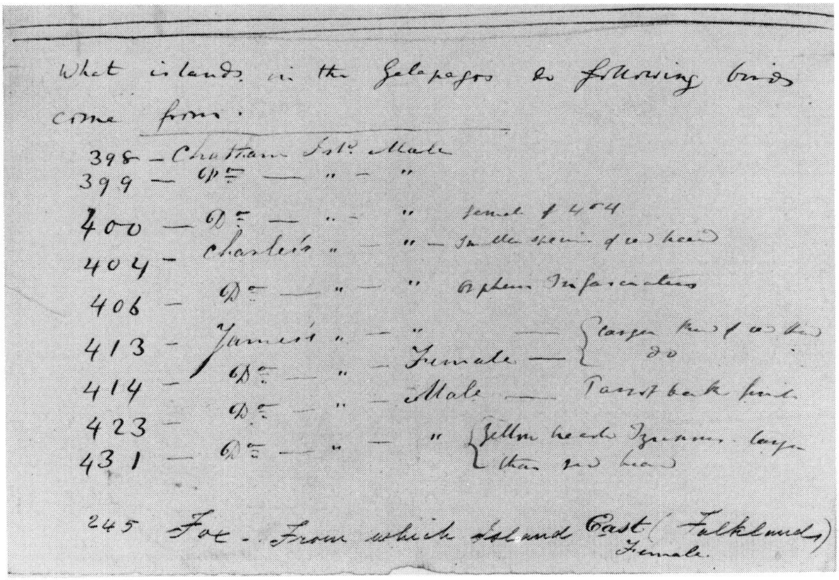
What islands in the Galapagos do following birds come from:
398 – Chatham Isl^d Male
399 – D^o — " — "
400 – D^o — " — "
404 – Charles's " — " —
406 – D^o — " — "
413 – James's " — " —
414 – D^o — " — Female
423 – D^o — " — Male
431 – D^o — " — "
245 Fox – From which Island East (Falklands) Female.

Figure 8. Darwin's request for information regarding the localities of Captain FitzRoy's Galapagos birds, with replies in the hand of an unidentified amanuensis. A second unidentified amanuensis, who is known to have worked for Darwin after the *Beagle* voyage, addressed the last question on the list, which was in turn answered by the first amanuensis. Additional memoranda, later added by Darwin, appear at the right of most of the entries. (Courtesy of the Syndics of Cambridge University Library.)

By mid-March of 1837, then, Darwin had finally come to accept the evolutionary implications inherent in his Galapagos collections; and he wasted little time in extending this evolutionary point of view to include patterns of geographic distribution among numerous other South American species, as well as the evidence of the fossil record. The *Red Notebook*, in which Darwin recorded most of these early evolutionary speculations, was completed by the end of June 1837 and was followed by a series of six notebooks on the transmutation of species (1837–42) and two additional notebooks devoted to man, mind, and comparative psychology (1838–39).[14] Although Darwin's famous Galapagos finches are not mentioned either in the *Red Notebook* or in any of the transmutation notebooks, it must have been during the spring of 1837 that Darwin realized what a great mistake he had previously made during his Galapagos visit when he had failed to separate his finches by island. This conclusion is borne out by two of the memoranda on Darwin's "Galapagos" list: "Capt Fitz Roy['s] Parrot beaked finch comes from James Isl^d", and "No specimens [*or* 'species'?] from James Island". This second observation, written in pencil and later erased, apparently refers to *Camarhynchus crassirostris* or to *Geospiza nebulosa* among the list of finches (Fig. 7). Darwin subsequently sought to rectify his failure to label his own Galapagos finch specimens by island, by collecting every scrap of available locality information from the accurately labeled collections of three other *Beagle* shipmates (Figs 8, 9), including that

[14]See Gruber & Barrett (1974) and Kohn, Smith & Stauffer (1982 : 419).

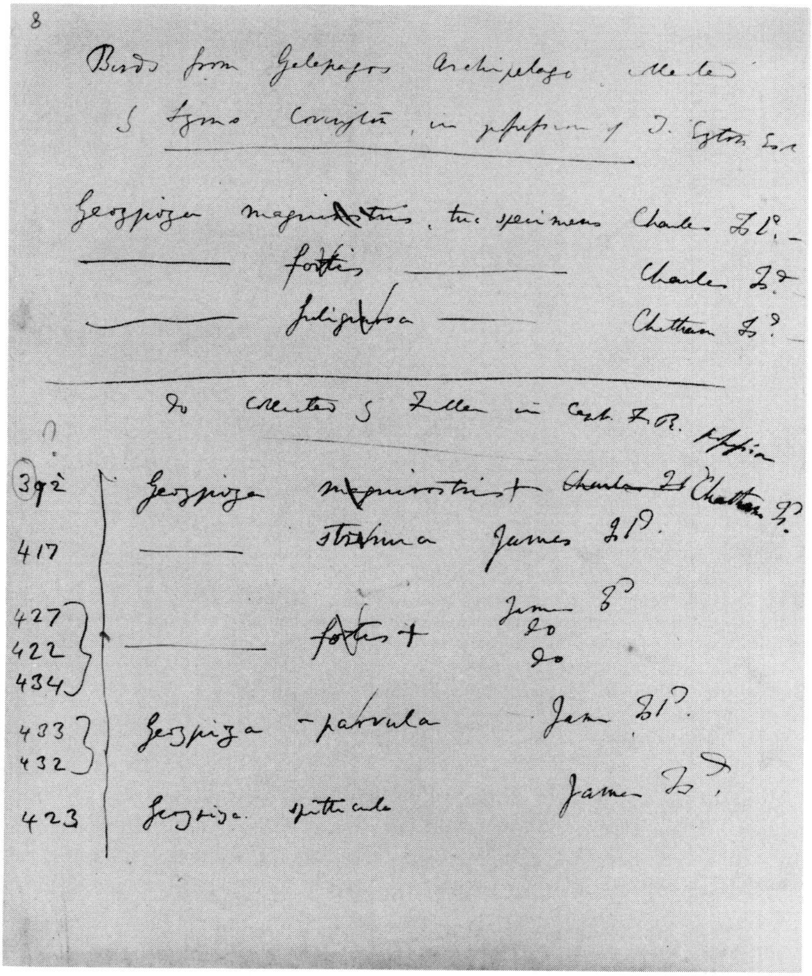

Figure 9. Darwin's notes on the island localities of his servant Covington's and another shipmate's Galapagos finches. (Courtesy of the Syndics of Cambridge University Library.)

of his own servant.[15] Darwin also tried to reconstruct, from memory, the localities of some of his own Galapagos finches (Fig. 10); but it is evident from bill and wing measurements of these specimens that, in at least two instances, he guessed incorrectly (Sulloway, 1982b:64).[16] Unfortunately, this dubious locality speculation—"¿Chatham Isd??"—for eight of his thirty-one finch specimens later became the basis for the official British Museum tags; and over the years Darwin's own question marks were dropped from most of these tags as the British Museum substituted various new labels for Darwin's originals. Still, Darwin, by collating locality information derived from the four separate *Beagle* collections of Galapagos finches (1841:100–106), was ultimately able to supply

---

[15]DAR 29.3 : 26, 30.
[16]See Darwin's voyage catalogue of specimens ("Printed Numbers Nos 1426 . . . 3342", Down House, Downe, Kent).

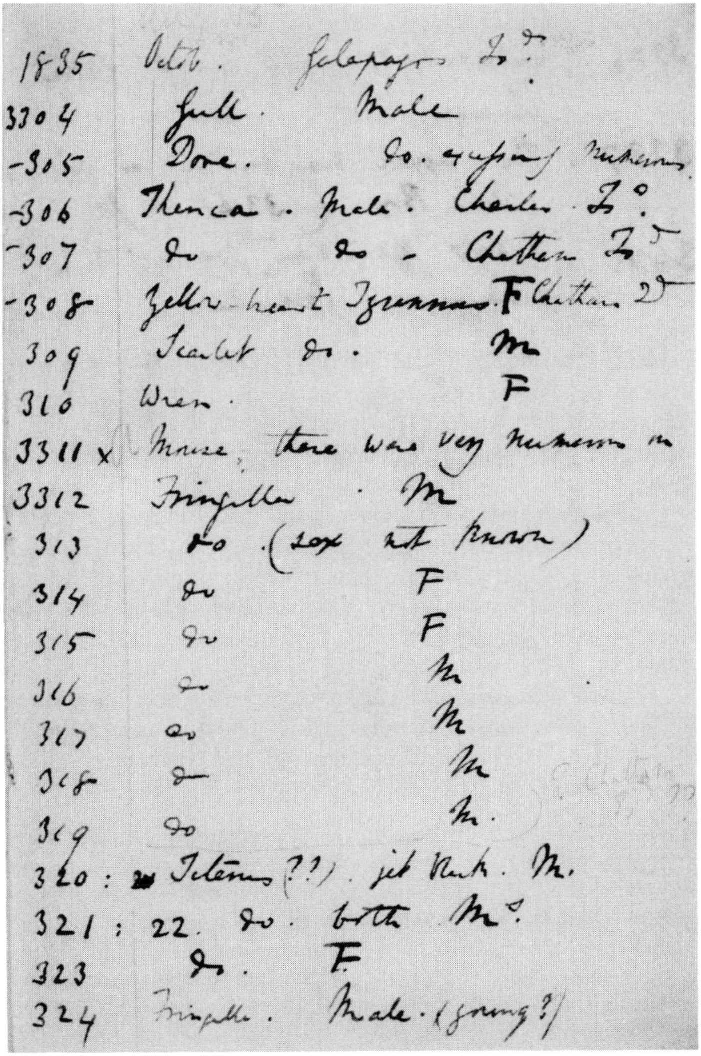

Figure 10. A page from Darwin's voyage specimen catalogue, recording some of the ornithological specimens collected in the Galapagos Archipelago. Specimen no. 3310 (*Certhidea olivacea*, mistakenly thought to be a "Wren") and nos. 3312–24 are Darwin's finches. Under the first eight "Fringilla", Darwin later drew a faint line in pencil and added the conjecture "?Chatham Is^d??" Because Darwin apparently remembered having first seen the largest-billed species of Galapagos finches (*Geospiza magnirostris magnirostris*) on Chatham Island and also remembered collecting specimens of the cactus finch—"Icterus (??)" = *G. scandens*—on James Island, he later guessed that the first eight "Fringilla" (mostly specimens of *G. magnirostris magnirostris*) must have come from Chatham. One must appreciate, however, that Darwin's entire cataloguing operation was done on the way to Tahiti, and that all of his finch specimens had already been thoroughly intermixed. What Darwin proceeded to do, then, was to catalogue his finches by general form rather than by island, beginning with the largest-billed birds. It is for this reason that he later confused two specimens of *G. magnirostris strenua*, collected on James Island, with specimens of *G. magnirostris magnirostris*, collected either on Chatham or Charles Island. (Courtesy of Down House and the Royal College of Surgeons of England.)

geographic distribution data for all but two of the thirteen species named by Gould.

FURTHER DEVELOPMENTS IN DARWIN'S UNDERSTANDING OF THE GALAPAGOS ISLANDS
(1837–1859)

In the first edition of Darwin's *Journal of Researches*, the text of which was completed in June of 1837 and sent to the printer in August, he aptly referred to the Galapagos Archipelago as "a little world within itself" and commented on how "very remarkable" its organic productions were (1839:454). But he limited himself to a detailed description of the islands and the various species he had collected there, making no statement about the heterodox evolutionary conclusions he had privately drawn from such evidence. In the second edition of his *Journal* (1845)—emboldened in part by Joseph Hooker's (1846, 1847a) findings of extensive island endemism among Darwin's Galapagos plants (Fig. 11)[17]—Darwin revised and greatly expanded, by nearly fifty percent, the text of his chapter on the Galapagos Islands. The finches of the Galapagos, four different species of which Darwin had illustrated in this second edition, were discussed more fully than in the first edition; and Darwin even hinted at his evolutionary interpretation of their closely allied nature when he remarked: "Seeing this gradation and diversity of structure in one small, intimately related group of birds, one might really fancy that from an original paucity of birds in this archipelago, one species had been taken and modified for different ends" (1845:380). Darwin also intimated, in several other veiled passages, the evolutionary conclusions that might be drawn from the Galapagos evidence.[18] It was in one of these interpolated passages that Darwin made perhaps his most famous pronouncement about the bearing of the Galapagos Islands on the question of the origin of species:

> The archipelago is a little world within itself, or rather a satellite attached to America, whence it has derived a few stray colonists, and has received the general character of its indigenous productions. Considering the small size of these islands, we feel the more astonished at the number of their aboriginal beings, and at their confined range. Seeing every height crowned with its crater, and the boundaries of most of the lava-streams still distinct, we are led to believe that within a period geologically recent the unbroken ocean was here spread out. Hence both in space and time, we seem to be brought somewhat near to that great fact—that mystery of mysteries—the first appearance of new beings on this earth.   (1845:377–78)

---

[17]Although not published until 1846 and 1847, the general results of Hooker's first two major papers on the plants of the Galapagos Islands were communicated to Darwin in early 1845, in time to be included in the second edition of Darwin's *Journal of Researches* (1845 : 395–97). George Robert Waterhouse's (1845) analysis of the 29 coleopterous insects collected by Darwin in the Galapagos also became available in time to be included in the second edition of the *Journal*. Of the 14 insects for which Darwin had supplied an island locality, Waterhouse found that each was confined to a single island. (Darwin was later able to provide island localities for some, but not all, of his coleoptera as an incidental result of having recorded habitat information in his zoology notes for certain of the specimens.)

[18]Perhaps the two most significant changes that were made in the second edition of Darwin's *Journal* were his great expansion of the Galapagos chapter (the text of the book as a whole had to be cut by about five percent) and the change that Darwin made in the book's title (he reversed the positions of the words "Geology" and "Natural History", to read *Journal of Researches into the Natural History and Geology of the Countries Visited during the Voyage of H.M.S. Beagle round the World . . .*). For further discussion of the changes between the first and second editions of Darwin's *Journal of Researches*, see Gruber (1981 : 259–99).

GALAPAGOS ARCHIPELAGO.

Name of Island.	Total No. of Species.	No. of Species found in other parts of the world.	No. of Species confined to the Galapagos Archipelago	No. confined to the one Island.	No. of Species confined to the Galapagos Archipelago, but found on more than the one Island.
James Island .	71	33	38	30	8
Albemarle Island	46	18	26	22	4
Chatham Island.	32	16	16	12	4
Charles Island .	68	39 (or 29, if the probably imported plants be subtracted)	29	21	8

Figure 11. Darwin's tabular presentation of Joseph Hooker's analysis of the flora of the Galapagos Islands. Darwin was particularly impressed by the frequency with which many endemic Galapagos genera, such as *Scalesia* and *Euphorbia*, are also endemic as distinct species on the separate islands (1845:396).

Darwin's publication of his solution to this "mystery of mysteries" had to wait another fourteen years, in part because Darwin himself still had not come to terms with certain unresolved issues in connection with his general theoretical argument. And the Galapagos case was an important part of the problem.

In particular, even when Darwin was making revisions for the second edition of his *Journal*, he had yet to appreciate the full significance of Joseph Hooker's botanical findings for explaining geographic speciation on the different islands of the Galapagos group. Thus he still maintained in 1845 that "neither the nature of the soil, nor height of the land, nor climate, nor the general character of the associated beings, and therefore their action on one another, can differ much in the different islands" (1845:397). Isolation, by implication, was the major cause of evolutionary change.[19] It was not until two years later, while reading Hooker's third major publication on the Galapagos flora (1847b), that Darwin finally grasped the full nature and significance of floral diversity between islands; namely, that this diversity is not just due to geographic speciation but is also caused by the frequent presence of plants on each island— random colonists—belonging to totally different genera and even families. In the margin of the relevant portion of Hooker's paper (1847b:239), Darwin wrote: "so the Flora of [the] different isl$^{d[s]}$ must be very different independently of representation".[20] This insight was to prove crucial to Darwin's subsequent

[19]This was the same position that Darwin had taken on the role of isolation in his Essay of 1844 (1909[1844]: 163, 168, 183, 189–90). See Sulloway (1979: 32–49) and Ospovat (1981: 193–200) for further discussion of Darwin's changing views on the role of geographic isolation in the speciation process.

[20]Darwin Reprint Collection, quarto item no. 36, Cambridge University Library. Curiously, Darwin had anticipated this insight on purely theoretical grounds as early as 1838, when he remarked in his second notebook on the transmutation of species: "N.B. (Isl^{ds} springing up [would be] more likely to have different species than those sinking. because arrival of any one plant might make condition[s] in any one isl^{d} different).—" (de Beer *et al.*, 1967: C 25e). Darwin again mentioned this idea in his Essay of 1844, although without altogether realizing its implications (1909[1844]: 187). It took Hooker's concrete and, even to Darwin, striking botanical evidence to allow this important insight to come to full fruition in 1847.

thinking about evolution in the Galapagos Islands. In fact, this idea contributed to a major shift in Darwin's evolutionary theorizing, a shift that eventually led to a significant new emphasis upon ecological factors as part of what Darwin later termed the "principle of divergence".[21]

Twelve years later (and twenty-four years after his brief visit to the Galapagos), Darwin's efforts to reconstruct the biological world within a comprehensive evolutionary framework culminated in the publication of the *Origin of Species* (1859)—"the book that shook the world" (Mayr, 1964:vii). In the *Origin,* Darwin devoted just 1.1 percent of the text to discussing the Galapagos Islands, barely more attention than he gave to New Zealand or the Madeiras. Pigeons, in comparison, merited three times as much text! As part of other supporting examples drawing upon oceanic islands, Darwin's Galapagos remarks were made in a number of different contexts—such as the organic bond of affinity between islands and the nearest mainland, the absence of certain classes of organisms on remote oceanic islands, and the high rate of species endemism on such islands. Perhaps more surprising than the seemingly minimal use Darwin made of the Galapagos case is that fully two-thirds of his Galapagos remarks were inspired by, and directly based on, Hooker's various botanical analyses of the islands' flora (1846, 1847a, 1847b). Why, Darwin asked in his chapter on geographical distribution, should islands "within sight of each other, having the same geological nature, the same height, climate, etc." be tenanted by different but closely related species? Could not these differences be seen, in fact, as an argument against his theory? To this issue, Darwin responded:

> This [problem] long appeared to me a great difficulty: but it arises in chief part from the deeply-seated error of considering the physical conditions of a country as the most important for its inhabitants; whereas it cannot, I think, be disputed that the nature of the other inhabitants, with which each has to compete, is at least as important, and generally a far more important element of success.   (1859 : 400–401)

Darwin employed this reasoning about the extensive biotic, and especially botanical, differences between the islands of the Galapagos to explain not only why natural selection would favor different varieties (or species) on the different islands but also why some forms, once established, would be able to exclude

---

[21]Ironically, Darwin's increased attention to ecological considerations in speciation—prompted as it was by Hooker's Galapagos publications—later led him to de-emphasize the role of isolation in the multiplication of species. Thus, while Darwin continued to believe that isolation was a primary factor in the evolution of island species, he mistakenly came to believe that speciation could occasionally occur on continents through partial (geographic) or ecological isolation (1859:103, 176–77). One consideration that apparently influenced Darwin's change in views was Hooker's finding that some plant genera, like *Scalesia,* possess more than one species on the different islands of the Galapagos group. In 1845 Darwin admitted to Hooker that such facts were "hostile" to his theory, and he wrote in a contemporary memorandum: "Several species of same genus w^d be apt to arise on same isl^d in proportion as that isl^d was badly placed for new colonists—for then they often w^d have to fill separate functions." Similarly, Darwin wrote at this time, "I must give up my crossing notions & advantage of Paucity of individuals. I must stick to new conditions & especially new groupings of organic beings" (DAR 205.4 : 19). Darwin seems subsequently to have reconciled this problem for the Galapagos biota by realizing that the large number of islands in the group would allow repeated interisland colonizations by species having achieved reproductive isolation on another island. To Hooker's 1847 observation that the Galapagos possessed 1.7 species per genus, compared with 1.0 species per genus for Keeling Island, Darwin responded in the margin of Hooker's paper: "But why so [many species] created. I can account for [this]. I think number of is$^{ld[s]}$ comes into play" (Hooker, 1847b: 247). See further Sulloway (1979 : 44–45) and Ospovat (1981 : 193–200).

other closely related forms from successfully colonizing neighboring islands. Thus Darwin correctly recognized that interisland colonizations within the Galapagos are not just a matter of chance, or of the distance between islands, but rather are dependent on major ecological considerations as well. He therefore correctly anticipated the theoretical view that both ecological conditions and interspecific competition are ultimately responsible for the patterns of geographic distribution observed among various Galapagos organisms. It is only in the last twenty years that Darwin's *Origin* insight—after considerable controversy—has been fully accepted (Grant 1981b), although Darwin himself is not always recognized as the first proponent of this view.[22]

CONCLUSION: THE DARWIN–GALAPAGOS LEGEND

The publication of the *Origin of Species* not only revolutionized the biological sciences, but it also made Darwin into a celebrated intellectual hero—a man thoroughly worthy of scientific deification and hence destined to become the subject of legend. And because myths and legends, above all else, gravitate toward the problem of origins, Darwin's discoveries increasingly became enshrouded by the typical misconceptions of reconstructed 'heroic' history. Accordingly, the true story of Darwin's conversion to the theory of evolution is a far cry from the Darwin–Galapagos legend that has arisen in the wake of Darwin's scientific triumph, and that adorns so many of the biology textbooks today.[23] In fact, the legend, which is composed of three major component myths, tends to obscure precisely what it pretends to explain, namely, the nature of scientific insight.

The first of these component myths is that of Darwin's 'eureka-like' conversion during his visit to the Galapagos Islands in 1835. It may appeal to our romantic conception of scientific discovery to imagine the lone voyager suddenly throwing off the shackles of creationist thinking when finally confronted, in the Galapagos, with a microcosmic paradigm of evolution in action. But this myth, for all of its inherent allure, is both wrong and misleading. What this myth especially tends to obscure is the fascinating question 'Why Darwin?' That is to say, why was it that Darwin, and no one else, was converted by evidence that was widely known to many other contemporary naturalists—naturalists who, like Richard Owen and John Gould, were often far superior to Darwin in their experience and abilities as systematists? The answer to this question is closely associated with the real nature of Darwin's genius as a scientist. Repeatedly the far-seeing amateur among specialized experts, Darwin exhibited his unique intellectual caliber in the pattern of 'gifted individualism'

---

[22]Grant (1981b : 660), for example, has mistakenly assumed that Darwin thought the habitats of the various islands in the Galapagos group were the same and, moreover, that Darwin considered interisland colonizations to be basically random. It is true that Darwin, in his Essay of 1844, believed colonizations, both of oceanic archipelagoes like the Galapagos and of the separate islands within such archipelagoes, to be largely random (1909[1844] : 185, 187; see also Sulloway, 1982a : 34). By the time he wrote the *Origin*, however, Darwin had completely changed his mind on this issue, primarily as a result of his greater appreciation of the extensive biotic differences among the various islands of the Galapagos group (1859 : 399–403). In this respect Darwin anticipated Bowman's (1961) views, although he did not, like Bowman, believe that large-scale adaptive radiation could proceed without subsequent interspecific competition (1859 : 111–30).

[23]See notes 1, 6 and 11.

that manifested itself in the process of his conversion. While other naturalists stood by and calmly rationalized the Galapagos evidence in creationist terms, Darwin—virtually alone—took up the heterodox challenge offered by that evidence.[24] Expressed another way, the Galapagos did not make Darwin; if anything, Darwin, through his superior abilities as a thinker and a theoretician, made the Galapagos; and, in doing so, he elevated these islands to the legendary status they have today.

The second of the three component myths associated with Darwin and the Galapagos is the myth that these islands provided him, at an early stage, with a basic paradigm for his theory of evolution by geographic isolation and natural selection. As I have shown in the case of Darwin's finches, nothing could be further from the truth; and the same conclusion applies to Darwin's Galapagos observations as a whole, which were only slowly incorporated into his final theory. Thus the *Origin of Species* was ultimately the product of twenty-two years of thinking and further research (1837–59), not the five weeks that Darwin spent in the Galapagos Islands or even the five years that he spent accompanying H.M.S. *Beagle* around the world. True, the Galapagos certainly provided Darwin with some crucial hints; but Darwin's full understanding of both evolution and the Galapagos case required almost as long as it took him to publish the *Origin of Species*. Moreover, much of Darwin's evolutionary argument, as finally presented in the *Origin*, had to be constructed from alternative sources, owing to Darwin's failure to appreciate, and to collect, the necessary Galapagos evidence in 1835. Other scientists have been collecting that 'necessary' Galapagos evidence ever since, which leads me to the third of the three component myths encompassing the Darwin–Galapagos legend.

This third and last myth involves the notion that Darwin singlehandedly discovered almost everything there is to know about evolution in the Galapagos—or at least everything of *basic* importance—and hence that subsequent research in these islands has merely been a sort of Kuhnian (1962) mopping-up operation characteristic of 'normal', postrevolutionary science. This myth, promulgated in the biology textbooks and especially in the popular literature about Darwin and the Galapagos, is largely a natural extension of the first two Darwin–Galapagos myths. As a typical manifestation of this third myth, Darwin is frequently credited with insights about his famous Galapagos finches that were actually the product of extensive post-Darwinian ornithological research. For example, in spite of Darwin's own famous *Journal* (1845 : 380) remark about one species of finch appearing to have been "modified for different ends", Darwin was by no means personally convinced that all thirteen species of Galapagos finches (especially the warbler finch) were indeed derived from a single ancestor (see also Darwin, 1841:105). Darwin's lingering

[24]There are apparently only two other nineteenth-century naturalists, besides Darwin, who appreciated the Galapagos evidence in an evolutionary light prior to 1859. Robert Chambers made use of Darwin's Galapagos findings in his *Vestiges of Creation* (1844); but Chambers was converted to the theory of evolution by other considerations, and he also placed a very different evolutionary interpretation on the Galapagos evidence than Darwin did (Hodge, 1972). Alfred Russel Wallace (1855) later called special attention to the Galapagos Islands in his first (and somewhat guarded) evolutionary publication. But Wallace, who was converted to the theory of evolution after reading Chamber's *Vestiges of Creation,* was also largely indebted to Darwin's *Journal* (1845) for his evolutionary interpretation of the Galapagos case. On Darwin's intellectual individualism, see Ghiselin's (1971) valuable treatment; and, for a developmental and psychological approach to this topic, Sulloway (in press).

doubts about the finches' possible common ancestry apparently contributed to his decision, when writing the *Origin of Species,* to omit any specific reference to this now famous biological paradigm of "evolution in action" (Sulloway, 1983:372). During the remainder of the nineteenth century, ornithologists generally believed Darwin's finches were descended from two or three different ancestors—a warbler, a ground finch, and a separate form that gave rise to the six species of *Camarhynchus.* This issue of ancestry was not resolved for more than half a century after the *Origin of Species* was published.

As I have shown elsewhere, David Lack's classic book *Darwin's Finches* (1947) did much to perpetuate this third aspect of the legend, even though Lack himself personally knew better. Indeed, Lack, in reversing his original position on the possible adaptive significance of the beaks among the different species of Darwin's finches (1945, 1947), went through much the same experience of *ex post facto* 'discovery' that Darwin himself did. For it was only after leaving the Galapagos Islands that Lack reached his new theoretical position and then realized the need for the kind of follow-up studies of the finches' feeding behavior that Bowman (1961, 1963), Abbott, Abbot & Grant (1975, 1977), and various other ornithologists have subsequently carried out.[25] Similar 'delayed discoveries' have undoubtedly characterized the work of numerous other Galapagos researchers, who—unlike Darwin—have often had the opportunity to return to the Galapagos Islands in order to collect crucial data, and to make observations, that previously seemed unimportant. Thus the history of research in the Galapagos Islands has been anything but the history of 'mopping up' the scientific tidbits that Darwin left behind. Rather, it is only after repeated expeditions by six generations of post-Darwinian scientists that the Galapagos Archipelago has yielded—with a seeming air of reluctance—many of its richest biological treasures to the world of science. And even today, after so much scientific progress, almost as many questions remain about evolution in the Galapagos as there are answers to the mysteries that Darwin and others have successfully resolved (Patton, 1984).

Of all the numerous scientists who have gone through the experience of making important discoveries in the Galapagos, only to realize sometime later that they have merely scratched the scientific surface and thereby created the need for further research, Charles Darwin perhaps expressed it best. In 1846, shortly after Joseph Hooker had so delighted him with the results of his analysis of Darwin's Galapagos plants, Darwin declared to his friend: "The Galapagos seems a perennial source of new things."[26] The Darwin–Galapagos legend notwithstanding, these famous islands will doubtless remain "a perennial source of new things" in science; and no one would be more disappointed than Charles Darwin if this were not the case.

ACKNOWLEDGEMENTS

I thank the following persons and institutions for their assistance and support in connection with this publication: the Charles Darwin Research Station, Isla

---

[25]On the subject of feeding behavior among Darwin's finches, see also Grant *et al.* (1976); Smith *et al.* (1978); Grant (1981a, 1981b, 1983); and Boag & Grant (1981). For a recent survey of the considerable literature (and debates) concerning the role of interspecific competition in nature, see Schoener (1982).
[26]DAR 114, letter no. 63, dated July 1846.

Santa Cruz, Galapagos Islands, where I was a visitor in 1968, 1970, and 1981–82; Commander Andrew David; Peter J. Gautrey; Ernst Mayr; Gordon C. Sauer; David W. Steadman; and the John Simon Guggenheim Memorial Foundation. I am grateful to the *Journal of the History of Biology* for permission to use materials that previously appeared in a different form in Sulloway, 1982c.

REFERENCES

ABBOTT, I., ABBOTT, L. K. & GRANT, P. R., 1975. Seed selection and handling ability of four species of Darwin's finches. *Condor, 77:* 332–335.

ABBOTT, I., ABBOTT, L. K. & GRANT, P. R., 1977. Comparative ecology of Galapagos Ground Finches (*Geospiza* Gould): Evaluation of the importance of floristic diversity and interspecific competition. *Ecological Monographs, 47:* 151–184.

BARLOW, N. (Ed.), 1933. *Charles Darwin's Diary of the Voyage of H.M.S. "Beagle".* Cambridge: Cambridge University Press.

BARLOW, N. (Ed.), 1945. *Charles Darwin and the Voyage of the Beagle,* with an Introduction by N. Barlow. London: Pilot Press.

BARLOW, N. (Ed.), 1963. *Darwin's Ornithological Notes.* With an Introduction, Notes, and Appendix by N. Barlow. *Bulletin of the British Museum (Natural History) Historical Series, 2,* no. 7.

BARLOW, N. (Ed.), 1967. *Darwin and Henslow: The Growth of an Idea. Letters 1831–1860.* Berkeley and Los Angeles: University of California Press.

BOAG, P. T. & GRANT, P. R., 1981. Intense natural selection in a population of Darwin's finches (Geospizinae) in the Galápagos. *Science, 214:* 82–85.

BOWMAN, R. I., 1961. Morphological differentiation and adaptation in the Galapagos finches. *University of California Publications in Zoology, 58:* 1–302.

BOWMAN, R. I., 1963. Evolutionary patterns in Darwin's finches. *Occasional Papers of the California Academy of Sciences,* no. 44: 107–140.

BROOM, R., 1929. On the extinct Galapagos tortoise that inhabited Charles Island. *Zoologica, 9:* 313–320.

CHAMBERS, R., 1844. *Vestiges of the Natural History of Creation.* London: J. Churchill.

DAMPIER, W., 1729. *A New Voyage Round the World.* 7th ed. In *A Collection of Voyages.* 4 vols. Vol.1. London: James and John Knapton.

DARWIN, C. R., 1837. [Notes upon the *Rhea Americana* and upon *Rhea Darwinii.*] *Proceedings of the Zoological Society of London, 5:* 35–36.

DARWIN, C. R., 1839. *Journal of Researches into the Geology and Natural History of the Various Countries Visited by H.M.S. Beagle, under the Command of Captain FitzRoy, R.N. from 1832 to 1836.* London: Henry Colburn.

DARWIN, C. R. (Ed.), 1841. *The Zoology of the Voyage of H.M.S. Beagle, under the Command of Captain FitzRoy, R.N., during the Years 1832–1836.* Part III: *Birds.* London: Smith, Elder & Co.

DARWIN, C. R., 1845. *Journal of Researches into the Natural History and Geology of the Countries Visited during the Voyage of H.M.S. Beagle Round the World, under the Command of Capt. Fitz-Roy, R.N.* 2nd ed. London: John Murray.

DARWIN, C. R., 1859. *On the Origin of Species by means of Natural Selection, or, The Preservation of Favoured Races in the Struggle for Life.* London: John Murray.

DARWIN, C. R., 1868. *The Variation of Animals and Plants under Domestication.* 2 vols. London: John Murray.

DARWIN, C. R., 1887. *The Life and Letters of Charles Darwin, Including an Autobiographical Chapter.* Edited by F. Darwin. 3 vols. London: John Murray.

DARWIN, C. R., 1903. *More Letters of Charles Darwin: A Record of His Work in a Series of Hitherto Unpublished Letters.* 2 vols. Edited by F. Darwin and A. C. Seward. New York: D. Appleton.

DARWIN, C. R., 1909[1844]. *The Foundations of the Origin of Species: Two Essays Written in 1842 and 1844.* Edited with an Introduction by F. Darwin. Cambridge: Cambridge University Press.

DARWIN, C. R., 1933. *Charles Darwin's Diary of the Voyage of H.M.S. "Beagle".* Edited by N. Barlow. Cambridge: Cambridge University Press.

DARWIN, C. R., 1958[1876]. *Autobiography: With Original Omissions Restored.* Edited with Appendix and Notes by his grand-daughter, N. Barlow. London: Collins.

DARWIN, C. R., 1963[1836]. *Darwin's Ornithological Notes.* Edited with an Introduction, Notes, and Appendix by N. Barlow. *Bulletin of the British Museum (Natural History) Historical Series, 2,* no. 7.

DARWIN, C. R., 1980[1836–37]. *The Red Notebook of Charles Darwin.* Edited by S. Herbert, with Introduction and Notes. *Bulletin of the British Museum (Natural History) Historical Series, 7;* Ithaca, N.Y.: Cornell University Press.

DARWIN, F., 1888. Charles Robert Darwin (1809–1882). *Dictionary of National Biography, 14:* 72–84.

DARWIN, F., 1909. Introduction to *The Foundations of the Origin of Species: Two Essays Written in 1842 and 1844, by Charles Darwin.* Edited by F. Darwin. Cambridge: Cambridge University Press.

DARWIN, F. (Ed.), 1903. *More Letters of Charles Darwin: A Record of His Work in a Series of Hitherto Unpublished Letters.* 2 vols. Edited by F. Darwin and A. C. Seward. New York: D. Appleton.

DAVIS, J. & MILLER, A. H., 1960. Family Mimidae. In E. Mayr & J. C. Greenway (Eds), *Check-List of Birds of the World: A Continuation of the Work of James L. Peters.* Vol. 9 : 440–458. Cambridge, Mass.: Museum of Comparative Zoology.

DE BEER, G., 1958. Foreword to *Evolution by Natural Selection*, by C. R. Darwin and A. R. Wallace. Cambridge: Cambridge University Press.

DE BEER, G. (Ed.), 1959. *Darwin's Journal. Bulletin of the British Museum (Natural History) Historical Series, 2,* no. 1.

DE BEER, G., 1962. The origins of Darwin's ideas on evolution and natural selection. *Proceedings of the Royal Society of London. Series B. Biological Sciences, 155:* 321–332.

DE BEER, G., 1963. *Charles Darwin: Evolution by Natural Selection.* London: Thomas Nelson and Sons; Garden City, N.Y.: Doubleday, 1964.

DE BEER, G., ROWLANDS, M. J. & SKRAMOVSKY, B. M. (Eds), 1967. *Darwin's Notebooks on Transmutation of Species.* Part VI: *Pages Excised by Darwin. Bulletin of the British Museum (Natural History) Historical Series, 3,* no. 5.

DOBZHANSKY, T., AYALA, F. J., STEBBINS, G. L. & VALENTINE, J. W., 1977. *Evolution.* San Francisco: W. H. Freeman.

DORST, J., 1974. *The Life of Birds.* 2 vols. Translated by I. C. J. Galbraith. London: Weidenfeld and Nicolson.

EIBL-EIBESFELDT, I., 1961. *Galapagos: The Noah's Ark of the Pacific.* Translated by A. H. Brodrick. Garden City, N.Y.: Doubleday.

EISELEY, L., 1961. *Darwin's Century: Evolution and the Men who Discovered It.* Garden City, N.Y.: Anchor Books/Doubleday.

FITZROY, R., 1839. *Narrative of the Surveying Voyages of His Majesty's Ships Adventure and Beagle, between the Years 1826 and 1836, Describing Their Examination of the Southern Shores of South America, and the Beagle's Circumnavigation of the Globe.* Vol. 2: *Proceedings of the Second Expedition, 1831–1836, under the Command of Captain Robert Fitz-Roy, R.N. With Appendix.* London: Henry Colburn.

GHISELIN, M. T., 1969. *The Triumph of the Darwinian Method.* Berkeley and Los Angeles: University of California Press.

GHISELIN, M. T. 1971. The individual in the Darwinian revolution. *New Literary History, 3:* 113–134.

GIFFORD, E. W., 1919. Field notes on the land birds of the Galapagos Islands and of Cocos Island, Costa Rica. *Proceedings of the California Academy of Sciences, 2:* 189–258.

GILLSÄTER, S., 1968. *From Island to Island: Oases of the Animal World in the Western Hemisphere.* Translated by J. Tate. London: George Allen and Unwin.

GOULD, J., 1837a. Remarks on a group of ground finches from Mr. Darwin's collection, with characters of the new species. *Proceedings of the Zoological Society of London, 5:* 4–7.

GOULD, J., 1837b. Observations on the raptorial birds in Mr. Darwin's collection, with characters of the new species. *Proceedings of the Zoological Society of London, 5:* 9–11.

GOULD, J., 1837c. On a new *Rhea (Rhea Darwinii)* from Mr. Darwin's collection. *Proceedings of the Zoological Society of London, 5:* 35.

GRANT, P. R., 1981a. The feeding of Darwin's finches on *Tribulus cistoides* (L.) seeds. *Animal Behavior, 29:* 785–793.

GRANT, P. R., 1981b. Speciation and the adaptive radiation of Darwin's finches. *American Scientist, 69:* 653–663.

GRANT, P. R., 1983. The role of interspecific competition in the adaptive radiation of Darwin's finches. In R. I. Bowman, M. Berson & A. E. Leviton (Eds), *Patterns of Evolution in Galapagos Organisms:* 187–199. Speciation Publication 1. San Francisco: AAAS, Pacific Division.

GRANT, P. R., GRANT, B. R., SMITH, J. N. M., ABBOTT, I. J. & ABBOTT, L. K., 1976. Darwin's finches: Population variation and natural selection. *Proceedings of the National Academy of Sciences, 73:* 257–261.

GRUBER, H. E., 1981. *Darwin on Man: A Psychological Study of Scientific Creativity.* 2nd ed. Chicago: University of Chicago Press.

GRUBER, H. E. & BARRETT, P. H., 1974. *Darwin on Man: A Psychological Study of Scientific Creativity.* Together with *Darwin's Early and Unpublished Notebooks.* Foreword by Jean Piaget. New York: E. P. Dutton.

GRUBER, H. E. & GRUBER, V., 1962. The eye of reason: Darwin's development during the *Beagle* voyage. *Isis, 53:* 186–200.

GRZIMEK, B., 1973. *Grzimek's Animal Life Encyclopedia.* Vol. 9: *Birds.* New York: Van Nostrand Reinhold Co.

HERBERT, S., 1974. The place of man in the development of Darwin's theory of transmutation. Part I. To July 1837. *Journal of the History of Biology, 7:* 217–258.

HERBERT, S. (Ed.), 1980. *The Red Notebook of Charles Darwin.* With an Introduction and Notes. *Bulletin of the British Museum (Natural History) Historical Series, 7;* Ithaca, N.Y.: Cornell University Press.

HIMMELFARB, G., 1959. *Darwin and the Darwinian Revolution.* New York: Doubleday; London: Chatto & Windus; Garden City, N.Y.: Doubleday, 1962. [I have cited from the 1962 edition.]

HODGE, M. J. S., 1972. The universal gestation of nature: Chamber's *Vestiges* and *Explanations. Journal of the History of Biology, 5:* 127–151.

HODGE, M. J. S., 1983. Darwin and the laws of the animate part of the terrestrial system (1835–1837): On the Lyellian origins of his zoonomical explanatory program. *Studies in the History of Biology, 7:* 1–106.

HOOKER, J. D., 1846. Enumeration of the plants in the Galapagos Islands, with descriptions of the new species. *Proceedings of the Linnean Society of London, 1:* 276–279.

HOOKER, J. D., 1847a. An enumeration of the plants of the Galapagos Archipelago; with descriptions of those which are new. *Transactions of the Linnean Society of London, 20:* 163–233.

HOOKER, J. D., 1847b. On the vegetation of the Galapagos Archipelago as compared with that of some other tropical islands and of the continent of America. *Transactions of the Linnean Society of London, 20:* 235–262.

HUXLEY, J., 1954. The evolutionary process. In J. Huxley, A. C. Hardy & E. B. Ford (Eds), *Evolution as a Process:* 1–23. London: George Allen and Unwin.

HUXLEY, J., 1960. The emergence of Darwinism. In S. Tax (Ed.), *Evolution after Darwin*. Vol. 1: *The Evolution of Life: Its Origins, History and Future:* 1–21. Chicago: University of Chicago Press.

HUXLEY, J., 1966. Charles Darwin: Galápagos and after. In R. I. Bowman (Ed.), *The Galapagos:* 3–9. Berkeley and Los Angeles: University of California Press.

IRVINE, W., 1955. *Apes, Angels, and Victorians: The Story of Darwin, Huxley, and Evolution.* New York: McGraw-Hill Book Co.

KIMBALL, J. W., 1978. *Biology.* 4th ed. Reading, Mass.: Addison-Wesley Publishing Co.

KOHN, D., SMITH, S. & STAUFFER, R., 1982. New light on *The Foundations of the Origin of Species:* A reconstruction of the archival record. *Journal of the History of Biology, 15:* 419–442.

KOTTLER, M. J., 1978. Charles Darwin's biological species concept and the theory of geographic speciation: the transmutation notebooks. *Annals of Science, 35:* 275–297.

KUHN, T., 1962. *The Structure of Scientific Revolutions.* Chicago: University of Chicago Press.

LACK, D., 1945. *The Galapagos Finches (Geospizinae): A Study in Variation. Occasional Papers of the California Academy of Sciences,* no. 31.

LACK, D., 1947. *Darwin's Finches.* Cambridge: Cambridge University Press.

LACK, D., 1969. Subspecies and sympatry in Darwin's finches. *Evolution, 23:* 252–263.

LEIGH, E. G., Jr., 1971. *Adaptation and Diversity: Natural History and the Mathematics of Evolution.* San Francisco: Freeman, Cooper, & Co.

LIMOGES, C., 1970. *La Sélection naturelle: Étude sur la premiere constitution d'un concept (1837–1859).* Paris: Presses Universitaires de France.

MAYR, E., 1964. Introduction to *On the Origin of Species: A Facsimile of the First Edition,* by C. Darwin. Cambridge and London: Harvard University Press.

MOODY, P. A., 1970. *Introduction to Evolution.* 3rd ed. New York, Evanston, and London: Harper & Row.

OSPOVAT, D., 1981. *The Development of Darwin's Theory: Natural History, Natural Theology, and Natural Selection, 1838–1859.* Cambridge and New York: Cambridge University Press.

PATTON, J., 1984. Genetical processes in the Galapagos. *Biological Journal of the Linnean Society, 21:* 97–111.

PAYNTER, R. A., Jr. & STORER, R. W., 1970. *Check-List of Birds of the World: A Continuation of the Work of James L. Peters.* Edited by R. A. Paynter, Jr. (in consultation with E. Mayr). Vol.XIII: *Emberizinae, Catamblyrhynchinae, Cardinalinae, Thraupinae, Tersininae.* Cambridge, Mass.: Museum of Comparative Zoology.

PETERSON, R. T., 1963. *The Birds.* Life Nature Library. New York: Time-Life Books.

PORTER, D., 1815. *Journal of a Cruise Made to the Pacific Ocean, by Captain David Porter, in the United States Frigate Essex, in the Years 1812, 1813, and 1814. Containing Descriptions of the Cape de Verd Islands, Coasts of Brazil, Patagonia, Chili, and Peru, and of the Galapagos Islands.* 2 vols. Philadelphia: Bradford and Inskeep; New York: Abraham H. Inskeep.

PORTER, D. M., 1980. The vascular plants of Joseph Dalton Hooker's *An enumeration of the plants of the Galapagos Archipelago; with descriptions of those which are new. Botanical Journal of the Linnean Society, 81:* 79–134.

RUSE, M., 1979. *The Darwinian Revolution: Science Red in Tooth and Claw.* Chicago and London: University of Chicago Press.

RUSE, M., 1982. *Darwinism Defended: A Guide to the Evolution Controversies.* Reading, Mass.: Addison-Wesley Publishing Co.

SALVIN, O., 1876. On the avifauna of the Galapagos Archipelago. *Transactions of the Zoological Society of London, 9:* 447–510.

SCHOENER, T. W., 1982. The controversy over interspecific competition. *American Scientist, 70:* 586–595.

SHARPE, R. B., 1888. *Catalogue of Birds in the British Museum.* Vol. 12: *Catalogue of the Passeriformes, or Perching Birds, in the Collection of the British Museum. Fringilliformes: Part III. Containing the Family Fringillidae.* London: Printed by Order of the Trustees.

SILVERSTEIN, A., 1974. *The Biological Sciences.* San Francisco: Holt, Rinehart, and Winston.

SMITH, S., 1960. The origin of "the Origin" as discerned from Charles Darwin's notebooks and his annotations in the books he read between 1837 and 1842. *The Advancement of Science, 16:* 391–401.

SMITH, J. N. M., GRANT, P. R., GRANT, B. R., ABBOTT, I. J. & ABBOTT, L. K., 1978. Seasonal variation in feeding habits of Darwin's ground finches. *Ecology, 59:* 1137–1150.

STEADMAN, D. W., 1981. Vertebrate fossils in lava tubes in the Galapagos Islands. In B. F. Beck (Ed.), *Proceedings of the Eighth International Congress of Speleology:* 549–550. Americus, Georgia: Georgia Southwestern College.

STEADMAN, D. W., 1982. *Fossil Reptiles, Birds, and Mammals from Isla Floreana, Galapagos Archipelago.* Ph.D. dissertation, University of Arizona, Department of Geosciences, Tuscon, Arizona.

STOPPARD, T., 1981. This other Eden. *Noticias de Galapagos, 34:* 6–7.

SULLOWAY, F. J., 1969. *Charles Darwin and the Voyage of the* Beagle. Senior honors thesis, Harvard University, Cambridge, Massachusetts.

SULLOWAY, F. J., 1979. Geographic isolation in Darwin's thinking: The vicissitudes of a crucial idea. *Studies in the History of Biology, 3:* 23–65.

SULLOWAY, F. J., 1982a. Darwin and his finches: The evolution of a legend. *Journal of the History of Biology, 15:* 1–53.

SULLOWAY, F. J., 1982b. *The* Beagle *Collections of Darwin's Finches (Geospizinae). Bulletin of the British Museum (Natural History) Zoology Series, 43,* no. 2.

SULLOWAY, F. J., 1982c. Darwin's conversion: The *Beagle* voyage and its aftermath. *Journal of the History of Biology, 15:* 325–396.

SULLOWAY, F. J., 1983. The legend of Darwin's finches. *Nature, 303:* 372.

SULLOWAY, F. J., in press. Darwin's early intellectual development: An overview of the *Beagle* voyage (1831–1836). In D. Kohn (Ed.), *The Darwinian Heritage: A Centennial Retrospect.* Wellington, New Zealand: Nova Pacifica; Princeton: Princeton University Press.

SWARTH, H. S., 1931. *The Avifauna of the Galapagos Islands. Occasional Papers of the California Academy of Sciences,* no. 18.

THORNTON, I., 1971. *Darwin's Islands: A Natural History of the Galápagos.* Garden City, N.Y.: The Natural History Press.

WALLACE, A. R., 1855. On the law which has regulated the introduction of new species. *Annals and Magazine of Natural History,* 2nd ser., *16:* 184–196.

WATERHOUSE, G. R., 1845. Descriptions of coleopterous insects collected by Charles Darwin, Esq., in the Galapagos Islands. *Annals and Magazine of Natural History, 16:* 19–41.

WICHLER, G., 1961. *Charles Darwin: The Founder of the Theory of Evolution and Natural Selection.* New York: Pergamon Press.

WILLIAMSON, M., 1983. Darwin's finches worth a mention. *Nature, 302:* 566.

Biological Journal of the Linnean Society (1984), *21:* 61–75. With 2 figures

Geology of Galapagos

TOM SIMKIN

National Museum of Natural History, Smithsonian Institution, Washington, D.C. 20560, U.S.A.

This paper reviews the origins and geological history of the Galapagos Islands. The islands arose from a 'hot-spot'. The oceanic crust on which the islands are built can be no more than 10 million years old, and the islands themselves have been in existence at least 3.3 million years.

The Galapagos are among the most active volcanic groups in the world, and the physical nature of the islands is dominated by lava structures of various ages, including lava tubes or tunnels, which have been of particular interest to biologists. Weathering of the rock to produce soil has generally been slow, particularly in the drier parts.

KEY WORDS:—Galapagos – geology – volcanism – lava flow – lava tunnels – lava weathering.

CONTENTS

INTRODUCTION

The Galapagos Islands—Herman Melville's "five-and-twenty heaps of cinders"—owe their existence above sea-level to their youthful volcanic nature. They are there because they are volcanically young. Furthermore, the dry equatorial climate slows the weathering of rock into soil and withholds the moisture that would soon cloak such volcanic features in vegetation if they were found in wetter climates. The result is a stark landscape quite alien to the

0024–4066/84/010061 + 15 $03.00/0

experience of most visitors. However, they quickly recognize variety in
this landscape, and sense that its history is inextricably tied to that of the wild
and wondrous organisms living on its surface. This paper summarizes what is
known of the archipelago's physical history, and reviews the active processes
that have shaped (and continue to shape) these special islands. It is a shortened
version of a chapter by the author in a book edited by Roger Perry (*Galapagos*
Oxford: Pergamon, 1984), which also discusses lava flows; compositions and
vents; explosive volcanism and tuff cones; volcano spacing, rifts and calderas;
volcano growth and form; and local seismicity.

SETTING

Plate tectonics—the concept that revolutionized the earth sciences in the
1960s—tells us that most of the world's volcanism takes place at plate
boundaries: either along the deep oceanic rift system, where plates are spreading
apart with the creation of new oceanic crust, or where plates are colliding, with
thinner crust being subducted under thicker at island arcs or continental
margins. But the Galapagos Islands are not explained by either setting. They lie
over 1000 km west of the magnificent continental margin volcanoes of the
Ecuadorian Andes, and equally far east of the point where the Pacific's major
rift, the East Pacific Rise, turns northward to form the Gulf of California. At this
major bend in the East Pacific Rise another rift system—the Galapagos
Spreading Centre—heads eastwards to pass north of the Galapagos Islands (Fig. 1).
Southward spreading from this centre is slow (3 cm/yr) relative to the fast
(> 7 cm/yr) eastward spreading from the East Pacific Rise, resulting in a net
crustal movement toward the east–south-east in the Galapagos Islands region.

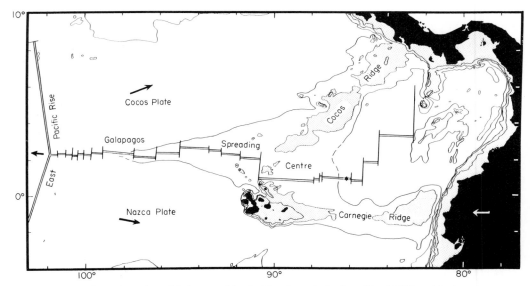

Figure 1. Tectonic setting of Galapagos Islands. Plate boundaries from Hey (1977) and bathymetry
from Chase (1968). Bathymetric contour intervals at 500 fathoms (914 m) and area above 1000
fathoms (1829 m) shown by stipple pattern. Site of deep-sea hydrothermal vent explorations on the
Galapagos Spreading Centre (Corliss *et al.*, 1979) marked by asterisk. Absolute plate motion vectors
(arrows) from Hey (1977).

Although the Galapagos Spreading Centre has provided the dramatic 1977 discovery of bizarre hydrothermal vent communities 2.5 km below the surfaces and subsequent observations on the oceanic rift system (Corliss et al., 1979; Ballard, van Andel & Holcomb, 1982), the site of this detailed work is over 400 km east of the Galapagos archipelago and shares only its name with the islands. To understand the origin of the islands we must go beyond the fundamental plate tectonics elements of spreading centres and subduction zones.

An early corollary of the new understanding of Earth's behaviour has been the idea of 'hot-spots', rooted deep in the mantle, supplying lava to and through the overlying crustal plates as they move slowly along. The Hawaiian Islands have been the type example of this process, with ages growing progressively older to the north-west (in the direction of plate movement) from the currently active hot-spot at the southeast end of the chain, but the Galapagos were listed as another example in Wilson's (1963) initial statement of the idea. The concentration of recent volcanism in the western islands defines the hot-spot, and plate movement to the east-south-east explains the older islands and submarine ridge in that direction as former products of the hot-spot. By the time of Morgan's (1971) extension of the concept, the presence of the Galapagos Spreading Centre was known and he suggested that ancient products of the Galapagos hot-spot had, through past millions of years, been deposited to the north as well as the south. Plate motion on that side of the spreading centre is to the north-east, explaining the submarine Cocos Ridge as another 'plume' of volcanic material being carried away from the Galapagos hot-spot. Thus the hot-spot idea explains not only the main features of island ages, but also the broader physiographic features of the region. The shallow submarine platform of the Galapagos archipelago has steep slopes to the west and south, but gentle slopes to the north-east and east, where it is essentially the intersection of the Cocos and Carnegie Ridges. This is exactly the physiographic pattern to be expected from a fixed mantle hot-spot that has, over the last 25 million years, supplied volcanic products to two overlying plates, one moving to the north-east and the other to the east-south-east (Hey, 1977). The Galapagos Spreading Centre that separates these two plates has received many detailed studies (see for example, Johnson et al., 1976, and Schilling, Kingsley & Devine, 1982), with particular attention to the striking petrochemical variation along its axis, and all have tended to strengthen the basic idea that a deep mantle hot-spot exists under the western Galapagos Islands. The regional gravity data of Case et al. (1973) and the isotope data of White & Hofmann (1978) are also compatible with a Galapagos hot-spot.

AGE

The questions that biologists constantly ask geologists in Galapagos are 'how old are the islands?' and 'how long has evolution had to work here?' Marine geophysical techniques, particularly the dating of sea-floor magnetic anomalies, indicate that the oceanic crust on which the islands are built can be no older than 10 million years. However, the hot-spot concept described above clearly implies that the islands are much younger than their underlying crust. Attempts to date the island rocks themselves have encountered several difficulties: (1) young lava flows tend to cover older ones, (2) erosion in most parts of these arid islands is too slow to cut through to the older interior, (3) potassium, the parent element whose radioactive decay to argon is most commonly measured in age determinations, is present in only very small quantities in most Galapagos rocks, and (4) the dry, rubbly lava flows that make up so much of the Galapagos have daunted much dilligent searching for the islands' oldest rocks (Herman Melville (1856) wrote of the islands "The interior of most of them is tangled and difficult of passage beyond description; the air is sultry and stifling; an intolerable thirst is provoked, for which no running stream offers its kind relief"). Nevertheless, much progress has been made, with the most recent radiometric work (Bailey, 1976) yielding confident ages around 3.3 million years for Espanola. Although Espanola lavas had been interpreted by McBirney & Williams (1969) as uplifted submarine flows, Hall, Ramon & Yepes (1980) have since found evidence that they are subaerial, confirming that true islands existed in Galapagos at least 3.3 million years (My) ago.

Allan Cox (1983) has tabulated radiometric and magnetic age determinations from the Galapagos. No rocks older than 0.7 My have been found on the historically volcanic islands of Fernandina, Isabela, Santiago, and Pinta, nor on the northern islands of Genovesa, Machena, or Culpepper. A 0.72 My age has been measured from Wenman. The central islands of Santa Cruz, Baltra, Rabida, Pinzon, and Floreana have rocks in the 0.7–1.5 My age range. Ages of 2.6–2.8 My have been measured on Santa Fe rocks and an uncertain 4.2 My ($\pm$ 1.8 My) age from Plazas, off the north-east coast of Santa Cruz. The 3.3 My age for Espanola has been discussed above. Cox estimates that the old volcano of southwest San Cristobal is older than 2.4 My (Colinvaux, 1968, has reported a 4 My date for this volcano, but the dated rock was not collected in place and my represent pre-island growth) and concludes (as did Hey, 1977) that the earliest Galapagos Islands probably emerged 3–5 My ago. There is no geological evidence that the islands have ever been connected, even by a chain of islands, with the mainland. These ages would seem to conflict with some fossil evidence from the islands—particularly the late Miocene (5–10 My) ages reported by Durham (1965) for the invertebrate fossils found on north-east Santa Cruz—however, Lipps & Hickman (1982) have recently concluded that most Galapagos fossils are represented by living species and that most island fossil deposits are not older than 2 My. Evolution has had only a few million years in which to work in the Galapagos, and these islands—in contrast to other still-active ocean archipelagos like the Azores, Canary, and Samoa Islands—are remarkably young.

The remainder of this paper will discuss the main geological features of Galapagos with emphasis on the processes that have formed them. Because the islands are so young, and their formative processes so vigorously continuing, the origin of most features is obvious without great leaps of imagination.

HISTORICAL VOLCANISM

The oldest documented eruption was in 1797 from Volcan Wolf*, although earlier visitors observed volcanism from unspecified parts of the archipelago.

---

*An arguably older eruption on Santiago is dated, but was not witnessed. The prominent lava flow in James Bay contains fragments of marmalade pots stashed by Buccaneers in 1683 (Heyerdahl, 1963). The flow was mentioned by Darwin, thus bracketing its age between 1683 and 1835: one of the rare uses of marmalade pots in volcano-chronology.

The historical volcanism of Galapagos has been catalogued by Richards (1962) and updated by Simkin et al. (1981). Fifty-three eruptions are known, from eight volcanoes, and another six volcanoes carry evidence of probably eruptions within the last few thousand years or less. Most eruptions have produced basaltic lava flows, commonly from linear fissures, rather than the violently exploding ash and pumice characteristic of the more gas-rich, siliceous volcanoes at colliding plate margins.

By any measure, the Galapagos are amongst the most active oceanic volcano groups in the world. During the last 15 years, when reporting of Galapagos volcanism has been reasonably complete, only Hawaii has been active in more years (10, compared with 7 each for Galapagos and Reunion, in the Indian Ocean; Iceland was active in 6 and no other deep ocean island group was active in more than 3). Simkin & Siebert (in press) have suggested measures of volcanic "vigour" that allow crude comparisons of different volcanic groups. These show Galapagos to be among the dozen most vigorous groups in the world; roughly comparable to Japan, and more active than regions such as the Philippines, Mexico, and the West Indies. It is certain, however, that the historical record for Galapagos is incomplete, probably even in recent years. The 1979 eruption of Cerro Azul, from a vent low on its east flank, was well recognized in Galapagos (Moore, 1980b). But it was not until two years later that Moore, upon climbing to the summit, saw lava within the caldera that she did not remember from pre-1979 visits. Inspection of photographs taken by other summit visitors in 1979 narrows the date of the summit flow to within a few months of the flank eruption, and they may well have been synchronous. Even more recently, an unnoticed eruption produced small lava flows visible on air photos taken on 26 March, 1982, that were not present when we walked past that area 15 months earlier (4 December, 1980). Normally the earliest phase of Galapagos eruptions are the most spectacular and activity is often virtually ended in a day or two— clouds or topography may obscure the activity, or there may simply be no human observers near enough to see the show.

ERUPTIVE STYLES AND PRODUCTS

People who drink bubbly liquids know that removing the bottle cap reduces pressure on the liquid, allowing gas to escape from solution. They also know that shaking the bottle, or holding it at the wrong angle, can disastrously affect the way in which the liquid leaves the bottle. Molten magma is also a bubbly liquid, and its gas content is an important factor affecting the way it disperses from its container. The degassing and consequent expansion of molten liquid as it nears the surface is a driving force behind most eruptions, and the varied eruptive products on Galapagos can be better appreciated if the (now-departed) gas phase is considered. In part because of the changes caused by degassing during eruption, geologists use the word 'magma' to refer to the gas-containing liquid before it reaches the surface, and 'lava' for the relatively gas-poor liquid that then flows onto the surface.

The other, much more obvious, factor to consider is that molten Galapagos lava, at temperatures above 1000°C, cools to the solid state at a wide variety of rates, depending on the temperature of the surrounding material and the volume of the liquid. Small fragments thrown high into the air above an erupting vent may chill instantly as spun glass fibres of 'Pele's hair'; larger clots of liquid splattered onto nearby cold rock may cool in seconds as a thin, hard, skin; and metres-thick lava flows may retain a molten liquid interior capable of movement and redistribution for months after the end of the eruption.

Of particular interest to biologists have been lava tubes, or tunnels, several of which have been mapped by Balazs (1975), Montoriol-Pous & de Mier (1977), and others. Commonly 5 m in diameter and hundreds of metres long, these natural caves have provided refuge for former Galapagos animals and protection for their remains. They have thus been fertile collecting sites for vertebrate palaeontologists (Steadman, 1981, 1982). The formation of lava tubes has been observed on Hawaii (Peterson & Swanson, 1974) both by the crusting over of lava streams near vents (such as the flow channels described above) and by the coalescence of pahoehoe toes at sluggish flow fronts on gentler, more distant, slopes. "As a pahoehoe toe budded from the flow front, a skin chilled around it. The skin inflated like a balloon as more lava oozed into it. Eventually the skin broke open owing to excess fluid pressure, and lava emerged as a new toe that rapidly became encased in its own skin. Repetitions of this budding process gradually lengthened the flow and developed a small tube, whose overlying crust thickened to form a rigid shell" (Peterson & Swanson, 1974: 215). As the tubes developed, they eroded downward from shallow depths to at least 13 m deep, cutting into underlying older flows. When the flow of lava ended at the vent, some or all of the lava drained out to leave a long, empty tube system. Because they transport large volumes of lava long distances with very little heat loss, lava tubes are recognized as important elements in the growth of large, broad, basaltic shield volcanoes.

Volcanism does not end with the end of an eruption. Deep lava flows (thicknesses of 50 m have been measured on Galapagos) may retain a liquid interior for a year or more, and continue to lose heat for decades. But huge magma chambers, filled with molten liquid and thermally blanketed by the overlying volcanic edifice, cool for many millions of years after their last eruption. The fumaroles and hot springs of many volcanic regions are surface reminders of these hot materials below.

Rainfall percolates downward through porous rocks, but if it encounters hot rock it is heated, becomes less dense, and rises back to the surface as steam. This may or may not contain volcanic gases, but the heat source is commonly still-cooling magma, and late gases from that magma commonly mix with the vapour on its way to the surface. These surface concentrations are called 'fumaroles', or 'solfataras' if sulphur precipitation is pronounced. They are located on major vertical fractures, like caldera boundary faults, that provide an easy passage to the surface and are often found near recent eruptive fissures, where still-cooling magma is very close to the surface.

In the submarine environment—where there is an abundance of water that is usually in short supply above sea-level in Galapagos—these hydrothermal systems work very efficiently, and submersible explorations of submarine volcanoes have recently revealed hydrothermal deposits, including metallic sulphides concentrated on caldera walls and recent vents (Lonsdale, Batiza & Simkin, 1982).

Galapagos fumaroles are most vigorous in climatically wet periods, when water is abundant, and are best seen in the early mornings, when the relative coolness of the air results in more condensation above the warm emanations. In Antarctica, such condensation results in the construction of huge ice towers over fumarole vents.

In Galapagos, fumaroles are known from all the historically active volcanoes, plus Volcan Ecuador (north-west Isabela), Marchena, Santa Cruz, and Santa Fe. They provide many warm, moist habitats for ferns, mosses, doves, and other moisture-loving organisms. Nordlie & Colony (1973) have described the fumarole that produced geyser-like water fountaining through the 1960s on the caldera wall of Volcan Alcedo. The gas flowed at a velocity around 6 m sec^{-1} and was mostly water vapour with large proportions of CO_2 and various sulphurous components. The associated pools, of mixed value to the large Alcedo wildlife population, carry a high boron content, abundant opal deposits, and temperatures up to 96°C in one.

ROCK COMPOSITION

Whether lava flowed quietly or was fragmented explosively, its composition is of considerable interest to geologists. Variations in the products of a volcano provide clues to processes within that volcano and, in Galapapos to the development of the archipelago as a whole. Some variations are obvious to the naked eye, such as the relative abundance of crystals that were already growing in the magma at the time it reached the surface; others are more cryptic. Indeed, most Galapagos rocks look very much the same, and most obvious colour variations—from black, to brown to red—are more the result of variation in oxidation than any fundamental variation in composition. Careful work with microscopes and chemical analyses is needed to recognize these more significant variations. The status of such detailed work (Simkin, Reeder & MacFarland, 1974) is not sufficiently advanced in most parts of the archipelago to allow satisfying generalizations, but some observations are clear. For example, the lavas of Fernandina are monotonously unchanging in composition, with specimens from the steep caldera walls and many locations on the flanks falling within a narrow range of tholeiitic basalt composition. This constancy contrasts dramatically with the wide range of compositions on the much smaller volcano of Pinzon. Nordlie, Thieben & Delaney (1982) found compositional variations that fit Nordlie's 1973 suggestion that the western volcanoes can be placed in a developmental sequence: Cerro Azul (youngest), Wolf, Fernandina, Darwin, Alcedo, Sierra Negra (oldest). They point out that chemical distinctions between the youngest and oldest volcanoes in this sequence— adjacent to each other and both recently active—shows that there is little connection between magma chambers at depth.

While the large western volcanoes are built of tholeiitic basalt, those of the central part of the archipelago (including San Cristobal) are mainly basalts of alkaline parentage. This pattern is reversed in the smaller volcanoes, with alkali basalts on Volcan Ecuador in the west, and tholeiitic rocks making up the central islands of Pinzon and Rabida. A broader complication is the finding (Baitis & Swanson, 1976; Bow, 1979) of low-K_2O tholeiitic basalts, comparable to those of the oceanic spreading centres, on the dominantly alkaline islands of Santiago and Santa Cruz. These young flows are related to the east-trending normal faults and aligned vents recognized by Swanson *et al.* (1974) on

Santiago and six other islands to the south-east. These features are believed to be related, in turn, to the east-trending Galapagos spreading centre just to the north of the archipelago (Fig. 1). While it is easy to say that the islands are dominantly basaltic with striking variation on some islands, there is much diversity, and more detailed petrological data are needed.

The ultimate origin of Galapagos rocks is the earth's mantle, many tens of kilometres below the surface. Recent attempts to understand the nature of this mantle include the petrologic work of Bow (1979), the isotope work of White (1979), and the trace element analysis of Shimizu, Masuda & Masui (1981).

FAULTING

A major process affecting the Galapagos is faulting—the movement of large blocks of the earth's crust with respect to one another.

All Galapagos travellers landing at the Baltra airstrip (Fig. 2) quickly see evidence of recent faulting: either driving north down the steep fault scarp to meet a boat at Caleta Aeolian or taking the ferry across the down-dropped block separating Baltra from Isla Santa Cruz to the south. Between the flat-lying lava flows that make up the island of Baltra are thin layers of limestone bearing shallow-water marine fossils (Hertlein, 1972). These lava flows, then, were once below sea level, and the east-trending faults so well shown on Fig. 2 record differential levels of uplift. Much of north-eastern Santa Cruz has also been uplifted, while the small, offshore islands of Plazas show tilting as well as uplift, like two parallel trap-doors lifted along east-trending faults that tilt their surfaces gently to the north. Similar east-trending faults have created Bahia Academy on the south coast of Santa Cruz. The bay and village of Puerto Ayora are in a down-dropped block bounded on one side by the north-facing cliff forming the inner harbour and on the other by the south-facing cliff, or barranco, behind the tortoise pens at the Darwin Station. These faults extend over 20 km to the east (Laruelle, 1967) where the uplifted submarine lavas of Santa Fe have been broken into subparallel, east-trending fault blocks. McBirney & Williams (1969) remarked that some of these faults formed within the last few thousand years; Tjitte De Vries reported fumaroles from them in 1971; and Bruce Nolf has found post-faulting, subaerial volcanism on the island.

Although faulting has been an important process throughout Galapagos history, all of it has apparently been 'normal faulting': one crustal block moving up or down, with respect to its neighbour, along a near-vertical fracture or fault plane. As would be expected from the extensional tectonic setting, no evidence is known for either thrust faulting (one block over or under another) of convergent plate boundaries or tear faulting (blocks shifted horizontally without vertical movement) of offsets along spreading plate boundaries.

Uplift

The vertical faulting described above has uplifted large areas of the central Galapagos. Swanson *et al.* (1974) regard this as a broad, tectonic uplift related to the Galapagos spreading centre to the north. The east-trending fault system in the islands supports this interpretation, but localized uplifts are also known in the islands, and more detailed geological mapping is needed for a full understanding

Figure 2. Isla Baltra, north-east Santa Cruz, and east-trending fault blocks. Vertical air photograph (by U.S. Air Force) taken in 1959 shows remnants of World War II military base and prominent fault scarp just north of the currently-used (eastern) airstrip. Parallel faults have isolated Baltra from Santa Cruz, to the south, and Seymour, to the north. The north arrow at lower right corresponds to a distance of two kms.

of the regional uplift pattern. On southern Isabela, inland from the coastal village of Villamil, are 100 ha of uplifted fossiliferous limestone, sandy beaches hundreds of metres from the coast, and a variety of shallow lagoons. Two historical examples of even more local uplift provide a good illustration of the dynamic nature of the Galapagos.

In 1954, a Disney film crew sailing along the west coast of Isabela noticed a remarkably white stretch of shoreline at Urvina Bay. Closer investigation showed that the white colour was shallow-water coral and calcareous algae exposed to 4 m above sea level. Also stranded were lobsters, marine turtles, and even a few fish, indicating that the uplift had been both swift and recent (Couffer, 1956). The uplift of this bay and nearly 6 km of adjacent shoreline, has been interpreted as a response to near-surface magma movement because an eruption from Volcan Alcedo, 15 km to the east, was reported later that year (Richards, 1957). However, K. A. Howard's careful comparison of 1946 and 1960 air photographs of Alcedo's finds no new lavas anywhere on the upper flanks, as reported for the 1954 eruption; a more likely location for the reported site of that eruption is Sierra Negra, over 40 km south of Urvina Bay.

More recently, uplift at Punta Espinoza on Fernandina, has resulted in a tourist landing dock that now stands fully out of the water at lower stages of the tide. Contrary to the situation at Urvina Bay, however, Punta Espinoza was raised in several steps. In mid-September, 1974, marine invertebrate studies by Jerry Wellington indicated approximately 30 cm of recent uplift at Espinoza. This was six weeks after a local earthquake, and more earthquakes—resulting in further uplift—were recorded during the following year. In February, 1976, Wellington estimated the total uplift at 80–90 cm and there has been no apparent movement since then. Although the total uplift was not large, some mangrove trees have died, along with many stranded barnacles, and M. P. Harris noticed that flightless cormorants moved their nests forward as the shoreline receded. Similar biological changes must have taken place during countless uplifts in the past.

Evidence of large-scale subsidence is not obvious in Galapagos, but as Darwin has pointed out (1839: 560) "the movement itself tends to conceal all evidence of it". Subsidence is to be expected as plate motion carries older volcanoes away from hot-spots and spreading centres (Menard, 1969) and this is a good reason to urge modern marine geological investigation of the archipelago. On the islands themselves, however, the abundant evidence of Galapagos uplift testifies to the dynamic, ongoing growth of these islands.

EROSION AND OTHER SURFACE PROCESSES

Volcanism and faulting alter the Galapagos landscape in dramatic, often spectacular, events of short duration. Other processes make equally major alterations in the Galapagos landscape by small, unspectacular, but relentless changes, over long periods of time. In the mild, dry climate and relatively calm seas of Galapagos, these processes are not as effective as they are in other environments, but they are nonetheless important and will be discussed briefly below.

Weathering and alteration

Lava flow surfaces decompose through time by a variety of mechanical and chemical processes, eventually forming soil. Differential expansion and contraction with temperature change helps to disintegrate rock; surface temperature measurements on Fernandina's 1968 ash ranged from 4° to 49°C (39–120°F). Plants break up rock with rootlets and their own organic decomposition products. Fumarole vents actively leach and alter basaltic rock as documented by Colony & Nordlie (1973) on Sierra Negra. But effective weathering in most areas requires moisture combining with elements from the atmosphere, the biosphere, and the rocks themselves to produce acids capable of sustained disintegration of rock. Such moisture is present in some parts of the islands, with an annual rainfall of 2.6 m having been measured in the highlands of Santa Cruz, but in most areas it is not. Weathering is slow in the drier parts of the islands. Darwin referred to the "general indecomposable character of the lava" (1844: 114), and Cox (1971) describes a San Cristobal lava flow that appeared to be no more than a few decades old, but is probably (on the basis of its palaeomagnetism) considerably more than several centuries old. Baitis & Lindstrom (1980: 372) call attention to "the lack of extensive soil development or erosion in the past several hundred thousand years following the final activity on Pinzon".

Unfortunately little work has been done on the soils of Galapagos since the tragic death of Jacques Laruelle in 1967. His work, from 1963 onwards, is summarized in his book (1967); Eswaran, Stoops & De Paepe (1973) have added to that work on Santa Cruz. Clearly, though, the soils of Galapagos depend heavily on the distribution of moisture—that essential ingredient in the disintegration of lava—and the distribution of fragmental deposits from explosive eruptions. Fragmental deposits may mantle a whole volcano, like Alcedo, providing an easy footing for vegetation that could not be created on a hard lava surface by thousands of years of normal weathering.

Water movement above sea level

Most lava flows are severely broken up during the cooling process. They are therefore notoriously porous: rainfall swiftly disappears into cracks, and permanent running streams are not found on Galapagos. However, volcanic ash deposits—particularly the fine ash that is the last to fall—can compact to form a very impervious surface, and heavy rains are not unknown even at lower elevations in Galapagos (the Darwin Station recorded 15 cm of rainfall in one 24-h period in 1976). Rain on a fresh ash surface quickly runs off down hill, rapidly eroding the soft, fragmental deposit in the process. Radial valleys with intervening ribs are formed by such erosion on the flanks of Galapagos tuff cones; the process is dramatically shown on the deposits of Fernandina's 1968 eruption. Two years after the eruption, erosion had already cut 6 m deep, V-shaped gulleys into the soft, fragmental materials on the caldera floor, and after four more years the underlying lava flows were exposed, producing a 6 m wide ephemeral stream bed, bounded by 7 m vertical walls of tephra.

Groundwater movement below the surface is also an important process in Galapagos. Groundwater dissolves parts of the material through which it passes and later may precipitate these salts when encountering changed conditions. Waters percolating through the 1968 fragmental deposits of Fernandina, for example, have precipitated sodium sulphate thenardite as a white coating on the steep walls cut by the erosion. Higher temperature movement of water and vapour through tuff cones hardens these initially soft layers, in a palagonitization process, making them better able to withstand the forces of erosion.

Redistribution of elements from the freshly fragmented glass actually weakens the older glass, but strengthens the total rock (palagonite tuff) by depositing these elements, particularly silica, as a strong cement binding the fragments together. Because the movement of water and vapour through tuff cones is likely to be irregular, the resulting cementation may also be irregular. Certainly the resistance to erosion of Galapagos tuff cones is often uneven, with segments like the much-photographed Pinnacle Rock of Bartolome standing isolated from the remainder of its cone. More work needs to be done on the palagonite tuffs of Galapagos, where they were first recognized by Darwin, and on the subsurface movement of groundwater so important to residents of the archipelago.

Marine erosion

The most dramatic continuing force on Galapagos is the pounding surf, and evidence of its power is found in islands such as Wenman, Culpepper, Espanola, and Roca Redonda which are mere erosional remnants of their former structures. The sea is rarely mirror-like in Galapagos, and travellers hoping to land on seemingly quiet shores often find crashing waves that relentlessly modify the coasts. The ability of coastal materials to withstand such pounding varies widely. However, little is known of the rates at which marine erosion works in Galapagos. Darwin described the large palagonite cone of Tortuga (Brattle) Island as "in a ruined condition, consisting of little more than half a circle open to the south" (1844: 109). Nearly 150 years later, the island remains "little more than half a circle", reminding us that well-cemented tuff cones are not destroyed overnight. In contrast, comparison of Darwin's comments on Fresh-water Bay with today's Buccaneer Cove, on Santiago, shows more substantial changes over the years.

Sedimentation and beaches

The other side of the erosion coin is sedimentation: material removed from one place being later deposited in another. The 1968 fragmental deposits eroded from the floor of Fernandina caldera have been redeposited (along with periodic additions of lava) on the floor of the caldera lake, while material eroded from Galapagos coastlines is redeposited (along with marine organisms and detritus) offshore. Johnson *et al.* (1976) report sea-floor sediment thicknesses of about 500 m over much of the shallow Galapagos Platform, although some areas are scoured clear of sediment by strong bottom currents.

Some material eroded from Galapagos coastlines, however, is simply shifted by long-shore currents and redeposited nearby as beach sands. Beautiful, long beaches are found on many of the islands, particularly near the fragmental deposits of explosive eruptions. The composition of these beach sands varies from place to place, depending on the available materials nearby. Altered volcanic glass is a common constituent, and some sands are rich in coral and shell fragments. The white sands of Marchena are rich in feldspar from the large feldspar crystals common in lavas from that island, and there is a much-visited green olivine sand beach at Punta Cormoran of Floreana. Olivine-rich xenoliths from the crust below Floreana are common in the lavas of that island, and were explosively fragmented, along with the liquid lava that carried them to the surface in the eruption which built the prominent cone at Punta Cormoran. The development of a large beach, though, is a long and complicated process. Galapagos beaches are not easily renewable resources, and it is a matter of concern that sand had recently been stripped from several beaches on Santa Cruz for use in building construction on that island.

CONCLUSIONS

The processes that shape geologically vigorous volcanic islands are the same as those that shape all parts of our planet, but they are particularly vivid on Galapagos. There are no examples, it must be admitted, of glacial sculpturing, and the huge changes wrought by man elsewhere have been fortunately slight in Galapagos. The major geological processes *not* displayed in Galapagos, are the deep burial, metamorphism, overthrusting, and compressive folding of the world's subduction zones. These will come but not for another 20 million years or so when (if current plate motions continue) the present islands will be buried under South America.

ACKNOWLEDGEMENTS

Students of Galapagos, both resident and migratory, are as remarkable, in many respects, as the more celebrated wildlife of the islands. I am fortunate in having met many fine teachers, helpers and friends through Galapagos and space does not permit mention of all those to whom I owe a substantial debt. Among those who have generously supplied information about Galapagos eruptions, though, I especially thank Tui De Roy Moore, Dagmar Werner, a succession of Darwin Station staff members, and ham radio operators Bud Devine and Forrest Nelson. Many friends have helped make field work enjoyable as well as educational and I particularly thank Bruce and Penny Nolf, Pete Hall, Tui Moore, and Bob Smith. This manuscript has benefitted from thoughtful reviews by Dick Fiske, Bill Melson, and Bruce Nolf. Mary McGuigan has conjured many manuscript drafts and in out of the word processor with skill and admirable good humour. Lastly, I thank my wife, Sharon, not only for help on Galapagos volcanoes but also for answering more 'is this paragraph clear to a non-geologist?' questions that any non-geologist should be called upon to answer.

BIBLIOGRAPHY

The essential review of Galapagos geology is the memoir by McBirney & Williams (1969). A chronological list of geological treatments of the whole archipelago, excluding those listed by McBirney & Williams, is: Wolf (1892), Lewis (1956), Sauer (1965), Williams (1966), McBirney & Aoki (1966), Laruella (1967), Case *et al.* (1973), Durham & McBirney (1975), Hall (1977), Moore (1980a), and Cox (1983). Since the 1969 memoir, geological treatments of the western islands include: Simkin & Howard (1970), Simkin (1972), Nordlie & Colony (1973), Colony & Nordlie (1973), Delaney *et al.* (1973), and Nordlie (1973). Discussions of the central islands include: Hertlein (1972), Swanson *et al.* (1974), Baitis & Swanson (1976), Bow (1979), and Baitis &

Lindstrom (1980). Other post-1969 discussions of individual islands include those of Cox (1971) on San Cristobal, and Beate (1979) on Genovesa. The list below omits many references when later related work by the same author is cited.

The interested reader is referred to the authoritative volcanological textbooks by Macdonald (1972) and Williams & McBirney (1979). The excellent photographic compilation by Carr & Greeley (1980) illustrates and discusses many Hawaiian examples of volcanic features common in Galapagos.

REFERENCES

BAILEY, K., 1976. Potassium–Argon ages from the Galapagos Islands. *Science, 192:* 465–466.

BAITIS, H. W. & LINDSTROM, M. M., 1980. Geology, petrography, and petrology of Pinzon Island, Galapagos Archipelago. *Contributions to Mineralogy & Petrology, 72:* 367–386.

BAITIS, H. W. & SWANSON, F. J., 1976. Ocean rise-like basalts within the Galapagos Archipelago. *Nature, 259:* 195–197.

BALAZS, D., 1975. Lava tubes on the Galapagos Islands. *National Speleological Society Bulletin, 37:* 1–4

BALLARD, R. D., VAN ANDEL, T. A. & HOLCOMB, R. T., 1982. The Galapagos Rift at 86°W: 5. Variations in volcanism, structure, and hydrothermal activity along a 30-kilometer segment of the rift valley. *Journal of Geophysical Research, 87:* 1149–1162.

BEATE, B., 1979. *Geologia y petrografia de la Isla Genovesa, Galapagos.* Thesis de Grado, Escuela Politecnica. Quito.

BOW, C. S., 1979. *The geology and petrogenesis of the lavas of Floreana and Santa Cruz Islands, Galapagos Archipelago.* Ph.D. Thesis. Univ. Oregon.

CARR, M. H. & GREELEY, R., 1980. *Volcanic Features of Hawaii.* Washington: NASA Scientific & Technical Information Branch, SP-403. 211pp.

CASE, J. E., RYLAND, S. L., SIMKIN, T. & HOWARD, K. A., 1973. Gravitational evidence for a low-density mass beneath the Galapagos Islands. *Science, 181:* 1040–1042 (see also *184:* 808–809).

CHASE, T. E., 1968. *Sea Floor Topography of the Central Eastern Pacific Ocean.* U.S. Dept. Interior, Fish and Wildlife Service, Bureau of Commercial Fisheries, Circular 291.

COLINVAUX, P. A., 1968. Reconnaissance and chemistry of the lakes and bogs of the Galapagos Islands. *Nature, 219:* 590–594.

COLONY, W. E. & NORDLIE, B. E., 1973. Liquid sulfur at volcan Azufre [Sierra Negra], Galapagos Islands. *Economic Geology, 68:* 371–380.

CORLISS, J. B., DYMOND, J., GORDON, L. I., EDMOND, J. M. *et al.*, 1979. Submarine thermal springs on the Galapagos Rift. *Science, 203:* 1073–1083.

COUFFER, J. C., 1956. The disappearance of Urvina Bay. *Natural History, 65:* 378–383.

COX, A., 1971. Paleomagnetism of San Cristobal Island, Galapagos. *Earth & Plant. Science Letters, 11:* 152–160.

COX, A., 1983. Ages of the Galapagos Islands. In R. I. Bowman, M. Berson & A. E. Leviton (Eds), *Patterns of Evolution in Galapagos Organisms:* 11–24. San Francisco: American Association for the Advancement of Science, Pacific Div.

DARWIN, C., 1844. *Geological Observations on Volcanic Islands.* London: Smith, Elder & Co., 175 pp.

DELANEY, J. R., COLONY, W. E., GERLACH, T. M. & NORDLIE, B. E., 1973. Geology of the Volcan Chico area on Sierra Negra Volcano. Galapagos Islands. *Geological Society of America Bulletin, 84:* 2455–2470.

DURHAM, J. W., 1965. Geology of the Galapagos. *Pacific Discovery, 18:* 3–6.

DURHAM, J. W. & McBIRNEY, A. R., 1975. Galapagos Islands. In R. W. Fairbridge (Ed.), *Encyclopedia of World Regional Geology:* 285–290. Stroudsburg, Pa.: Dowden, Hutchinson & Ross.

ESWARAN, H., STOOPS, G. & DE PAEPE, P., 1973. A contribution to the study of soil formation on Isla Santa Cruz, Galapagos. *Pedologie, 23:* 100–122.

HALL, M. L., 1977. *El Volcanismo en el Ecuador.* Quito: Biblioteca Ecuador, 120 pp. (17–37 Galapagos).

HALL, M. L., RAMON, P., & YEPES, H., 1980. The subaerial origin of Española (Hood) island and the age of terrestrial life in the Galapagos. *Noticias de Galapagos, 31:* 21.

HERTLEIN, L. G., 1972. Pliocene fossils from Baltra (South Seymour) Island, Galapagos Islands. *Proceedings of the Californian Academy of Sciences Fourth Series XXXIX, (3):* 25–46.

HEY, R., 1977. Tectonic evolution of the Cocos-Nazca spreading center. *Geological Society of America Bulletin, 88:* 1404–1420.

HEYERDAHL, T., 1963. Archaeology in the Galapagos Islands. *Occasional Papers of the California Academy of Science, 44:* 45–51.

JOHNSON, G. L., VOGT, P. R., HEY, R., CAMPSIE, J. & LOWRIE, A., 1976. Morphology and structure of the Galapagos Rise. *Marine Geology, 21:* 81–120.

LARUELLE, J., 1967. *Galapagos.* Gent: Natuurwetenschappelijk Tijdschrift.

LEWIS, G. E. 1956. Galapagos Islands (Archipielago de Colon) Province. *Geological Society of American Memoir 65:* 289–291. (Handbook of South American Geology, edited by W. F. Jenks).

LIPPS, J. H. & HICKMAN, C. S., 1982. Paleontology and geologic history of the Galapagos Islands. *Geological Society of America, Abstracts for Annual Meeting:* 548.

LONSDALE, P., BATIZA, R. & SIMKIN, T., 1982. Metallogenesis at seamounts on the East Pacific Rise. *Marine Technology Society Journal, 16 (3):* 54–61.

LONSDALE, P. & SPIESS, F. N., 1979. A pair of young cratered volcanoes on the East Pacific Rise. *Journal of Geology, 87:* 157–173.

MACDONALD, G. A., 1972. *Volcanoes.* Englewood Cliffs, N. J.: Prentice-Hall, 510 pp.

McBIRNEY, A. R. & AOKI, K., 1966. Petrology of the Galapagos Islands. In R. I. Bowman (Ed.), *The Galapagos:* 71–77. Berkeley: University of California Press.

McBIRNEY, A. R. & WILLIAMS H., 1969. Geology and petrology of the Galapagos Islands. *Geological Society of America Memoir, 118:* 197 pp.

MELVILLE, H., 1856. The Encantadas. In *Piazza Tales.* New York: Dix & Edwards, 66 pp.

MENARD, H. W., 1969. Growth of drifting volcanoes. *Journal of Geophysical Research, 74:* 4827–4837.

MONTORIOL-POUS, J. & DE MIER, J., 1977. Contribucion al conocimiento vulcano-espeliologico de la Isla de Santa Cruz (Galapagos, Ecuador). *Speleon, 23:* 75–91.

MOORE, T. DE ROY 1980a. *Galapagos—Islands Lost in Time.* New York: Viking.

MOORE, T. DE ROY, 1980b. The awakening volcano. *Pacific Discovery, 33 (4):* 25–31.

MORGAN, W. J., 1971. Convection plumes in the lower mantle. *Nature, 230:* 42–43.

NORDLIE, B. E., 1973. Morphology and structure of the western Galapagos volcanoes and a model for their origin. *Geological Society of America Bulletin, 84:* 2931–2956.

NORDLIE, B. E. & COLONY, W. E., 1973. Fumarole with periodic water fountaining, Volcan Alcedo, Galapagos Islands. *Geological Society of America Bulletin, 84:* 1709–1720.

NORDLIE, B. E., THIEBEN, S. E. & DELANEY, J. R., 1982. Chemical composition of the basalts of Cerro Azul, Volcan Wolf, and Sierra Negra, Western Galapagos Islands. *IAVCEI-IAGC Scientific Assembly, Reykjavik. Iceland, Abstract 41.*

PETERSON, D. W. & SWANSON, D. A., 1974. Observed formation of lava tubes. *Studies in Speleology, 2:* 209–222.

RICHARDS, A. F., 1957. Volcanism in Eastern Pacific Ocean Basin: 1945–1955. *Cong. Geol. Internatl., 20 sess,* 19–31.

RICHARDS, A. F., 1962. Archipelago de Colon, Isla San Felix and Islas Juan Fernandez. *Catalog of Active Volcanoes of the World 14,* Rome: IAVCEI. 50 pp.

SAUER, W., 1965. *Geologia del Ecuador.* Quito: Talleres Graficos del Ministerio de Educacion, 383 pp. (344–360 on Galapagos).

SCHILLING, J.-G., KINGSLEY, R. H. & DEVINE, J. D., 1982. Galapagos hot spot sprading center system. 1. Spatial petrological and geochemical variations (83°W–101°W). *Journal of Geophysical Research, 87:* 5593–5610.

SHIMIZU, H., MASUDA, A. & MASUI, N. 1981. Rare-earth element geochemistry of volcanic and related rocks from the Galapagos Islands. *Geochemical Journal, 15:* 81–93.

SIMKIN, T., 1972. Origin of some flat-topped volcanoes and guyots. *Geological Society of America Memoir, 132:* 183–193.

SIMKIN, T. & HOWARD, K. A., 1970. Caldera collapse in the Galapagos Islands, 1968. *Science, 169:* 429–437.

SIMKIN, T., REEDER, W. G. & MACFARLAND, C. (Eds), 1974. *Galapagos Science: 1972 status and needs.* Published by Smithsonian Institution for Charles Darwin Foundation, Wash., D.C., 87 pp.

SIMKIN, T. & SIEBERT, L., (1984). Explosive eruptions in space and time: durations, intervals and a comparison of the world's active volcano belts. In F. R. Boyd (Ed.), *Explosive Volcanism:* 110. Washington: National Academy of Sciences.

SIMKIN, T., SIEBERT, L., McCLELLAND, L., BRIDGE, D., NEWHALL, C., & LATTER, J. H., 1981. *Volcanoes of the World.* Stroudsburg, PA: Hutchinson Ross, 240 pp.

STEADMAN, D. W., 1981. Vertebrate fossils in lava tubes in the Galapagos Islands. *Proceedings of the 8th International Congress of Speleology:* 549–550.

STEADMAN, D. W., 1982. The origin of Darwin's Finches (Fringillidae, Passeriformes). *Transactions of the San Diego Society of Natural History, 19 (19):* 279–296.

SWANSON, F. J., BAITIS, H. W., LEXA, J. & DYMOND, J., 1974. Geology of Santiago, Rabida, and Pinzon Islands, Galapagos. *Geological Society of America Bulletin, 85:* 1803–1810.

WHITE, W. M., 1979. Pb isotope geochemistry of the Galapagos Islands. *Carnegie Institute of Washington, Washington Yearbook, 78:* 331–335.

WHITE, W. M. & HOFMANN, A. W., 1978. Geochemistry of the Galapagos Islands: implications for mantle dynamics and evolution. *Carnegie Institute of Washington, Washington Yearbook, 77:* 596–606.

WILLIAMS, H., 1966. Geology of the Galapagos Islands. In R. I. Bowman (Ed.), *The Galapagos:* 65–70. Berkeley: University of California Press.

WILLIAMS, H. & McBIRNEY, A. R., 1979. *Volcanology.* San Francisco, Freeman, Cooper, 397 pp.

WILSON, J. T., 1963. Hypothesis of earth's behaviour. *Nature, 198:* 925–929.

WOLF, T., 1982. *Geografia y Geologia del Ecuador.* Quito: Casa de la Cultura Ecuatoriana (1975 republished), 798 pp. (517–542 on Galapagos).

Biological Journal of the Linnean Society (1984), *21:* 77–95. With 2 figures

A new look at evolution in the Galapagos: evidence from the late Cenozoic marine molluscan fauna

M. J. JAMES

Department of Paleontology, University of California, Berkeley, California 94720, U.S.A.

Endemism is not as common in the marine invertebrate fauna of the Galapagos Islands region as in the adjacent terrestrial biota. Marine invertebrates in the Galapagos are largely cosmopolitan species from the Panamic, Indo-Pacific, Californian, or Peruvian faunal provinces. However, an endemic component is also present in the fauna. The observed pattern among marine invertebrate organisms can be accounted for by at least two processes: (1) genetic continuity between mainland and island populations mediated through planktonic larvae; and (2) lower rates of intrinsic evolutionary change. The evolutionary scenario standardly applied to terrestrial organisms in the Galapagos, namely, adaptive radiation and speciation in reproductive isolation from mainland source populations, does not apply to all marine invertebrates. Evidence in support of the alternative scenario for marine invertebrates comes from both published records of species occurring in the islands and recent studies of fossil-bearing deposits on several islands in the archipelago. Two misconceptions—considering the islands and sedimentary deposits to be older than now thought, and equating the rate of evolution of the terrestrial biota with the marine biota—can lead to an incorrect interpretation of evolution in the Galapagos. Contrasts between marine invertebrate and terrestrial organisms serve to illustrate some fundamental differences which have important evolutionary implications. Some of these are: endemism; dispersal; taxonomic relationships; island definitions; rates of evolutionary change; and age of fossils. In terms of Darwin's evolutionary scenario, terrestrial organisms represent the paradigm and marine organisms represent the paradox.

KEY WORDS:—Galapagos Islands – Mollusca – marine invertebrates – fossils – Cenozoic – evolution – biogeography.

CONTENTS

0024–4066/84/010077 + 19 $03.00/0

INTRODUCTION

The Galapagos Islands have been of great interest to evolutionary biologists for many years. This is primarily due to the unique terrestrial biota that stands as a paradigm to the work of Charles Darwin. Adaptive radiation and speciation in reproductive isolation from mainland source populations are the processes invoked to account for the pattern of endemic insular taxa. Unfortunately, the marine invertebrate fauna has not received as much attention as the well-known terrestrial biota (Hedgpeth, 1969). The purpose of this paper is to review what is known about evolution of marine invertebrates (particularly molluscs) in the Galapagos, using examples from Neogene and Recent taxa. Investigation of evolutionary rates and patterns among Galapagos organisms will result in an understanding of evolutionary mechanisms among the studied species.

In February 1982, I participated in a palaeontological expedition to the Galapagos Islands in which the primary goal was to collect both palaeontological and stratigraphic information. Previous research expeditions to the islands have collected fossils either incidentally or as a secondary objective, usually while studying a neontological topic. A systematic and thorough field sampling programme was undertaken (see Fig. 1 for all island localities mentioned in text). The main objective was to gain an improved understanding of the taxonomy, stratigraphy, palaeoecology, palaeobiogeography, and evolutionary implications of previously-reported Late Cenozoic fossil faunas of the Galapagos.

Also participating in the 1982 expedition were William D. Pitt (trip organizer, Sacramento, California) and Lois J. Pitt; Carole S. Hickman (University of California, Berkeley); and Jere H. Lipps (University of California, Davis). The results of this expedition (which represent the combined efforts of all five participants) and additional information assembled from the literature by this author have shed new light on the taxonomic, biogeographic, and evolutionary relationships of the Late Cenozoic marine invertebrate fauna of the Galapagos Islands.

GEOGRAPHIC SETTING

The Galapagos Islands are true oceanic islands situated astride the equator spread over a wide geographic area ranging from longitude 89° 15′ 30″W to 92°

01′ 00″ W and from latitude 1° 40′ 00″ N to 1° 36′ 00″ S (Fig. 1). Geographically, these islands are not a compact, coherent group, as they extend both above and below the equator, and are influenced by different oceanographic conditions in the north and south (Abbott, 1966; Wooster and Hedgpeth, 1966). However, the Galapagos are by no means the most isolated island group in the world, being surpassed in this regard by the Hawaiian Islands.

BIOGEOGRAPHIC SETTING

The marine invertebrate biota (particularly the molluscs) of the Galapagos consists of three distinct components: (1) a tropical eastern Pacific element consisting of species from the Panamic faunal province (Keen, 1971); (2) true

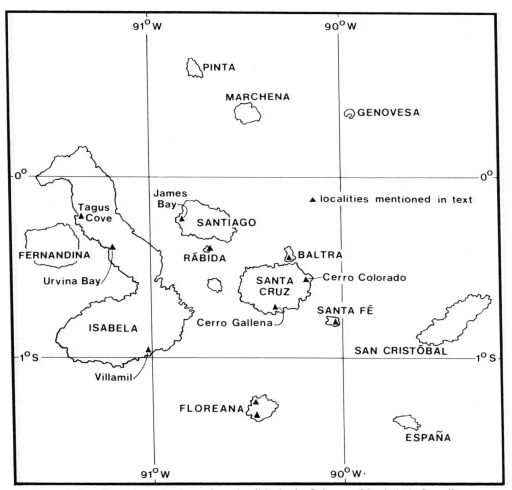

Figure 1. Late Cenozoic marine invertebrate localities in the Galapagos Islands (▲). Co-ordinates of fossil locality on San Cristobal metioned by Charles Darwin (see text for discussion) are unknown.

endemic shallow-water species that are unique to the islands, only a few of which are higher taxa (Keen, 1971), and (3) vagrant species from the Indo-Pacific, Californian, and Peruvian faunal provinces (Emerson, 1967, 1978; Marincovich, 1973; Lindberg & McLean, 1981). Species in the first component occur in both the Galapagos and mainland populations. The second component consists of valid species, based on neontological or palaeontological definitions, that are either extant or extinct in the islands today. These taxa arrived in the islands, diverged sufficiently to be considered valid species and, by definition, do not occur elsewhere in the world. Populations of the third component have become established on the islands through dispersal mechanisms, such as oceanic current transport. A fourth, and as yet unrecognized, faunal element appears to be a deep-water highly endemic species group (J. W. Valentine, pers. comm., 29 January 1983; currently under investigation by J. W. Valentine & D. Jablonski). Deep-water (i.e. bathyal and abyssal) highly endemic gastropod faunas are not unknown or unusual. Hickman (1976) has documented a fauna of bathyal gastropods of the family Turridae from Oligocene deposits in Oregon.

Although unlikely, biogeographic affinities between the Galapagos fossil fauna and species in the Caribbean faunal province cannot be ruled out completely. Keigwin (1978) has stated that the Panama Land Bridge closed at 3.1 My B.P., effectively preventing marine faunal interchange between the two areas after that time. Cox (1983) and Simkin (in press) have summarized information about ages of the Galapagos, and Hall, Ramon & Yepes (1980) have presented evidence suggesting that subaerial islands existed in the Galapagos Archipelago at least 3.3 My ago. The difference of about 0.2 My (or 200 000 years) would be the 'window' during which deposits (as yet undiscovered) in the Galapagos could potentially contain taxa with direct affinities to the Caribbean faunal province. However, Geister (1977) has presented evidence that several seaways existed between the eastern Pacific and Caribbean during late Pleistocene interglacial high sea-level stands. Using occurrences of the reef coral *Pocillopora* in late Pleistocene Caribbean deposits, Geister presented arguments for either survival of *Pocillopora* from the Tertiary Caribbean coral fauna or a hypothetical re-introduction from the eastern Pacific. Durham (1980) considered the appearance of *Pocillopora* in Pleistocene deposits on Guadalupe Island, Mexico (eastern Pacific), to be synchronous with some of the Caribbean occurrences. Vermeij (1978: 258–266) has reviewed problems of the direction and magnitude of biotic interchange between the eastern Pacific and Caribbean if a sea-level connection occurred. If the Pleistocene openings of the Isthmus of Panama were of sufficient duration, then biotic contact between the Caribbean and eastern Pacific (and thus Galapagos) could have occurred intermittently.

An interesting pattern has emerged from the study of Indo-Pacific faunal elements occurring in the eastern Pacific (Emerson, 1967, 1978). Occasional Indo-Pacific vagrant species occur on true oceanic islands in the eastern Pacific. However, very few species successfully transverse the relatively short distance to the tropical American mainland and, to date, no Panamic species are known to have undergone the reverse pattern of establishing populations in the Indo-Pacific faunal province (Emerson, 1967). This pattern of unequal interchange between the two provinces is not adequately explained by the lack of distributing mechanisms from the Panamic province to the Indo-Pacific province (Abbott, 1966).

Factors influencing the biogeographic affinities of the Galapagos marine invertebrate fauna have been summarized by Abbott (1966). These are: (1) distributing mechanisms and access routes; (2) physical barriers; (3) temperature controls on colonization; and (4) incidence of pelagic larvae and timing of reproductive cycles. Distributing mechanisms and access routes consist of ocean currents that carry planktonic larvae to and from the islands. To a lesser degree, adult marine invertebrates can also be transported on drifting debris. A more remote possibility is transport to the islands as part of the fouling community on ship bottoms. Physical barriers to distribution consist of the Panama Land Bridge and Ekman's East Pacific Barrier (Ekman, 1953: 72–77), both of which obstruct east–west exchanges, and various ocean currents that obstruct north–south exchanges. Temperature regimes in inshore waters around the islands control the successful colonization by species of particular temperature tolerances. The incidence of pelagic larvae and the timing of reproductive cycles affect the 'raw material' available for oceanic current dispersal.

In terms of numbers of species, the Galapagos invertebrate fauna is depauperate (Table 1) with respect to the Panamic faunal province (Emerson, 1967). Kay (1979: 14) comments on the similarly depauperate or attenuate nature of the Hawaiian mollusc fauna in comparison to the Indo-Pacific faunal province. Several physical factors may explain the reduced species diversity of the Galapagos: (1) the islands are relatively isolated (925 km west of Cabo San Lorenzo, Ecuador) from mainland source populations; (2) during part of the year oceanographic circulation brings cold water from the Humbolt (Peru Coastal) and the Peru Oceanic Currents to the islands (Abbott, 1966), which can operate as a selection mechanism preventing the establishment of stenothermic species; and (3) ocean currents in the eastern Pacific responsible for transporting planktonic larvae to the islands can be weak, indirect, or inconsistent. El Nino events (Wyrtki, Stroup, Patzert, Williams, & Quinn, 1976), i.e., the appearance of warm, low salinity water in the eastern Pacific, increase growth rates of some coral species in the Galapagos (Druffel, 1981). The absence of true coral reef environments in the Galapagos and the Panamic province in general, accounts in part for the depauperate nature of the fauna (Emerson, 1967). However, substantial localized coral growth does occur in the Galapagos (Wellington, 1978; Glynn, Wellington, & Birkeland, 1979). The implications of the geological nature of the islands for the shallow marine fauna and reef development have been investigated by Rosen (in prep.). A factor that does not limit the development of species populations in the islands is their geologic age, now considered to be Pliocene or Pleistocene based on K-Ar dating techniques (Cox, 1983, Simkin, in press). Both the probabilities of successful transport (dispersal) and successful population establishment (colonization) must be considered when evaluating the taxonomic and biogeographic affinities of species in the Galapagos.

An Atlantic Ocean analog to the Galapagos Islands may be St Paul's Rocks, an isolated group of barren islets lying on the mid-Atlantic ridge just north of the equator (Edwards & Lubbock, 1983). However, important differences exist, particularly concerning the degrees of endemicity seen in the two areas. In the case of Galapagos molluscs, about 42% of the species are endemic (Table 1), but at St Paul's Rocks only 5% of the shallow water marine fauna is endemic (four species of fishes) (Edwards & Lubbock, 1983). Many biotic differences observed between the Galapagos and St Paul's Rocks are accounted for by components of MacArthur & Wilson's (1967) theory of island biogeography, such as isolation, source areas, and area-diversity patterns. Williamson (1981: 82–92, 1983) has presented a critique of MacArthur & Wilson (1967) pointing out that the "popular but obsolete" (Williamson, 1983) theory is being replaced as new theories are developed. Berry (1979) has suggested a cautious approach to the application of island biogeography theory, preferring instead a species-by-species analysis to more accurately determine causal mechanisms.

THE GALAPAGOS MALACOFAUNA

Table 1 presents a compilation of information taken mainly from Keen (1971) on the occurrence of molluscs in the Panamic province and the Galapagos. This is only a first approximation of the malacofauna of the islands and should be interpreted as such. Much additional information must be gathered from field studies and the literature before complete knowledge of Galapagos molluscs is possible. When additional distributional data becomes available, the exact values in Table 1 will change but the general pattern will likely remain.

Two conclusions can be made from the data in Table 1. First, the Galapagos islands contain only a small percentage (12%) of the molluscan species found in the Panamic province (323 of 2803 species, or 12%). For those taxonomic groups that occur in the Galapagos (i.e. excluding Aplacophora and Monoplacophora), only between 6% and 31% of the known Panamic species are found in the islands. Thus, the Galapagos can be considered a depauperate outpost of the Panamic province. Second, overall species endemism is moderately high (mean 42%) using the criteria of Briggs (1966). If one considers only the prosobranch gastropods (taxonomically the most intensely studied group of molluscs), then the degree of species endemism is higher (45%). However, the pattern of molluscan endemism established here does not conform to the predictions of Briggs (1966) concerning degrees of endemism in relation to changes in marine palaeotemperatures during the Pleistocene glaciations. Lindberg, Roth, Kellogg & Hubbs (1980) consider the distance of an island (or island group) from the mainland as an ameliorating factor in the influence of Pleistocene glaciations on marine palaeotemperatures. Given these conditions, high endemism will be predicted, contrary to Briggs (1966), for the Galapagos. Other groups of Galapagos marine organisms exhibit lower endemism (shore fishes 27% (Walker & Rosenblatt, 1961) and brachyuran crabs 15% (Garth, 1946)) and conform to the predictions of Briggs (1966).

For the molluscan groups presented in Table 1 that occur in the islands, endemism ranges from 0% (Scaphopoda) to 75% (Cephalopoda). The classes Aplacophora and Monoplacophora are apparently unknown from the Galapagos according to Keen (1971). Only 6% of the Panamic Bivalvia occur in the Galapagos. This low figure is not completely explained by the dispersal abilities of bivalves (see Jablonski & Lutz, 1983). Perhaps it is because bivalves have been overlooked by collectors in the area. The Scaphopoda are not represented by endemic species in the Galapagos.

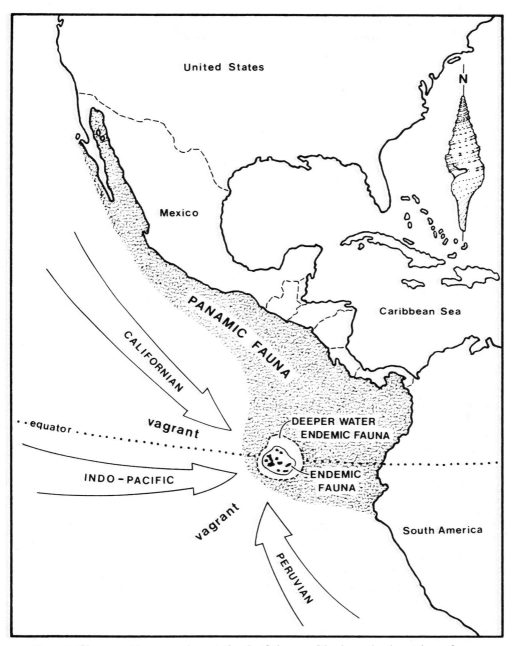

Figure 2. Biogeographic source elements for the Galapagos Islands marine invertebrate fauna, relying particularly on molluscan distributional data. Large arrows indicate faunal elements dispersed to the Galapagos from the Indo-Pacific, Peruvian, and Californian faunal provinces. The bulk of the Galapagos marine biota originates in the Panamic province. Near-shore and deeper water endemic faunas are Panamic subsets.

Table 1. Composition of the Galapagos Islands malacofauna. Data from Keen (1971), with additional information from Lindberg & McLean (1981), Smith & Ferreia (1977), and Ferreia (1978)

Taxonomic group	No. of Panamic spp.*	No. of Galapagos spp.†	Percentage Panamic in Galapagos‡	No. of Galapagos endemics§	Percentage endemic¶
Prosobranchia					
Archeogastropoda	182	37	20	20	54
Mesogastropoda	797	88	11	34	39
Neogastropoda	893	127	14	59	47
Bivalvia	803	44	6	10	23
Polyplacophora	54	12	22	8	67
Scaphopoda	25	5	20	0	—
Cephalopoda	13	4	31	3	75
Aplacophora	1	0	0	0	—
Monoplacophora	3	0	0	0	—
Marine Pulmonata	32	6	19	1	17
Total	2803	323	12	135	42

* Includes all species in Panamic and Galapagos
† Includes both widespread and endemic species
‡ Column 2 divided by column 1 × 100
§ Occur only in Galapagos Islands
¶ Column 4 divided by column 2 × 100

ALTERNATIVE EVOLUTIONARY SCENARIOS

There is an apparent paradox between the two *a priori* biogeographic or evolutionary scenarios that, stated implicitly or explicitly, have occasionally appeared in the literature on marine invertebrate organisms in the Galapagos. The first scenario assumes that, because the islands are isolated in a cold water regime where currents are of temporal uncertainty, the marine invertebrate fauna has undergone adaptive radiation and speciation to a comparable degree as the well-known terrestrial biota. As a result of this line of reasoning, some taxonomists have accentuated morphological differences rather than similarities between species. Although difficult to pinpoint, this perspective seems to exist occasionally in the taxonomic literature as an unspoken assumption. One would be implying near heresy, in an evolutionary sense, to conclude that evolution of the marine biota has *not* been as spectacular as the evolution of the terrestrial biota. A recent example of this attitude has been clearly stated by Bratcher & Burch (1971: 537) in a paper on the taxonomy of marine neogastropods: "Since Charles Darwin's day, the Galapagos Islands have fascinated naturalists as being the 'cradle of new species' for both vertebrate and invertebrate fauna, so it is not surprising that some of the *Terebra* species should prove to be new". Mayr (1978) described similar roadblocks against the advancement of evolutionary thinking as "silent assumptions, never fully articulated". Given the available modern evidence, I think it is time to re-evaluate some of the taxonomic interpretations of invertebrate organisms in the Galapagos.

The second *a priori* scenario considers some of the same evidence as in the first scenario but arrives at a different conclusion. This alternative view considers ocean currents in the eastern Pacific as dispersal routes to the islands for planktonic larvae. This would result in greater faunistic similarity to mainland populations and hence high taxonomic diversity when compared to island size.

The conflicting notions of isolation, dispersal, and rate of evolutionary change (or species longevity, see below) are difficult to reconcile. Stanley (1979: 170) suggested that radiation of shallow-water marine invertebrates around island archipelagos may be more common than generally thought. However, two of the properties of these animals, namely, high dispersal ability and low rate of speciation (Stanley, 1979: 260–265 and 229–260, respectively) suggest otherwise.

Latitudinal environmental gradients may exist for marine organisms in the Galapagos owing to differences in water-mass characteristics brought by currents differentially affecting the northern or southern islands in the archipelago (Abbott, 1966). These gradients can be greater than climatic differences on land. For marine invertebrates, the pattern of population distribution may be quite complex due to ocean current patterns within the islands themselves. Further studies of single species are needed to determine patterns of geographic variation between islands for comparison to differences with mainland populations.

PALAEONTOLOGICAL CONSIDERATIONS

Previous investigations

Accounts of the occurrence of fossil-bearing deposits in the Galapagos extend back to observations made by Charles Darwin in 1835. He noted the presence of molluscan shells on Chatham Island (San Cristobal) "embedded several hundred feet above the sea, in the tuff of two craters, distant from each other" (Darwin, 1891: 130). It is not known if Darwin collected specimens from these tuff cones, and I found no material in the British Museum (Natural History) amongst the invertebrate fossils collected by Darwin during the *Beagle* voyage. Although not observed directly, the mechanism of fossilization for beds observed by Darwin on San Cristobal is likely to be similar to that described from Cerro Gallena and Cerro Colorado on Santa Cruz Island by the 1982 expedition (MS in prep.), and is considered to be a widespread phenomenon in the Galapagos. Observations by the 1982 expedition with a discussion of the mechanism of formation of this little-known and poorly-understood mode of fossilization will be presented at a later date (MS in prep.). Berthiaume (1938) and Durham (1942) have described other examples of Cenozoic marine fossils preserved in volcanic tuffs. Wolf (1895) described "sparse shell remains" in palagonite tuff 100 m above sea level in the Galapagos but did not discuss the exact location of these deposits.

Dall & Ochsner (1928) presented the first extensive treatment of Cenozoic fossils from the Galapagos. In addition to describing fossil localities from three islands and commenting on the likely geologic ages of the deposits, they described 39 new molluscan species that were then considered endemic, extinct species. Many of these species have since been shown to be synonyms of Recent Panamic species (Keen, 1971). The perspective of Dall & Ochsner (1928) seems to have been one of emphasizing taxonomic and morphological differences rather than similarities. Additional references to Galapagos fossils were made by Dall (1924). However, in contrast to their taxonomic opinions, Dall & Ochsner

(1928) correctly considered the marine molluscan affinities of the Galapagos to be with the Panamic faunal province, an opinion shared by Hertlein & Strong (1939) in their in-depth treatment of fossils from the islands. Hertlein & Strong (1939) also commented on the occasional affinity between Galapagos species and the Caribbean and Indo-Pacific faunas. Hertlein (1972) documented the presence of 30 bivalve and 77 gastropod species from fossil deposits on Isla Baltra. Emerson (1967, 1978), and Zinsmeister & Emerson (1979) have reviewed occurrences of Recent molluscs with Indo-Pacific affinities in the Galapagos. Sufficient quantities of information (such as that presented in the recent studies cited above) are just becoming available to allow important biogeographic conclusions to be drawn. These can be made by comparing the temporal composition of the molluscan fauna through analysis of both living and fossil species occurrences.

Previous age assignments

Dall & Ochsner (1928) attribute fossil deposits at Vilamil (Isla Isabela), Isla Baltra, and Cerro Colorado (Isla Santa Cruz) to Tertiary and Pleistocene ages without being specific about age-determination criteria. Based on unspecified fossil evidence, the beach deposit at Vilamil is determined to be younger than either of the deposits found at Cerro Colorado and Baltra. Hertlein & Strong (1939) assign Late Pleistocene or subfossil ages of deposits from James Bay (Isla San Salvador), Isla Baltra, Tagus Cove (Isla Isabela), and Isla Rabida. Hertlein (1972) re-evaluated the age of fossil deposits on Isla Baltra (previously considered Late Pleistocene by Hertlein & Strong, 1939) and assigned a Pliocene age to the deposits. However, he did not attempt a more specific refinement to Upper, Middle, or Lower Pliocene. Durham (1965a, b), and Durham & McBirney (1975) assigned the oldest age to any fossil deposit in the Galapagos by considering fragmental blocks of fossiliferous marine limestone from near Cerro Colorado (Isla Santa Cruz) to be Late Miocene. Unfortunately, specific age determination criteria were not reported by Dall & Ochsner (1928), Hertlein & Strong (1939), Durham (1965a, 1965b), Hertlein (1972), or Durham & McBirney (1975). However, Durham (1979) stated that his late Miocene age-determination was based on "a large pecten that is difficult to differentiate from *Lyropecten crassicardo* (Conrad, 1857) of the California upper Miocene".

RESULTS OF THE 1982 EXPEDITION

Results of the 1982 expedition have been summarized briefly by Lipps & Hickman (1982), and Pitt & James (1983). Pitt (in press) has presented a brief narrative description of the 1982 expedition. A summary and a brief elaboration of results to date are presented below. Results of the 1982 expedition are used here to show how fossil evidence bears on the question of evolutionary rates and patterns among Galapagos organisms.

Types of deposits

Lipps & Hickman (1982) have described and classified six types of marine deposits bearing remains of organisms. These deposits include: (1) tuff cones with marine fossils; (2) limestone and sandstone interbedded with basalt flows; (3) terrace deposits above sea level; (4) beach rock; (5) supratidal talus debris, and (6) recently uplifted (3 m in 1954 (Richards, 1957)) tidal and subtidal rocks and sand.

Revised age assignments

Previous workers have reported that the ages of these deposits (with the exception of (6) above, Urvina Bay) were Miocene, Pliocene and/or Pleistocene. Lipps & Hickman (1982) stated that these ages cannot be reconciled with presently-available geologic evidence (such as Cox, 1983 and Simkin, in press). Taxonomic work in progress in Pitt & James (1983) has shown that the majority of fossils are represented by living species. All fossil specimens found embedded in tuff cones at Cerro Gallena and Cerro Colorado on Isla Santa Cruz are assignable to recent species known to be living in the near-shore waters of the Galapagos or elsewhere in the Panamic faunal province (MS in prep.). This is a conservative estimation of the taxonomic affinities of these marine invertebrate fossils, because many of the standard taxonomic characters, such as shell colour pattern and internal soft anatomy (e.g. radular morphology), are not preserved. Lipps & Hickman (1982) suggested that the tuff cones, interbedded sedimentary rocks, and terrace deposits on most islands are no older than about two million years, and that beach rock and talus deposits are at most a few hundred years old. They relied mainly on the information in Cox (1983) for age-determination criteria (J. H. Lipps and C. S. Hickman, pers. comm. 1982). These interpretations are compatible with plate tectonic and volcanologic interpretations of the young age of the Galapagos Islands (Cox, 1983; Simkin, in press; see also Pitt, in press, fig. 2).

Stratigraphic correlation

The dynamic processes of volcanic island formation and incorporation of sedimentary fossil deposits appear to have been quite variable throughout the island chain. Each deposit in the six-part classification scheme of Lipps & Hickman (1982) can alternatively be considered unique temporally, lithologically, and palaeontologically (C.S. Hickman, pers. comm., 27 March 1982). This would reduce the likelihood of ever successfully making an archipelago-wide stratigraphic correlation without the aid of additional evidence, such as isotopic dates or biostratigraphy. Work in progress (by JHL) on the foraminifera of several deposits will hopefully clarify and contribute to our understanding of the age, environmental, and stratigraphic relationships of the deposits.

Biogeographic affinities

Basic taxonomic decisions ultimately determine the biogeographic (and evolutionary) affinities of a fauna. To date, most fossil specimens collected by the 1982 expedition have been assigned to taxa with living representatives in the Panamic molluscan faunal province. However, one fossil from Isla Baltra appears to be a representative of the Indo-Pacific molluscan faunal province, a specimen tentatively assigned to an Indo-Pacific species of the widespread archeogastropod genus *Nerita* (W. D. Pitt, pers. comm., 7 April 1983). This represents the first known example of a fossil representative of the vagrant species component of the Galapagos (Emerson, 1967, 1978). Fallaw (1983) has recently shown that narrowing of the Pacific Ocean basin during Mesozoic and Cenozoic time aided migration of marine invertebrates between the Americas and the Indopacific–Asian landmasses.

Having presented background information on the Galapagos and a discussion of the results of the 1982 expedition, some of the differences between terrestrial and marine organisms are considered below. These serve to illustrate why different evolutionary scenarios might apply to the two groups.

TERRESTRIAL–MARINE COMPARISONS

Biological contrasts between the land and sea biotas of the Galapagos have generally not been made. However, this is unfortunate because some interesting differences emerge suggesting important evolutionary implications.

Endemism and higher taxa

A review of the taxonomic literature of the Galapagos provides two observations concerning endemism and the incidence of higher taxa. These suggest different evolutionary scenarios for terrestrial and marine organisms. Firstly, endemism is observed to be lower in the marine invertebrate fauna than in the terrestrial biota (Hedgpeth, 1969). Secondly, very few endemic marine invertebrate higher taxa have been described from the islands. Kay (1979: 15) noted a similar situation with respect to endemicity at higher taxonomic levels for marine molluscs in the Hawaiian Islands. Thus, when evolutionary divergence has occurred it has been manifested mainly at the species level. Recognition of higher taxa is a subjective process dependent on the opinions of individual taxonomists specializing in particular groups. As a consequence, some taxonomic groups will have higher taxonomic levels (usually genera) described as endemic from the Galapagos while others will not. Stanley (1979: 138) provided one of the few coherent generic concepts I have read. He stated that a typical genus forms by one or a small number of *markedly divergent* speciation steps (emphasis mine). By highlighting not only the number of speciation events but also a high degree of divergence per speciation, Stanley has emphasized both the quantity and magnitude of evolutionary change required to provide a new genus. Separating out biological reality from taxonomic artifact is of supreme importance when trying to understand the evolutionary history of organisms in the Galapagos.

Dispersal stage and mechanism

One important attribute of marine invertebrate reproductive biology, in contrast to that of terrestrial organisms, lies in the fact that many groups, including some molluscs, exchange genetic material *via* planktonic larvae (Thorson, 1950). Terrestrial vertebrates possess no such readily-dispersed stage (Matthew, 1930). Furthermore, the mechanism of transport for marine invertebrates is largely through drifting in ocean currents in contrast to being blown by the wind (e.g. birds, plants), rafting (e.g. plants, reptiles), or attachment to birds (e.g. plants) for terrestrial organisms (see Carlquist, 1974: 45–96). MacArthur & Wilson (1967: 157–159) have pointed out that the common loss of dispersal abilities in many terrestrial island organisms (such as birds, insects, and plants) is directly attributable to alteration of dispersal mechanisms *via* "various combinations of several morphological changes". A fundamental difference exists in the biological stage of development at which dispersal occurs in marine invertebrates and terrestrial vertebrates. In the former, dispersal occurs at the larval stage very early in the life history of the organism, and in the latter, dispersal is by way of gravid females (or more than one individual of separate sexes). The establishment of the fact of larval dispersal in marine molluscs has been thoroughly reviewed by Lutz & Jablonski (1980). Plant dispersal in and between islands usually involves both a specialized seed adaptation and a bird vector (see Carlquist, 1974: 45–96).

Probability of dispersal

Given the differences in dispersal stage and mechanism in marine invertebrate and terrestrial organisms, dispersal of terrestrial organisms is likely to be an accidental event of low probability (with the possible exception of plants), whereas in marine organisms dispersal is a life history event of higher (almost mandatory) probability. The probability of dispersal for an organism has important evolutionary consequences (see Stanley, 1979: Ch. 9), and hence, the differences between marine invertebrate and terrestrial organisms are of great value.

Dispersal and the founder principle (Mayr, 1942) have been critically examined by Rotondo, Springer, Scott & Schlanger (1981) in relation to the establishment of endemic biotas. Island integration (Rotondo, *et al.*, 1981) is suggested as a possible alternative mechanism in the formation of Hawaiian Islands endemism. Application of this concept to the Galapagos Islands is uncertain because the Galapagos are not a linear island chain (Byerly, 1980).

Degree of isolation

The higher probability of genetic contact with mainland source populations through planktonic larvae means that marine invertebrates experience a lower degree of isolation than the majority of terrestrial organisms. Isolation of the island group by sheer distance from the mainland has been important evolutionarily and now has major biogeographic consequences for the terrestrial biota. However, this pattern is not observed to the same degree in the marine biota due to the dispersal of planktonic larvae by ocean currents.

Taxonomic hierarchy

Consideration of the evolutionary relationships of the Galapagos biota usually involves several taxonomic comparisons, which can be ranked in a hierarchical fashion. Firstly, when investigating ancestor–descendant relationships for a particular lineage, taxonomists usually compare island species with South American mainland or Caribbean species. At this level the greatest taxonomic or morphological differences are seen. It is standard for comparisons of this sort to be made for both marine invertebrate and terrestrial organisms. Secondly, inter-island populations of a species can be compared and, in these cases, smaller morphological differences are usually observed. These comparisons, although quite common in terrestrial vertebrates and plants, are almost absent in marine invertebrates. I know of no inter-island population differences in marine invertebrates that are comparable to those of terrestrial organisms in the Galapagos. A multivariate study of one gastropod species (*Conus nux*) throughout the eastern Pacific did not reveal any taxonomically significant morphological differences between five populations in the Galapagos and 18 other populations located between the Galapagos and the Gulf of California (James, 1982). Lastly, intra-island taxonomic comparisons reveal even more subtle morphological differences (usually at the subspecies or race level) but have only been noted in terrestrial organisms. This hierarchy of taxonomic differences (mainland to archipelago, inter-island, and intra-island) is what makes the terrestrial biota of the Galapagos of such great evolutionary interest. These differences are not significant for marine invertebrates. However, valid taxonomic comparisons can be made between mainland populations (or otherwise in the Panamic province) and island populations of endemic species. Endemism at higher taxonomic levels is conspicuously absent from the fauna. Thus, the hierarchy of taxonomic differences as seen in terrestrial organisms is not seen in marine invertebrate organisms.

Island definitions

The taxonomic hierarchy described above assumes island populations are clear-cut and can be easily compared with other island or mainland populations elsewhere. A problem concerning the study of marine organisms associated with islands involves the bathymetric definition of what is actually *in* the Galapagos (Shumway & Chase, 1963). Many marine species (particularly molluscs) have been described from very deep water adjacent to the islands and are standardly considered part of the marine biota (e.g., Keen, 1971). The subject of deeper water marine faunas associated with the Galapagos Islands has recently been considered by Valentine (as discussed above). However, to date, this fauna remains unstudied and consequently unpublished. Until it can be examined more thoroughly it is best not considered further in this discussion. For comparative purposes, restricting (by definition) the marine fauna to those species occurring in relatively shallow water may be necessary. In this way inter-island populations can be more readily compared for morphological and evolutionary differences. This situation is in contrast to the terrestrial biota where problems of defining the geographic limits of particular islands are not difficult. For the terrestrial biota, island populations are clear-cut: i.e., an island is defined by its coastline. Marine birds and fish form an exception to this rule, however.

Intrinsic rate of evolutionary change

The rate at which organisms diverge evolutionarily will have an impact on the degree of differentiation among taxa in an area such as the Galapagos. The data presented by Stanley (1978, 1979) for bivalved molluscs and mammals, although still highly controversial, indicate that these groups of organisms undergo differential rates of evolutionary change. If this is true for invertebrate and vertebrate organisms in general, then another line of reasoning can be added to the processes accounting for different patterns of evolutionary change among the two groups. Lower rates of intrinsic evolutionary change will probably result in a reduced amount of external morphological change through time. External morphology is the basis of most taxonomic decisions in fossil and living molluscs (and other shelled invertebrates). The lack of morphological change through time is the raw data (translated into taxonomic similarity) for suggesting a different evolutionary scenario among marine invertebrate organisms.

Age of fossils

In addition to the taxonomic hierarchy based on present-day geographic differences among taxa, another valuable comparison, divulging information on rates of evolutionary change lies in the temporal aspect of comparing fossil remains with living organisms in the Galapagos. In this regard, marine invertebrate fossils from the islands are older than the known vertebrate fossil record (Steadman, 1982; Steadman & Ray, 1982; Lipps & Hickman, 1982; Pitt & James, 1983). These older marine invertebrate fossils provide a more complete temporal picture from which we may infer evolution in the islands than can be obtained for terrestrial organisms. Comparisons (taxonomic or morphologic) of present-day island populations run the risk of using incorrect character choices as the basis for establishing biogeographic or evolutionary relationships. Fossil evidence will more clearly indicate ancestor-descendant relationships. In this respect the palynological record provides valuable information about plant species occurrences in the past as well as climatic changes (Colinvaux, 1969, 1972).

Taxonomic relationships

Because marine invertebrate organisms have not diverged as far (morphologically) from mainland source populations as have many of the Galapagos terrestrial organisms, the taxonomic (and thus phylogenetic) relationships of living and fossil species are not as blurred as in, for example, Darwin's Finches (Steadman, 1982). In contrast, however, very little work has been carried out investigating the evolutionary relationships of marine invertebrate organisms, beyond the purely descriptive level. Biochemical studies on the giant tortoises *Geochelone elephantopus* (Marlow & Patton, 1981), Galapagos finches (Geospizinae) (Yang & Patton, 1981), and introduced rodent populations of *Rattus rattus* (Patton, Yang & Myers, 1975) have provided information on the taxonomic and phylogenetic relationships of terrestrial organisms. This provides a complement to purely morphological studies. Unfortunately, taxonomic decisions can be biased by the reputation of the spectacular terrestrial biota of the Galapagos. Additional work is needed beyond the level of alpha taxonomy for marine invertebrate organisms in the Galapagos to determine more precisely their evolutionary relationships.

PATTERN AND PROCESS REVISITED

Different patterns are seen in the evolution of marine and terrestrial organisms in the Galapagos. In terms of Darwin's evolutionary scenario, terrestrial organisms represent the paradigm and marine organisms represent the paradox. Fortunately, there are two processes that account for these patterns. Firstly, as mentioned above, many marine organisms have the ability to exchange genetic material between island and mainland populations through planktotrophic (or merely planktonic, *sensu* Lutz & Jablonski, 1980) larvae. Terrestrial organisms do not have this ability. Secondly, molluscs undergo intrinsically slower rates of evolutionary change than mammals (Stanley, 1979: 101–142). If this is a general indication of the difference between vertebrates and invertebrates, then it is not surprising that, when combined with the ability to exchange genetic material mentioned above, marine invertebrate organisms in the Galapagos do not exhibit a large degree of evolutionary divergence from mainland source populations. The differential rate of evolutionary divergence then comprises this 'new look at evolution in the Galapagos Islands'.

CONCLUSIONS

Evolution of the marine invertebrate biota of the Galapagos Islands, (as evidenced both by the living species, and by recent studies of fossil-bearing deposits) has proceeded at a much slower rate than the evolution of the terrestrial biota. The geologically young age of the islands suggests that terrestrial evolution, producing the spectacular terrestrial biota, has occurred comparatively quickly. The evolutionary scenario for terrestrial organisms is usually one of dispersal followed by adaptive radiation and speciation into unoccupied niche space. All of this is a consequence of the low probability of further influxes of genetic material from mainland source populations, thus maintaining a restricted gene pool. This scenario does not, however, generally hold for marine invertebrate organisms where rates of dispersal and genetic exchange are much higher. Consideration of this alternative evolutionary scenario should be useful when making taxonomic decisions relating to the marine invertebrate fauna of the Galapagos Islands.

ACKNOWLEDGEMENTS

For aiding the 1982 expedition I thank Dr Friedemann Koster for making available the services and facilities of the Charles Darwin Research Station (CDRS); the Galapagos National Park Service for expediting our field work; Dr David Duffy, former Director CDRS, for extending to one of us (WDP) an invitation to investigate the marine paleontology of the islands; the Navy and Air Force of Ecuador for access to fossil localities on Isla Baltra; and the De Roy family (Jaqueline, Andre, and Gil) for valuable locality information. Sincere thanks are extended to Mr Patrick F. Fields for providing thorough reviews of multiple drafts of the manuscript and for making numerous valuable suggestions that greatly improved its content. Ms Cynthia S. Leung copyedited the manuscript. Figures were prepared by Mary Taylor (UCMP). I appreciate the courtesy extended by Professor R. J. Berry for inviting the submission of this paper. Critical reviews of the manuscript were kindly provided by Drs J. Wyatt Durham and David R. Lindberg resulting in the clarification of several points. However, neither necessarily agreed with all aspects of this paper, and consequently, the author bears full responsibility for everything discussed herein.

REFERENCES

ABBOTT, D. P., 1966. Factors influencing the zoogeographic affinities of Galapagos inshore marine fauna. In R. I. Bowman (Ed.), *The Galapagos: Proceedings of the Symposia of the Galapagos International Scientific Project:* 108–122. Berkeley: University of California Press.

BERRY, R. J., 1979. The Outer Hebrides: where genes and geography meet. *Proceedings of the Royal Society of Edinburgh, 77B:* 21–43.

BERTHIAUME, S. A., 1938. Orbitoids from the Cresent Formation (Eocene) of Washington. *Journal of Paleontology, 12*(5): 494–497.

BRATCHER, T. & BURCH, R. D., 1971. The Terebridae (Gastropoda) of Clarion, Socorro, Cocos, and Galapagos Islands. *Proceedings of the California Academy of Sciences Fourth Series* Vol. 37, No. 21: 537–566.

BRIGGS, J. C., 1966. Oceanic islands, endemism, and marine paleotemperatures. *Systematic Zoology 15*(2): 153–163.

BYERLY, G., 1980. The nature of differentiation trends in some volcanic rocks from the Galapagos Spreading Center, *Journal of Geophysical Research, 85*(B7): 3797–3810.

CARLQUIST, S. J., 1974. *Island Biology.* New York: Columbia University Press. 660 pp.

COLINVAUX, P. A., 1969. Vegetation of a Galapagos island before and after an ice age. *Annals of the Missouri Botanical Garden, 56* (3): 419.

COLINVAUX, P. A., 1972. Climate and the Galapagos Islands. *Nature, 240:* 17–20.

COX, A. 1983. Age of the Galapagos Islands. In R. I. Bowman, M. Berson & A. E. Leviton (Eds), *Patterns of Evolution in Galapagos Organisms:* 11–23. San Francisco: American Association for the Advancement of Science, Pacific Division.

DALL, W. H., 1924. Note on fossiliferous strata of the Galapagos Islands explored by W. H. Ochsner of the Expedition of the California Academy of Sciences in 1905–6. *Geological Magazine, 61*(723): 428–429.

DALL, W. H. & OCHSNER, W. H., 1928. Tertiary and Pleistocene Mollusca from the Galapagos Islands. *Proceedings of the California Academy of Sciences, Fourth Series* Vol. 17, No. 4: 89–138.

DARWIN, C. R., 1891. *Geological Observations on the Volcanic Islands and parts of South America visited during the Voyage of H.M.S. 'Beagle'* (3rd edition), London: Smith, Elder, & Co., xiii+648 pp.

DRUFFEL, E. M., 1981. Radiocarbon in annual coral rings from the eastern tropical Pacific Ocean. *Geophysical Research Letters, 8*(1): 59–62.

DURHAM, J. W., 1942. Eocene and Oligocene coral faunas of Washington. *Journal of Paleontology, 16*(1): 84–104.

DURHAM, J. W., 1965a. Geology of the Galapagos Islands. *Pacific Discovery, 18*(5): 3–6.

DURHAM, J. W., 1965b. The Galapagos Islands Expedition of 1964. *Annual Reports for 1964 of the American Malacological Union:* 53.

DURHAM, J. W., 1979. A fossil *Haliotis* from the Galapagos Islands. *The Veliger, 21*(3): 369–372.

DURHAM, J. W., 1980. A new fossil *Pocillopora* (Coral) from Guadalupe Island, Mexico. In D. M. Power (Ed.), *The California Islands: Proceedings of a Multidisciplinary Symposium:* 63–70. Santa Barbara, California: Santa Barbara Museum of Natural History.

DURHAM, J. W. & McBIRNEY, A. R., 1975. Galapagos Islands. In: R. W. Fairbridge (Ed.), *The Encyclopedia of World Regional Geology*, Part 1: 285–290. Stroudsburg: Dowden, Hutchison and Ross, Inc.

EDWARDS, A. & LUBBOCK, R., 1983. Marine zoogeography of St. Paul's Rocks. *Journal of Biogeography 10*(1): 65–72.

EKMAN, S., 1953. *Zoogeography of the Sea.* London: Sidgwick and Jackson, Ltd. xiv+417 pp.

EMERSON, W. K., 1967. Indo-Pacific faunal elements in the tropical eastern Pacific, with special reference to the mollusks. *Venus, 25*(3–4): 85–93.

EMERSON, W. K., 1978. Mollusks with Indo-Pacific faunal affinities in the eastern Pacific Ocean. *The Nautilus, 92*(2): 91–96.

FALLAW, W. C., 1983. Trans-Pacific faunal similarities among Mesozoic and Cenozoic invertebrates related to plate tectonic processes. *American Journal of Science, 283:* 166–172.

FERREIA, A. J., 1978. A new species of chiton (Neoloricata: Ischnochitonidae) from the Galapagos Islands. *Bulletin of the Southern California Academy of Sciences 77*(1): 36–39.

GARTH, J. S., 1946. Distributional studies of Galapagos Brachyura. *Allan Hancock Pacific Expeditions, 5:* 603–648.

GEISTER, J., 1977. Occurrence of Pocillopora in late Pleistocene Caribbean coral reefs. In: *Second Symposium international sur les coraux et recifs coralliens fossiles, Paris, September 1975, B.R.G.M. (Paris) Memoires* No. 89: 378–388.

GLYNN, P. W., WELLINGTON, G. M. & BIRKELAND, C., 1979. Coral reef growth in the Galapagos: Limitation by sea urchins. *Science, 203:* 47–49.

HALL, M. L., RAMON, P. & YEPES, H., 1980. The subaerial origin of Espanola (Hood) Island and the age of terrestrial life in the Galapagos. *Noticias de Galapagos, 31:* 21.

HEDGPETH, J. W., 1969. An intertidal reconnaissance of rocky shores of the Galapagos. *Wasmann Journal of Biology, 27:* 1–24.

HERTLEIN, L. G. & STRONG, A. M., 1939. Marine Pleistocene mollusks from the Galapagos Islands. *Proceedings of the California Academy of Sciences, Fourth Series* Vol. 23, No. 24: 367–380.

HERTLEIN, L. G., 1972. Pliocene fossils from Baltra (South Seymour) Island, Galapagos Islands. *Proceedings of the California Academy of Sciences, Fourth Series* Volume 34, No. 3: 25–46.

HICKMAN, C. S., 1976. Bathyal gastropods of the family Turridae in the Early Oligocene Keasey Formation in Oregon, with a review of some deep-water genera in the Paleocene of the eastern Pacific. *Bulletins of American Paleontology, 70* (292): 1–119.

JABLONSKI, D. & LUTZ, R. A., 1983. Larval ecology of marine benthic invertebrates: paleobiological implications. *Biological Reviews, 58*(1): 21–90.

JAMES, M. J., 1982. Analysis of morphometric variation in *Conus nux:* Biogeographic patterns in the eastern Pacific. *Western Society of Malacologists, Annual Report 14:* 12–13.

KAY, E. A., 1979. *Hawaiian Marine Shells.* Reef and Shore Fauna of Hawaii, Section 4: Mollusca, Bernice P. Bishop Museum Special Publication 64(4): xviii + 653 pp.

KEEN, A. M., 1971. *Sea Shells of Tropical West America* (2nd edition), Palo Alto: Stanford University Press, xiv + 1064 pp.

KEIGWIN, L. D., JR., 1978. Pliocene closing of the Isthmus of Panama, based on biostratigraphic evidence from nearby Pacific Ocean and Caribbean Sea cores. *Geology, 6:* 630–634.

LINDBERG, D. R. & MCLEAN, J. H., 1981. Tropical eastern Pacific limpets of the family Acmaeidae (Mollusca, Archeogastropoda): Generic criteria and descriptions of six new species from the mainland and the Galapagos Islands. *Proceedings of the California Academy of Sciences* Vol. 42, No. 12: 323–339.

LINDBERG, D. R., ROTH, B., KELLOGG, M. G. & HUBBS, C. L., 1980. Invertebrate megafossils of Pleistocene (Sangamon Interglacial) age from Isla Guadalupe, Baja California, Mexico. In D. M. Power (Ed.), *The California Islands: Proceedings of a Multidisciplinary Symposium:* 41–62. Santa Barbara, California: Santa Barbara Museum of Natural History.

LIPPS, J. H. & HICKMAN, C. S., 1982. Paleontology and geologic history of the Galapagos Islands. *GSA Abstracts with Programs 14*(7): 548.

LUTZ, R. A. & JABLONSKI, D., 1980. Molluscan larval shell morphology: Ecological and paleontological applications. In D. C. Rhoads & R. A. Lutz (Eds), *Skeletal Growth of Aquatic Organisms: Biological Records of Environment Change:* 323–377. New York: Plenum Press.

MACARTHUR, R. H. & WILSON, E. O., 1967. *The Theory of Island Biogeography.* Princeton: Princeton University Press. 203 pp.

MARINCOVICH, L. N., JR., 1973. Intertidal mollusks of Iquique, Chile. *Los Angeles County Natural History Museum Science Bulletin 16:* 1–49.

MARLOW, R. W. & PATTON, J. L., 1981. Biochemical relationships of the Galapagos Giant tortoises (*Geochelone elephantopus*). *Journal of Zoology, London 195:* 413–422.

MATTHEW, W. D., 1930. The dispersal of land animals. *Scientia,* July: 33–42.

MAYR, E., 1942. *Systematics and the Origin of Species.* New York: Columbia University Press. 334 pp.

MAYR, E., 1978. The nature of the Darwinian revolution. In S. L. Washburn & E. R. McCown (Eds), *Human Evolution: Biosocial Perspectives:* 11–31. Menlo Park: Benjamin/Cummings Publishing Co.

PATTON, J.L., YANG, S.Y. & MYERS, P., 1975. Genetic and morphologic divergences among introduced rat populations (*Rattus rattus*) of the Galapagos Archipelago, Ecuador. *Systematic Zoology, 24* (3): 296–310.

PITT, W. D. (in press). Late Cenozoic marine paleontology of the Galapagos Islands. *Annual Report of the Charles Darwin Research Station.*

PITT, W. D. & JAMES, M. J., 1983. Late Cenozoic marine invertebrate paleontology of the Galapagos Islands. *Western Society of Malacologists, Annual Report, 15:* 14–15.

RICHARDS, A. F., 1957. Volcanism in eastern Pacific Ocean basin: 1945–1955. In: *International Geological Congress 20th Session, Mexico City, 1956. Section 1, Volcanology of the Cenozoic,* Volume 1, pp. 19–31.

ROSEN, B. R., (in prep). The geological nature of the Galapagos Islands and implications for its shallow marine fauna and reef development.

ROTONDO, G. M., SPRINGER, V. G., SCOTT, G. A. J. & SCHLANGER, S. O., 1981. Plate movement and island integration—A possible mechanism in the formation of endemic biotas, with special reference to the Hawaiian Islands. *Systematic Zoology, 30*(1): 12–21.

SHUMWAY, G. & CHASE, T. E., 1963. Bathymetry in the Galapagos Islands. *Occasional Papers of the California Academy of Sciences, 44:* 11–19.

SIMKIN, T., (in press). Geology of Galapagos. In R. Perry (Ed.), Chapter 2 in Pergamon Press book on Galapagos (in new "Key Environments" series edited by John Treherne).

SMITH, A. G. & FERREIA, A. J., 1977. Chiton fauna of the Galapagos Islands. *The Veliger, 20* (2): 82–97.

STANLEY, S. M., 1978. Chronospecies' longevities, the origin of genera, and the punctuational model of evolution. *Paleobiology, 4:* 26–40.

STANLEY, S. M., 1979. *Macroevolution, Pattern and Process.* San Francisco: W. H. Freeman and Co. 332 pp.

STEADMAN, D. W. 1982. The origin of Darwin's finches (Fringillidae, Passeriformes). *Transactions of the San Diego Society of Natural History, 19*(19): 279–296.

STEADMAN, D. W. & RAY, C. E., 1982. The relationships of *Megaoryzomys curioi*, an extinct Cricetine rodent (Muroidea: Muridae) from the Galapagos Islands, Ecuador. *Smithsonian Contributions in Paleobiology* No. 51, 23 pp.

THORSON, G., 1950. Reproductive and larval ecology of marine bottom invertebrates. *Biological Reviews, 25:* 1–45.

VERMEIJ, G. J., 1978. *Biogeography and Adaptation.* Cambridge, Massachusetts: Harvard University Press. 332 pp.

WALKER, B. W. & ROSENBLATT, R. H., 1961. The marine fishes of the Galapagos Islands. *Abstracts of Symposium Papers, Tenth Pacific Science Congress, Honolulu, Hawaii:* 470–471.

WELLINGTON, G. M., 1978. Undersea wonders of the Galapagos. *National Geographic, 154*(3): 362–381.

WILLIAMSON, M., 1981. *Island Populations.* Oxford: Oxford University Press. 286 pp.

WILLIAMSON, M., 1983. Variations in population density and extinction. *Nature, 303:* 201.

WOLF, T., 1895. Die Galpagos-Inseln. *Verhandlungen der Gesellschaft fur Erdkunde zu Berlin, 22:* 246–265.

WOOSTER, W. S. & HEDGPETH, J. W., 1966. The oceanographic setting of the Galapagos. In R. I. Bowman (Ed.), *The Galapagos: Proceedings of the Symposia of the Galapagos International Scientific Project:* 100–107. Berkeley: University of California Press.

WYRTKI, K., STROUP, E., PATZERT, W., WILLIAMS, R. & QUINN, W., 1976. Predicting and observing El Nino. *Science 191:* 343–346.

YANG, S. Y. & PATTON, J. L., 1981. Genic variability and differentiation in the Galapagos finches. *The Auk, 98:* 230–242.

ZINSMEISTER, W. J. & EMERSON, W. K. 1979. The role of passive larval dispersal in the distribution of hemipelagic invertebrates, with examples from the tropical Pacific Ocean. *The Veliger, 22*(1): 32–40.

Biological Journal of the Linnean Society (1984), *21:* 97–111. With 4 figures

Genetical processes in the Galapagos

JAMES L. PATTON

*Museum of Vertebrate Zoology, University of California,
Berkeley, California 94720, U.S.A.*

Current understanding of the mode and tempo of evolutionary divergence within the major groups of terrestrial vertebrates of the Galapagos Archipelago is summarized from three perspectives: (1) the number and relative timing of introductions for each group; (2) the influence of historical factors and present-day population demography of patterns and amounts of genetic diversity; and (3) possible mechanisms of adaptive radiation, or macroevolution, within the tortoise and finch groups.

Native and introduced rats, lava lizards, geckos, and iguanas most likely had more than one episode of introduction from already differentiated mainland stocks. The finches and tortoises appear to have originated from but a single respective radiation. The influence of these differences in invasion histories is clearly evident in patterns of within-island genetic biochemical diversity and between-island or between-species differentiation. It is argued that much of the pattern of within-species geographic differentiation in genetic characters results from the temporal history of populations, including the amount of migration between islands, as well as differential selection pressures. Major morphological differentiation within both the finch and tortoise radiations has occurred with minimal change at structural gene loci. Studies of, for example, developmental heterochrony and character heritability analyses are needed to understand this apparent paradox.

KEY WORDS:—Evolution – Galapagos – rodents – lava lizards – geckos – tortoises – finches – iguanas.

CONTENTS

INTRODUCTION

The biota of the Galapagos Islands, as viewed through the eyes of Charles Darwin and countless subsequent biologists of somewhat lesser stature, has

0024–4066/84/010097 + 15 $03.00/0

4

played a major role in the growth of evolutionary biology far beyond the prominence of the archipelago itself in size or geographic position. This is not to say, however, that we do not still have much to learn about basic evolutionary processes, particularly during this current period of polarized fermentation of evolutionary paradigms. The Galapagos have much to teach us of gradual *v.* punctuational tempos of evolutionary change, dispersalist *v.* vicariant views of biogeography, neutralism *v.* selection for maintenance of variation, and genetic *v.* epigenetic modes of macro-evolutionary change. I believe that continued evolutionary studies of island biota, particularly those of complex archipelagoes such as the Galapagos, will continue to provide us with fundamental insights into the workings of evolution, whether it be with regard to some of these issues or to others.

What I wish to achieve in this brief review is to examine three general areas in the realm of evolutionary studies in the Galapagos, by focusing on selected members of the terrestrial vertebrate fauna. I have chosen this group both because I know it best and because we have the best data base available on the genetic aspects of population structure and differentiation for these taxa.

These three areas of concern are as follows; firstly, a description of the number of separate introductions per group, with an estimation of the timing of these introductions relative to the history of the archipelago. While these are not direct components of the genetic processes operating in the evolution of the fauna or flora, a consideration of these two fundamental aspects of evolutionary history is extremely important to any understanding of genetical processes. This affects our interpretation of rates of evolutionary change and thus, depending upon one's biasses, our interpretations of the causes of that change. Secondly, an examination of evolutionary processes at the population level which have affected the extent and pattern of genetic differentiation within taxa in the archipelago, and thirdly, a contemplation of the processes which may be involved in the radiation of morphotypes within adaptive lineages; in other words, processes of macro-evolutionary change.

NUMBER AND TIMING OF INTRODUCTIONS

Data of widely different quality are available for several major groups of the terrestrial vertebrate fauna (namely the native and introduced rats, finches, tortoises, lava lizards, geckos, and iguanas) which can be used to examine patterns of colonization, including both numbers and timing of such events. For the most part, these data are sufficient to determine patterns of relationships among island populations and species, but in no case has a thorough parallel study been completed on the likely mainland direct ancestor of any particular Galapagos taxon. Unfortunately, this is a crucial lack of information, leaving us for the moment with more speculation than firm conclusions regarding genetical processes in operation during the evolution of these taxa.

The native rats

Only for the native rats has a direct comparison been made with a mainland form identified as a possible relative. There are two native rat groups currently extant in the islands: the genus *Nesoryzomys,* an endemic with no close relatives among the mainland South American rodent complex to which it belongs (the oryzomyine cricetine group, *sensu* Hershkovitz, 1962), and the genus *Oryzomys*, the rice rats which are a highly diverse and speciated group of sylvan tropical and subtropical rodents of the same basic stock as *Nesoryzomys*. In addition to these two genera, a third genus of giant but now extinct rats, *Megaoryzomys*, is known from the fossil record in the archipelago. This group is a presumed derivative of the thomasomyine complex of sylvan cricetid rodents of mainland South America, and thus unrelated to both *Nesoryzomys* and *Oryzomys* (Steadman & Ray, 1982). These conclusions suggest three separate introductions, two to represent the endemic genera *Nesoryzomys* and *Megaoryzomys*, both of which are strongly divergent from all mainland relatives (see Patton & Hafner, 1983; Steadman & Ray, 1982). However, the single extant species of *Oryzomys* in the islands, *O. bauri* from Isla Santa Fe, is virtually identical to a living species from the west coast of South America, *O. xantheolus*, in both morphological and biochemical (electromorphic allozyme) parameters (see Patton & Hafner, 1983). This similarity is so great as to suggest that *O. bauri* (and probably *O. galapagoensis*, a now extinct similar taxon from Isla San Cristobal) represents an introduction that may have taken place within the last few hundred years (with aboriginal travellers to the islands a likely dispersal agent!).

For the remainder of the fauna, we must infer the number and timing of introductions from the pattern and degree of measureable differentiation within the existing radiations in the archipelago. For the most part, such an inference can best be made from biochemical measures of differentiation, since such can be related to time after divergence on a rather linear basis (see reviews by Nei, 1975; Sarich, 1977; Wilson, Carlson & White, 1977; but see also Fitch, 1976).

The introduced rat Rattus rattus

Three types of analysis provide the same picture as to patterns of interisland divergence among the populations of introduced rats in the archipelago (see Fig. 1). Each of these patterns is derived from a different data source, two are morphological character sets (multivariate discriminant function analysis of cranial mensural data and qualitative 'epigenetic' character complexes; see Berry, 1970) and the third is an electromorphic analysis of 37 presumptive gene loci (see Patton, Yang & Myers, 1975). The three sets of data are remarkably concordant in indicating similar patterns of island population groupings (see Fig. 1 and Patton *et al.*, 1975), suggesting that the observed patterns of inter-island relationships result from distinct periods of historical introductions of black rats into the archipelago. Each of these is hypothesized to follow major temporal episodes of human occupation involving different groups of islands. The individual islands or island groups so involved and the time and pattern of extralimital and interisland movements of black rats are identified in Fig. 1 and discussed more fully in Patton *et al.* (1975).

The divergence patterns of the introduced rats thus imply three periods of major introductions, with each affecting different parts of the archipelago. This effect is still seen today in the observable traits of the extant populations. Since the history of introductions of black rats corresponds to major periods of historical human involvement in the islands, the divergence patterns observed appear to result primarily from historical accidents of the timing of introductions and the source populations of origin.

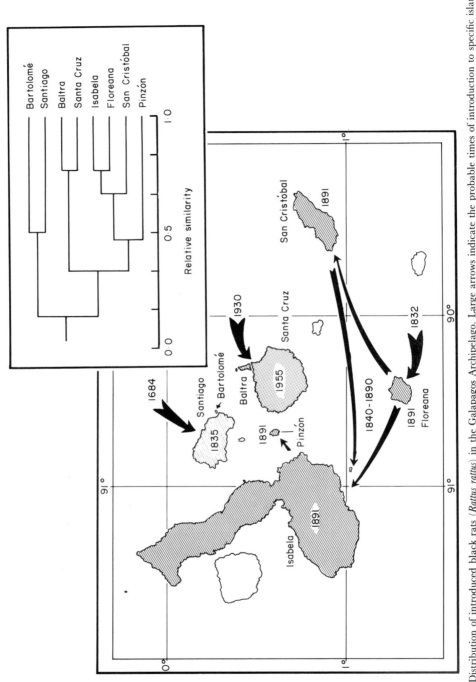

Figure 1. Distribution of introduced black rats (*Rattus rattus*) in the Galapagos Archipelago. Large arrows indicate the probable times of introduction to specific islands or island groups; smaller arrows indicate periods and directions of major interisland movement patterns. Dates with each island identify the time when black rats were initially recorded on that island. Inset: composite phenogram of interisland similarity among introduced rat populations, based on morphological and genetical analyses (Patton *et al.*, 1975).

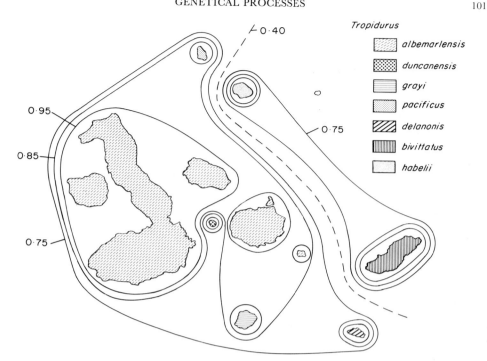

Figure 2. Taxonomic distribution of lava lizards, genus *Tropidurus*, in the Galapagos. Contour intervals group islands based on a phenogram of Rogers' genetic similarity values derived from electromorphic analyses (Wright, 1983).

Lava Lizards

Seven species of lava lizards of the genus *Tropidurus* are currently recognized, each restricted to a separate island or group of islands (Van Denburgh & Slevin, 1913; Wright, 1983). Since no sets of taxa are truly sympatric in the archipelago, the test of sympatry cannot be used to argue for multiple invasions for this taxon. Instead, we must rely on inferences derived from patterns and levels of interisland differentiation. In the present case, divergence in electromorphic biochemical characters gives us some idea as to the rate of population change, since these characters are largely divergent from the time of population splitting and accumulate changes at an approximately similar rate (see Nei, 1975; Wilson *et al.*, 1977). Interisland patterns of genetic similarity for *Tropidurus* are shown in Fig. 2. Here, a phenogram of Rogers' genetic similarity values is presented as a contour map of inter-island similarity levels. Most islands cluster at moderately high genetic similarities (above the 75% level), but there are two groups of populations and taxa differentiated at about twice that level (at 40% similarity). This dual set of divergence levels among populations of *Tropidurus* in the archipelago suggests that either: (a) there were multiple (probably two) introductions from differentiated mainland taxa; or (b) a single introduction followed by periods of radically different rates of biochemical divergence, very rapid at first and then abruptly slowed. Since molecular change occurs at a rather regular albeit somewhat 'bumpy' rate (see Fitch,

1976), the first of these two possibilities is the more likely. It is favoured by Wright (1983) in his evaluation of the evolutionary divergence of this taxon.

Geckos

Some seven species of geckos, genus *Phyllodactylus*, are present in the archipelago, but, unlike *Tropidurus*, there is at least one example of species sympatry (that of *P. darwini* and *P. leei* on Isla San Cristobal). These taxa, differentiated at about the 40% genic similarity level (Wright, 1983), clearly represent separate introductions, as *darwini* belongs to a different mainland species group of the genus (see Van Denburgh, 1912). It is significant to note that the level of differentiation between *P. darwini* and other geckos in the islands is equivalent to that seen maximally between *Tropidurus* taxa (40%). That the geckos represent two separate introductions yet have the same maximal divergence level as *Tropidurus* supports the hypothesis that a double invasion characterized the evolution of the lava lizards in the archipelago as well.

Tortoises

The giant tortoises show extreme similarity among the examined island races with a maximal level of genic differentiation of 8%. This is only one-fifth of the level of average divergence observed between gecko or lava lizard populations (see Marlow & Patton, 1981; Wright, 1983). *Geochelone elephantopus* on the islands is, however, markedly divergent from the extant South American mainland lineages suggesting but a single introduction from an ancestral stock, now extinct, from coastal western South America. The likely time of divergence of the *elephantopus* lineage from the extant mainland ones must predate the origin of the Galapagos archipelago by a considerable time span (see Marlow & Patton, 1981, for further discussion).

Darwin's finches

In a situation similar to that of the tortoises, the finches in the archipelago would appear to have resulted from but a single introduction. Maximal genic divergence levels between taxa are only about 15%, and all taxa share an inordinately high proportion of the eletromorphic 'alleles' detected (see Yang & Patton, 1981). The complex patterns of island sympatry of various finch species clearly have resulted from intra-archipelago speciation, regardless of the mode of that divergence, be it classical allopatric or sympatric (see Lack, 1947; Grant & Grant, 1979).

PATTERNS OF PHYLOGENESIS

Data available for the major groups of terrestrial vertebrates of the Galapagos indicate three basic and distinct patterns of phylogenetic divergence (Fig. 3).

Firstly are the groups with very low levels of interspecific differentiation but with related taxa showing divergence times of considerable length and clearly predating the origin of the archipelago. This pattern is exemplified by the land

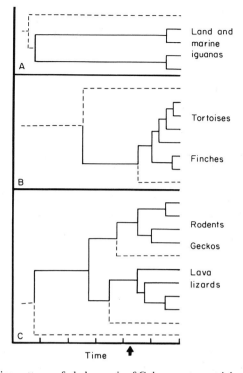

Figure 3. Three major patterns of phylogenesis of Galapagos terrestrial vertebrates suggested by electromorphic and/or immunological molecular data. The time scale is arbitrary; the arrow indicates approximate time of origin of the archipelago and/or initial introduction of each group; the solid lines in each phenogram indicate the radiations of the Galapagos taxa *per se*, the dashed lines indicate the relationship of these taxa to their closest living mainland sister group (modified from Wyles & Sarich, 1983). A, Pattern of both the land and marine iguanas which show minimal within species differentiation but must have shared a common ancestor considerably earlier than the origin of the islands; B, both the giant tortoises and Darwin's finches appear to be endemic radiations of the archipelago, stemming from but a single ancestral introduction. In the case of the tortoises this ancestral stock is now extinct, while the finches presumably have a close relative still extant on the South American mainland (following Steadman, 1982); C, both native and introduced rats, geckos, and probably lava lizards owe their diversity in the islands to multiple introductions from separate mainland stocks already differentiated to some degree.

and marine iguanas (Fig. 3A) based on both immunological and electrophoretic levels of molecular differentiation (see Wyles & Sarich, 1983). Virtually no measureable genetic differentiation has occurred within either taxon for the island populations examined (this includes several local populations of marine iguanas from Isla Santa Cruz and both species and several island populations of land iguanas; R. W. Marlow, unpubl. data). Hence, the two iguana groups must have had their time and place of origin outside of the archipelago, despite the fact that they are found only in that region today. A more thorough study of patterns of geographic variation among island populations of both taxa, particularly among the marine iguanas which display some marked island differentiation in morphological features such as size and colour pattern, is required from a biochemical standpoint.

The second pattern of phylogenesis is exemplified by the native rodents and presumably both lava lizards and geckos. This is a pattern of a high level of interisland genetic divergence (and morphological divergence, judging from the classical taxonomy of each group) complicated by the likely multiple introductions within each of these groups from already differentiated mainland taxa (Fig. 3C). In other words, for these three groups much of the current pattern of differentiation of taxa and populations within the Galapagos is a function, at least in part, of divergence in mainland stocks before colonization of the archipelago. The total extent of this historical effect cannot be ascertained with current data.

The third pattern of differentiation of Galapagos terrestrial vertebrates is that represented by both the finches and the tortoises. This is one of a high level of interisland divergence and speciation stemming from but a single original introduction into the archipelago by the progenitor stock. For the finches, this mainland stock is probably still extant (following Steadman, 1982), while for the tortoises the direct mainland ancestor is now extinct and the surviving mainland relatives are only quite distantly related (see Marlow & Patton, 1981).

This summary brings me to a consideration of those processes which have affected the rate and degree of interisland genetic differentiation observed in the taxa just examined, subsequent to their introduction into the islands.

EVOLUTIONARY PROCESSES AND ISLAND POPULATIONS

I wish to address two major questions in this section, both of which are of fundamental importance for any consideration of island evolution. Firstly, do island taxa show reduced levels of variability (in either or both genetic and morphological traits) when compared to mainland forms, as might be expected as a consequence of founder effects? If so, can we come to any conclusions about levels of variability and potential for radiation? Secondly is there any relationship between inter-island patterns of variability and (a) extra-island introductions, (b) population and/or island size, and (c) interisland movement patterns? If we can exclude these, and other demographic/stochastic variables, then we can focus more appropriately on specific questions derived from the selectionist paradigm.

For only two groups of Galapagos taxa is there sufficient information to approach both of these questions. These are the finches and the introduced rats. Morphological variability measures have been made for both taxa (see, for example, Lack, 1945; Bowman, 1961; Patton et al., 1975; Grant et al., 1976), but will not be dealt with here in substance. Rather, I would like to focus on patterns of genetic variability, as demonstrated by electromorphic analysis. These characters are perhaps less subjective in the assumptions of underlying heritability, and therefore easier to relate to the major evolutionary forces of selection, genetic drift, gene flow, and so forth.

Individual heterozygosity and population polymorphism levels for electromorphic variables are given in Table 1 for the terrestrial vertebrate taxa for which such measures are available. In general, these measures are lower than similar ones for mainland taxa of the same class (Nevo, 1978; Kilpatrick, 1981), but the level of statistical significance between these measures cannot be appropriately determined. Moreover, in no case is a direct comparison available for measures of genic variability between a Galapagos taxon/population and at least a marginally related mainland taxon/population. For a realistic discussion

Table 1. Genic variability in Galapagos terrestrial vertebrate taxa, as measured by electromorphic analysis

Taxon	Ns	Np	Ni	Nl	H	P
Nesoryzomys	1	1	10	37	0.0162	2.70
Oryzomys	1	1	10	37	0.0029	2.86
Geospiza	6	51	202	27	0.0576	17.50
Geochelone	1	9	200	20	0.0280	6.40
Phyllodactylus	7	17	222	20	0.0339	16.06
Tropidurus	7	18	231	23	0.0272	14.74
Conolophus	2	4	36	25	0.0216	16.00
Amblyrhynchus	1	3	39	25	0.0033	4.00
Island populations, all vertebrates*					0.0470	17.30
Non-island populations, all vertebrates*					0.0870	28.70

*For species ranging in both mainland and islands; see Nevo (1978). Ns, number of species examined; Np, number of populations surveyed; Ni, number of individuals sampled; Nl, number of loci in each analysis; H, mean percentage heterozygosity per individual per population; and P, mean percentage polymorphism per population.

of levels of island taxon variability, this latter comparison is an absolute requirement. Estimates of the extent of founder effects, for example, can only be made by contrasting data from island and mainland populations of the same closely related species.

Intraspecific variation in Rattus

There is an excellent correlation between the level of population heterozygosity and island size for the introduced black rat populations (Patton *et al.* 1975; see Fig. 4). There are two basic arguments in the literature to

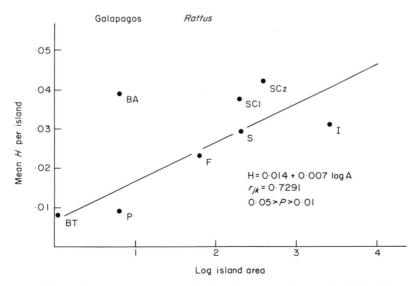

Figure 4. Relationship between average individual heterozygosity and island size for samples populations of introduced rats (*Rattus rattus*). Island abbreviations are: BT, Bartolome; P, Pinzon; BA, Baltra; F, Floreana; S, Santiago; SCl, San Cristobal; SCz, Santa Cruz; and I, Isabela.

explain this, and other similar observations: (1) the selectionist position that such a relationship exists because island size provides an index of niche availability (genic variation/niche width hypothesis; see Soulé, 1976; Nevo, 1978), and (2) that of the stochastic school which argues that the relationship is simply a result of population size, as measured by island size (Soulé, 1976; Schnell & Selander, 1981). For *Rattus*, the latter position is certainly a realistic one, although data are not available to test adequately either hypothesis. For example, the relationship between heterozygosity and island size is quite strong ($r=0.729$; Fig. 4) with only the population sample from Baltra showing significant deviation. This population is nearly four times as variable as would be predicted by the size of the island (and the size of its rat population?). However, Baltra represents the major port in the archipelago, both for ships and aeroplanes, and thus represents the most likely place for continued introduction of rats from outside the archipelago. Such continued immigration would be expected to increase levels of population variability beyond that maintained by intrapopulational demography. Thus, the higher than expected level of variability in the Baltra population supports the gene flow/population size, but not the niche width variability hypothesis (Patton *et al.*, 1975; Schnell & Selander, 1981).

Heterozygosity and island size relationships in Darwin's Finches

Sufficient samples are available for three species of finches to examine a similar relationship between island size and population genic variability level, as was summarized above for the introduced black rats. These data are presented in Table 2 (see also Yang & Patton, 1981). For both species of *Geospiza* (*fuliginosa* and *fortis*), there is no observable relationship between measured population genic variability (heterozygosity) and island size, while for *Certhidea olivacea* a rather strong relationship between these variables is present ($r=0.244$, 0.160, and 0.996, respectively; Table 2). Since the samples of *Certhidea* come from the same islands as those of the two *Geospiza* species, these data do not

Table 2. Population mean individual heterozygosity (H) values for three species of Galapagos finches (*Geospiza fuliginosa*, *G. fortis*, and *Certhidea olivacea*), with their relationship to island size given

Island	Log. Area	*fuliginosa*	*fortis*	*Certhidea*
Isabela	3.352	0.0635	0.0617	—
Santa Cruz	2.995	0.0509	0.0456	0.0556
Fernandina	2.389	0.0741	—	—
Santiago	2.308	0.0671	0.0611	—
San Cristobal	2.290	0.0593	—	—
Floreana	1.806	0.0741	0.0556	—
Marchena	1.653	0.0741	0.0750	0.0370
Pinta	1.301	0.0337	0.0875	—
Espanola	1.255	0.0535	—	0.0185
Santa Fe	0.875	0.0556	0.0370	—
Genovesa	0.643	—	0.0370	0.000
	$r=$	0.244ns	0.160ns	0.996*

*$P<0.05$; ns, non-significant.

follow the expectations of the niche width hypothesis. On the other hand, comparisons of morphological differentiation and population movements in *Certhidea* and the two *Geospiza* species suggest that interisland movements (gene flow) may be largely responsible for the differences in heterozygosity and island size patterns (Yang & Patton, 1981). *Geospiza fortis* and *G. fuliginosa* are both characterized by a high level of morphological similarity (all island populations of both species are part of single, monotypic assemblages; Lack, 1969), high between population levels of genetic similarity, and numerous examples of extralimital population movements. These three features, singly and in combination, suggest a high level of interisland gene flow which is sufficient to maintain interisland population homogeneity, morphological and genetical. With *Certhidea*, however, the pattern is the reverse; a high degree of between-island genetic divergence (S = 0.897 *v*. 0.956 and 0.957 for the two *Geospiza* species), many morphologically distinct island races, and no records of inter-island movements. This is not to say, however, that the island nature of finch populations has no effect on population differentiation. Indeed, the island populations of finch species are, on average, two to four times more distinct than are populations of mainland species of birds (Table 3).

These data for introduced rats and for finches suggest that much of the short-term patterns of variability and differentiation result from the direct interaction of demographic and historical processes such as population size, migration rate, and population history. This does not mean that selection is not operating or is unimportant (see, for example, Boag & Grant, 1981 for clear evidence of directional selection on morphological features in one finch population); it does mean that we cannot eliminate population processes such as total population size and gene flow rates as significant mechanisms of evolutionary differentiation within taxa of the Galapagos archipelago.

It is interesting that, while Table 1 does show generally reduced levels of island genetic variability, the level of heterozygosity for the finches is quite high

Table 3. Wright's standardized variance in allele frequency for population samples of Galapagos finches, adjusted for sample size (see Yang & Patton, 1981), compared to similar data for mainland bird species from Barrowclough (1980)

Galapagos species	Fst
G. fortis	0.0652
G. scandens	0.0197
G. fuliginosa	0.0544
G. difficilis	0.0570
G. magnirostris	0.0461
C. parvulus	0.0574
P. crassirostris	0.0338
C. olivacea	0.1248
Fst all taxa	0.0573 ± 0.0310
Fst *Geospiza* sp.	0.0485 ± 0.0175
Mainland species	
Fst non-colonial	0.0105 ± 0.0129
Fst colonial	0.0276 ± 0.0299

(5.76%; see Yang & Patton, 1981). This is significant because birds are notorious for low levels of biochemical differentiation and variability, and also the finches are fully as variable as continental populations of the average vertebrate taxon (Selander, 1976). This suggests (a) that the original number of colonizers for the eventual finch radiation was rather large, (b) that any variability lost as a result of the bottleneck of founding was quickly recovered by rapid population buildup (Nei, Maruyama & Chakraborty, 1975), or (c) that there were actually several, serial introductions of the finch ancestor(s) from the mainland. Each of these possibilities has significance for the subsequent radiation of the finches in the archipelago, and of our interpretation of that radiation.

ADAPTIVE RADIATION—MORPHOLOGICAL *V.* GENETIC EVOLUTION

Much has been argued in the recent literature about the patterns of adaptive radiation among animal groups and the genetic or epigenetic basis for major morphological change (Løvtrop, 1974; Gould, 1977; Stanley, 1979; among others). It is no longer a novel observation that morphological and genetical measures of differentiation are often discordant within a given group (e.g., Wilson, 1976; Wilson *et al.*, 1977), and it is therefore of no real surprise to record similar observations for Galapagos organisms. Two groups of the terrestrial vertebrates provide the most obvious cases for the lack of a correlation between these two measures of differentiation, the tortoises and the finches.

Both groups show well-known and marked morphological differentiation, yet the respective radiations have been accompanied by minimal genetic divergence, as least at the structural gene loci examined by electrophoresis. While morphological divergence has not been objectively measured in either the tortoises or finches (objective in the sense of Cherry, Case & Wilson, 1978; Cherry *et al.*, 1982), the wide spectrum of carapace and bill morphologies, respectively, for these two groups is well recognized, and has been since Darwin (Van Denburgh, 1914; Lack, 1947; Bowman, 1961). For the tortoises, the saddle-backed and domed carapace shapes do not form monophyletic assemblages within themselves (Marlow & Patton, 1981), so that the saddle-backed form must have been derived independently several times (assuming that the domed condition is primitive). This morphological differentiation has not been accompanied by any genetic change, suggesting that the mechanism of differentiation was one which produced significant morphological novelty in a small number of generations. Similarly with the finches, the diversity of beak, and therefore feeding, morphologies was accomplished without substantial genic divergence. The average level of genetic divergence among the five genera of finches is only about 8%, less than one-half the level for parulid warblers for example (Barrowclough & Corbin, 1978). Indeed, even within the single genus *Geospiza*, whose species display a wide range of bill morphologies and ecological roles (e.g., Lack, 1947; Bowman, 1961; Abbott, Abbot & Grant, 1977), there is only a very small amount of genetic differentiation (mean genetic divergence is only 4.4%; Yang & Patton, 1981).

The finch and tortoise data thus present a paradox: great morphological change with little appreciable structural gene change (i.e. high rates of morphological differentiation but very low rates of molecular evolution). As mentioned above, this is a common pattern (e.g., Wilson, *et al.*, 1977). It highlights the current issue in evolutionary biology regarding the modes or mode of major morphological transformations, or macro-evolution (see general statements and positions by Gould, 1980; Stanley, 1979; Mayr, 1982; and others). I would argue that examination of the patterns of morphological differentiation as exhibited by the tortoise and finch taxa of the Galapagos, could, for example, provide us with some of the keys to unlock this paradox. Both of these radiations are relatively uncomplicated, both sets of taxa can be examined under the experimental design of the developmental heterochrony paradigm, which permits substantial morphological change via minimal genetic alteration (Albrech, Gould, Oster & Wake, 1979); and the morphologies exhibited by these taxa, while of seemingly complex shape patterns, are ripe for examination by the integrative methods of shape analysis currently being developed (e.g. Strauss & Bookstein, 1982). Indeed, Grant (1981) has already approached this issue with some success.

Finally, for both sets of taxa, because of captive breeding programmes (tortoises) and long-term population analyses (finches), the morphological characters upon which the classical views of their respective radiations rest can be examined for heritability levels. In this vein, Boag & Grant (1978) have made an important initial step for the finches, but considerably more data are needed for this group, for example both with respect to beak shape and to plumage colour pattern. These kinds of data are not as yet available for the tortoises. Until such data are gathered with precision and integrative analyses made, the underlying mechanisms of genetic differentiation during adaptive radiation cannot be elucidated or understood.

SUMMARY

The Galapagos fauna has played a major role in the development of the evolutionary paradigm, yet we have barely scratched the surface in both understanding the patterns and processes of evolutionary divergence for this fauna and utilizing the examples provided by the Galapagos fauna to test major questions and issues of current evolutionary biology.

The fundamentally important issues of the relationships among the Galapagos taxa of given faunal elements and of these to mainland relatives still needs to be addressed for the majority of taxa. Commonality of patterns will help to distinguish between alternative hypotheses of island biogeography (Endler, 1982), evolutionary processes of differentiation (Gould, 1980), and modes of speciation (Bush, 1975). Focus on comparative intertaxon ontogenetic developmental sequences and character heritability estimates will help in examining the issues of adaptive radiation and macroevolution.

There is much left to do, and the Galapagos provide a selective set of very tractable organisms to further a more fundamental understanding of evolutionary dynamics. Darwin's evolutionary paradigm was initiated, at least in part, by the Galapagos fauna; it is only appropriate that this fauna and flora continue to be active participants in the further refinement of this paradigm.

ACKNOWLEDGEMENTS

Professor R. J. Berry invited me to contribute this effort and Dr J. W. Wright kindly allowed me to cite his unpublished data on lava lizards and geckos; to these colleagues I am sincerely grateful. I also especially acknowledge the fertile minds of Drs J. C. Daly and G. F. Barrowclough during discussions of the subject matter of this paper, and of Dr P. R. Grant in his continuing intellectual advancement of our knowledge and understanding of patterns of evolutionary change within Galapagos organisms.

REFERENCES

ABBOTT, I., ABBOT, L. K. & GRANT, P. R., 1977. Comparative ecology of Galapagos ground finches (*Geospiza* Gould): evaluation of the importance of floristic diversity and interspecific competition. *Ecological Monographs, 47:* 151–178.

ALBERCH, P., GOULD, S. J., OSTER, G. F. & WAKE, D. B., 1979. Size and shape in ontogeny and phylogeny. *Paleobiology, 5:* 296–317.

BARROWCLOUGH, G. F., 1980. Gene flow, effective population sizes, and genetic variance components in birds. *Evolution, 34:* 789–798.

BARROWCLOUGH, G. F. & CORBIN, K. W., 1978. Genetic variation and differentiation in the Parulidae. *Auk, 95:* 691–702.

BERRY, R. J., 1970. Covert and overt variation, as exemplified by British mouse populations. In R. J. Berry & H. N. Southern (Eds.), *Variation in Mammalian Populations:* 3–26. Symposium Zoological Society London No. 26. London: Academic.

BOAG, P. T. & GRANT, P. R., 1978. Heritability of external morphology in Darwin's finches. *Nature, London, 274:* 793–794.

BOAG, P. T. & GRANT P. R., 1981. Intense natural selection in a population of Darwin's finches (Geospizinae) in the Galapagos. *Science, N.Y., 214:* 82–84.

BOWMAN, R. I., 1961. Morphological differentiation and adaptation in the Galapagos finches. *University of California Publications in Zoology, 58:* 1–302.

BUSH, G. L., 1975. Modes of animal speciation. *Annual Review of Ecology and Systematics, 6:* 339–364.

CHERRY, L. M., CASE, S. M. & WILSON, A. C., 1978. Frog perspective on the morphological difference between humans and chimpanzees. *Science, N.Y., 200:* 209–211.

CHERRY, L. M., CASE, S. M., KUNKEL, J. G., WYLES, J. S. & A. C. WILSON., 1982. Body shape metrics and organismal evolution. *Evolution, 36:* 914–933.

ENDLER, J. A., 1982. Problems in distinguishing historical from ecological factors in biogeography. *American Zoology, 22:* 441–452.

FITCH, W. M., 1976. An evaluation of molecular evolutionary clocks. In F. J. Ayala (Ed.), *Molecular Evolution:* 160–178. Sunderland, Mass: Sinauer Association.

GOULD, S. J., 1977. *Ontogeny and Phylogeny:* 501. Cambridge, Mass: Harvard Univ. Press.

GOULD, S. J., 1980. Is a new and general theory of evolution emerging? *Paleobiology, 6:* 119–130.

GRANT, P. R., 1981. Patterns of growth in Darwin's finches. *Proceedings of the Royal Society of London* B, *212:* 403–432.

GRANT, B. R. & GRANT, P. R., 1979. Darwin's finches; population and sympatric speciation. *PNAS (USA), 76:* 2359–2363.

GRANT, P. R., GRANT, B. R., SMITH, J. N. M., ABBOT, I. & ABBOTT, L. K., 1976. Darwin's finches: population variation and natural selection. *PNAS (USA) 73:* 257–261.

HERSHKOVITZ, P., 1962. Evolution of Neotropical cricetine rodents (Muridae) with special reference to the phyllotine group. *Fieldiana: Zoology, 46:* 1–524.

KILPATRICK, C. W., 1981. Genetic structure of insular populations. In M. H. Smith & J. Joule (Eds.), *Mammalian Population Genetics:* 28–59. Athens, Georgia: Univ. Georgia Press.

LACK, D., 1945. The Galapagos finches (Geospizinae): A study in variation. *Occasional Papers of the California Acadamy of Science, 21:* 1–59.

LACK, D., 1947. *Darwin's Finches.* Cambridge: University Press.

LACK, D., 1969. Subspecies and sympatry in Darwin's finches. *Evolution, 23:* 252–263.

LOVTROP, S., 1974. *Epigenetics. A Treatise on Theoretical Biology:* 548 pp. London: Wiley.

MARLOW, R. W. & PATTON, J. L., 1981. Biochemical relationships of the Galapagos giant tortoises (*Geochelone elephantopus*). *Journal of Zoology, London, 195:* 413–422.

MAYR, E., 1982. Speciation and macroevolution. *Evolution, 36:* 1119–1132.

NEI, M., 1975. *Molecular Population Genetics and Evolution:* 288 pp, Amsterdam: North-Holland.

NEI, M., MARUYAMA, T. & CHAKRABORTY, R., 1975. The bottleneck effect and genetic variability in populations. *Evolution, 29:* 1–10.

NEVO, E., 1978. Genetic variation in natural populations: patterns and theory. *Theoretical Population Biology, 13:* 121–177.

PATTON, J. L. & HAFNER, M. S., 1983. Biosystematics of the native rodents of the Galapagos Archipelago, Ecuador. In R. I. Bowman, M. Berson & A. E. Leviton (Eds), *Patterns of Evolution in Galapagos Organisms.* Amer. Assoc. Adv. Sci., Pacific Div.; San Francisco.

PATTON, J. L., YANG, S. Y. & MYERS, P., 1975. Genetic and morphologic divergence among introduced rat populations (*Rattus rattus*) of the Galapagos Archipelago, Ecuador. *Systematic Zoology, 24:* 296–310.

SARICH, V. M., 1977. Rates, sample sizes, and the neutrality hypothesis for electrophoresis in evolutionary studies. *Nature, Lond., 265:* 24–28.

SCHNELL, G. D., & SELANDER, R. K., 1981. Environmental and morphological correlates of genetic variation in mammals. In M. H. Smith & J. Joule (Eds), *Mammalian Population Genetics:* 60–99. Athens, Georgia: Univ. Georgia Press.

SELANDER, R. K., 1976. Genic variation in natural populations. In F. J. Ayala (Ed.), *Molecular Evolution:* 21–25. Sunderland, Mass: Sinauer Assoc.

SOULÉ, M., 1976. Allozyme variation: its determinants in space and time. In F. J. Ayala (Ed.), *Molecular Evolution:* 60–77. Sunderland, Mass: Sinauer Assoc.

STANLEY, S. M., 1979. *Macroevolution: Pattern and Process.* San Francisco: Freeman.

STEADMAN, D. W., 1982. The origin of Darwin's finches (Fringillidae, Passeriformes). *Transactions of the San Diego Society of Natural History, 19:* 279–296.

STEADMAN, D. W. & RAY, C. E., 1982. The relationships of *Megaoryzomys curioi*, an extinct cricetine rodent (Muroidea: Muridae) from the Galapagos Islands, Ecuador. *Smithsonian Contributions to Paleobiology,* No. 51, 23 pp.

STRAUSS, R. E. & BOOKSTEIN, F. L., 1982. The truss: body form reconstruction in morphometrics. *Systematic Zoology, 31:* 113–135.

VAN DENBURGH, J., 1912. Expedition of the California Academy of Sciences to the Galapagos Islands, 1905–1906. VI. The geckos of the Galapagos Archipelago. *Proceedings of the California Academy of Sciences,* (Ser. 4) *1:* 405–430.

VAN DENBURGH, J., 1914. The gigantic land tortoises of the Galapagos Archipelago. *Proceedings of the California Academy of Sciences* (Ser 4) *2:* 203–374.

VAN DENBURGH, J. & SLEVIN, J. R., 1913. The Galapagoan lizards of the genus *Tropidurus*; with notes on the iguanas of the genera *Conolophus* and *Amblyrhynchus*. *Proceedings of the California Academy of Sciences* (Ser. 4) *2:* 133–202.

WILSON, A. C., 1976. Gene regulation in evolution. In F. J. Ayala (Ed.), *Molecular Evolution:* 225–234. Sunderland, Mass: Sinauer Assoc.

WILSON, A. C., CARLSON, S. C. & WHITE, T. J., 1977. Biochemical evolution. *Annual Review of Biochemistry, 46:* 573–639.

WRIGHT, J. W., 1983. The evolution and biogeography of the lizards of the Galapagos Archipelago: evolutionary genetics of *Phyllodactylus* and *Tropidurus* populations. In R. I. Bowman, M. Berson & A. E. Leviton (Eds), *Patterns of Evolution in Galapagos Organisms.* Amer. Assoc. Adv. Sci., Pacific Div., San Francisco.

WYLES, J. S. & SAEICH, V. M., 1983. Are the Galapagos iguanas older than the Galapagos? Molecular evolution and colonization models for the archipelago. In R. I. Bowman & A. E. Leviton (Eds), *Patterns of Evolution in Galapagos Organisms.* Amer. Assoc. Adv. Sci., Pacific Div., San Francisco.

YANG, S. Y. & PATTON, J. L., 1981. Genic variability and differentiation in Galapagos finches. *Auk, 98:* 230–242.

Biological Journal of The Linnean Society (1984) *21:* 113–136. With 9 figures

Recent research on the evolution of land birds on the Galapagos

P. R. GRANT

Division of Biological Sciences, University of Michigan, Ann Arbor, Michigan 48109–1048, U.S.A.

A decade of research on the evolution of Galapagos land birds is reviewed. and outstanding questions to be answered are highlighted. Evolutionary studies have been restric 1 almost entirely to the four species of mockingbirds and the 13 species of Darwin's finches. Long-term field studies have been initiated on representatives of both groups. Co-operative breeding has been discovered in the mockingbirds (and hawks).

Lack's (1945, 1947) monographic treatment of Darwin's finches has been largely upheld and extended by morphological, ecological, behavioural and biochemical studies. While the phylogenetic origins of Darwin's finches still remain uncertain, the major groupings of the finches have been confirmed by the results of protein polymorphism analysis. Fossils of Darwin's finches have been discovered recently: their potential for illuminating evolutionary change has not yet been realized. Three other major developments are (1) quantitative confirmation of the role of interspecific competition in the adaptive radiation, (2) experimental confirmation of the role of morphological and song cues in species recognition, and experimental evidence of their evolution in the speciation process, and (3) direct study of natural selection on heritable quantitative traits in a population, and identification of its causes. Continuing studies of population variation are likely to reveal the contemporary importance of selection, migration and hybridization, and thereby help us to more fully understand the causes of the adaptive radiation of Darwin's finches.

KEY WORDS:—Darwin's finches – mockingbirds – speciation – competition – reproductive isolation – phylogeny – fossils – natural selection.

CONTENTS

Introduction 113
Mockingbirds 115
Darwin's finches 117
    What are their systematic origins? 117
    What are their ancestor–descendant relationships? 118
    How many species evolved, how many became extinct and over what period of time? . 119
    Why have no more species evolved? 122
    How is the adaptive radiation explained? 125
    Some current and future developments 131
This is the house that Lack built 132
Acknowledgements 132
References. 133

INTRODUCTION

There are 28 species of land birds resident on the Galapagos: all but six are endemic (Table 1). Until recently most of our knowledge about them has come

113

0024–4066/84/010113+24 $03.00/0
©1984 The Linnean Society of London

Table 1. The resident species of Galapagos land birds (Harris, 1973a; Grant, 1983a). *Indicates an endemic species. *G. difficilis* was originally named *G. nebulosa* (Sulloway, 1982a, 1982b)

English name	Scientific name
Galapagos Hawk	*Buteo galapagoensis*
Galapagos Rail	*Laterallus spilonotus*
Paint-billed Crake	*Neocrex erythrops*
Galapagos Dove	*Zenaida galapagoensis*
Dark-billed Cuckoo	*Coccyzus melacorhyphus*
Barn Owl	*Tyto alba*
Short-eared Owl	*Asio flammeus*
Vermillion Flycatcher	*Pyrocephalus rubinus*
Large-billed Flycatcher	*Myiarchus magnirostris*
Galapagos Martin	*Progne modesta*
Galapagos Mockingbird	*Nesomimus parvulus*
Charles Mockingbird	*Nesomimus trifasciatus*
Hood Mockingbird	*Nesomimus macdonaldi*
Chatham Mockingbird	*Nesomimus melanotis*
Yellow Warbler	*Dendroica petechia*
Small Ground Finch	*Geospiza fuliginosa*
Medium Ground Finch	*Geospiza fortis*
Large Ground Finch	*Geospiza magnirostris*
Sharp-beaked Ground Finch	*Geospiza difficilis*
Cactus Ground Finch	*Geospiza scandens*
Large Cactus Ground Finch	*Geospiza conirostris*
Vegetarian Finch	*Platyspiza crassirostris*
Small Tree Finch	*Camarhynchus parvulus*
Medium Tree Finch	*Camarhynchus pauper*
Large Tree Finch	*Camarhynchus psittacula*
Woodpecker Finch	*Cactospiza pallida*
Mangrove Finch	*Cactospiza heliobates*
Warbler Finch	*Certhidea olivacea*

from studies of museum specimens. Ten years ago it could be said that, in terms of breeding behaviour and population ecology, sea-birds on the Galapagos were known better than the land-birds. The situation is now reversed, as a result of extensive field studies of hawks, mockingbirds and Darwin's finches. Nevertheless it is still true that most of the species of land birds are poorly known.

In this article I shall review advances in our understanding of the evolution of mockingbirds and finches. Evolutionary studies of the other species have lagged behind. Although, in the last few years, we have learned more about the behaviour and ecology of the hawk (de Vries, 1975, 1976; Faaborg *et al.*, 1980), rail (Franklin, Clark & Clark, 1979), dove (Grant, P. R. & Grant, K. T., 1979; Grant, P. R. & Grant, B. R., 1980a) and short-eared owl (de Vries, 1975; Grant, P. R. *et al.*, 1975; Grant, P. R. & Grant, B. R., 1980a), the evolutionary forces on the Galapagos that have shaped their features of interest have not been the prime focus of studies. For example, it has been established that the hawk has a co-operative, polyandrous, breeding system (deVries, 1975, 1976). The incidence of polyandry varies among island populations. The study of such variation can lead to a better understanding of the maintenance of the co-operative breeding habit (Faaborg *et al.*, 1980, Faaborg & Patterson, 1981). But since the related continental species, *Buteo harrisi*, also exhibits this breeding system (Mader, 1979), it is unlikely to have evolved on the Galapagos. Therefore, I will omit it from the review, and concentrate instead on the broad questions of evolution on the Galapagos islands that have been illuminated by research conducted in the last ten years.

MOCKINGBIRDS

Mockingbirds pose some interesting evolutionary problems. They have differentiated into four forms recognizable by plumage patterns, size and eye colour. The pattern of differentiation is not understood, and we are not even certain that the four forms are different species as currently recognized (Harris, 1973, 1974). Bowman & Carter (1971) kept *Nesomimus parvulus*, *N. trifasciatus* and *N. macdonaldi* in captivity and reported without further detail that "all attempts to hybridize them failed". These negative results are at least consistent with the current practice of treating the forms as separate species.

Figure 1 shows their distribution in the archipelago. Each one of three species is restricted to one large southern island and its satellites: the fourth is widely distributed through the archipelago, but its populations are not strongly differentiated. No two species occur on the same island. They may be the product of one or more colonizations from the continent. Their distribution, like their differentiation, has yet to be explained satisfactorily (see Swarth, 1931).

Mockingbirds present a taxonomic puzzle too. Unless the ancestral species on the continent has become extinct, all island species are derived from the ancestors of the single species in western South America, *Mimus longicaudatus*. This species is distinctive in wing, tarsus and bill dimensions but not especially so in relation to variation among Galapagos species. For this reason Abbott & Abbott (1978) place the island species in the same genus (*Mimus*) as the mainland species, just as Rothschild & Hartert (1899) had done many years earlier. However the continental species is also distinctive in plumage, so whether one or two genera should be adopted, in recognition of the attainment of a particular level of evolutionary divergence in the archipelago, is a difficult question to resolve. Results of captive breeding could shed light on this question, but if anything they add confusion. Having failed to interbreed the island species, all unquestionably congeneric, Bowman & Carter (1971) accidentally brought about a cross between a male *M. longicaudatus* and a female *N. parvulus*! One of the progeny reached adulthood. The generic status of Galapagos mockingbirds is therefore uncertain.

Aside from classification problems, another evolutionary problem has come to light in the last few years. Like the hawks, Galapagos mockingbirds breed co-operatively (Grant, P. R. & Grant, N., 1979). Unlike the hawks, the breeding system is monogamous, with young males, almost invariably sons, helping the parents to feed their nestlings. Kinnaird & Grant (1982) reasoned that the co-operative breeding habit was possessed by mockingbirds at the time they colonized the Galapagos. This appears to be wrong. A detailed, unpublished, breeding study of *M. longicaudatus* in Peru (M. D. Williams, pers. comm.) produced no evidence of co-operative breeding. Therefore it seems likely that the habit evolved on the islands: if so, why? Why is it a regular feature of mockingbird breeding? In contrast, helping at the nest appears to be only an

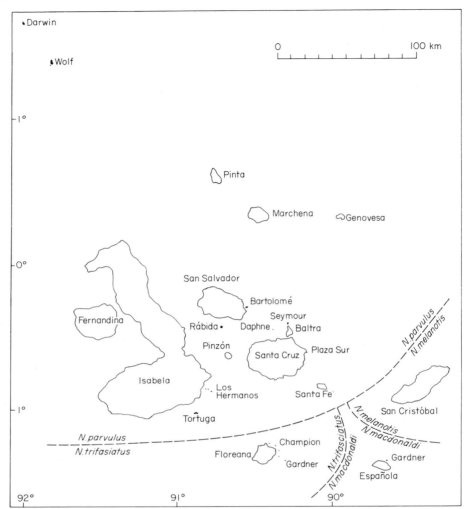

Figure 1. The distribution of four species of mockingbirds (*Nesomimus* spp.) in the Galapagos.

occasional and accidental feature of finch breeding (Price, Millington & Grant, 1983).

Two approaches are being used to answer these questions (R. L. Curry, pers. comm.). One is to apply general models for the evolution of co-operative breeding in birds (Brown, 1974, 1978; Emlen, 1978, 1982a, 1982b) to Galapagos mockingbirds. The models are modified to allow for the ecological differences between continent and islands, such as the different risks of predation (Marchant, 1960; Grant, P. R. & Grant, B. R., 1980a).

The second approach is to make a comparative study of the breeding structure of populations on different islands in order to identify the most influential factors and their effects. Breeding structure is certainly likely to vary among populations. For example, on the 10 ha island of Champion near I. Floreana there were only 49 mockingbirds (*N. trifasciatus*) in 1980 (unpubl. pers.

obs.). This probably inbred population had a female-biased sex ratio. In the same year the population of at least 3000 *N. parvulus* on I. Genovesa had a male-biased sex ratio (Kinnaird & Grant, 1982). In the five years this population has been studied, the maximum size of a group participating in territorial defence in the non-breeding season was nine individuals. In contrast, Hatch (1966) observed a group of up to 40 individuals in a single season's study of the population (*N. macdonaldi*) on I. Española. Studies of these populations by R. L. Curry and myself are continuing.

DARWIN'S FINCHES

Monographs by Lack (1945, 1947) and Bowman (1961) dealt comprehensively with the evolution of this group of birds. I shall summarize recent additions to our knowledge in answers to five major questions. Where did they come from? Which species gave rise to which? How many species evolved and over what period of time? Why have no more species evolved? Why did they evolve in the directions they did?

What are their systematic origins?

There is general agreement that the closest relatives are emberizine finches in South or Central America. There is also general agreement that they are all more closely related to each other than any one is to a living species on the continent. There is no agreement on the most closely related continental species. Bowman (1961) favoured a species, *Melanospiza richardsoni*, living not on the continent but on the West Indian island of St Lucia; see also Cutler (1970). Harris (1972) stressed the similarities between Darwin's Finches and *Coereba flaveola*. Steadman (1982a) has made the most comprehensive attempt to come to grips with this problem and proposed that two races of *Volatinia jacarina*, one in Central America and the other in South America, provided two separate groups of colonists to Cocos Island and the Galapagos respectively, and from these all Darwin's finches are derived.

None of these attempts to answer the question lay all doubts to rest. Steadman (1982a) exposed and discussed weaknesses of preceding efforts to identify the continental descendant of the ancestral species. However his candidate, *V. jacarina*, differs conspicuously from all of Darwin's finches by a highly characteristic vertical display flight performed by males, and by an open cup-shaped nest—all Darwin's finches build domed nests. The climatic environments of *V. jacarina* and Darwin's finches on the Galapagos are not very different, therefore it is not clear why such marked behavioural differences would have evolved on the Galapagos if *V. jacarina* was close to the ancestral stock.

Thus the ancestors of Darwin's finches have not been unambiguously identified by studies of phenotypic resemblance. The solution to the problem may be reached by studies of genetic resemblance. Sibley & Ahlquist (in press) and C. G. Sibley (pers. comm.) have used the technique of DNA–DNA hybridization to assess the systematic relationships of many species of passerine birds. Some of their findings are surprising. Many S American species currently classified as emberizine finches are in fact more closely related to members of the

Thraupidae (tanagers) than to emberizines. Conceivably then, Darwin's finches are not finches, but like many finch-like species on the continent they are tanagers. If so they should be called Darwin's finch–tanagers. Extension of the DNA–DNA hybridization technique to Darwin's finch material holds the promise of a resolution of the problem of identifying the systematic origins of the group and the most closely-related, extant, continental species (C. G. Sibley, pers. comm.).

What are their ancestor–descendant relationships?

Lack (1947) used the similarities and differences among species in plumage, size and shape, especially of the bill, to construct a tentative phylogenetic tree (Fig. 2). The main features of the tree are (a) an early differentiation of warbler-like finches (*Certhidea olivacea* on the Galapagos and *Pinaroloxias inornata* on Cocos Island) from the ancestral stock, (b) subsequent separation of the tree finches from the remainder (ground finches), and (c) an even later differentiation of the tree finch and ground finch groups, with *G. difficilis* being closest to the ancestor.

The main features of this scheme have been confirmed by the results of an analysis of protein polymorphisms (Yang & Patton, 1981; Polans, 1983; initiated by Ford, Ewing & Parkin, 1974). First, the warbler finch is the most distinctive species (Fig. 3), biochemically, and hence diverged from the rest at the earliest time (Yang & Patton, 1981). Second, the six ground finches cluster together, and four of the tree finches cluster together: material was lacking for the other two species. Barrowclough (1983) has integrated the results of the morphological and biochemical studies.

A small discrepancy with Lack's tree is that the tree finches are found to have differentiated more recently than the beginning of the ground finch differentiation (Yang & Patton, 1981). Also by this analysis *G. difficilis* is not

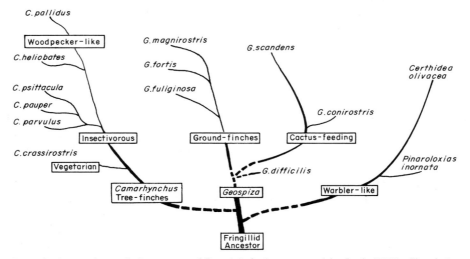

Figure 2. A tentative evolutionary tree of Darwin's finches suggested by Lack (1947). *Pinaroloxias inornata* occurs on Cocos Island, all other species occur on the Galapagos.

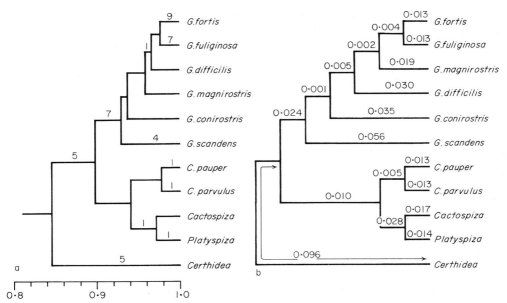

Figure 3. Relationships among Darwin's finch species based on an analysis of protein polymorphisms (from Yang & Patton, 1981). Numbers along branch lengths of the dendrogram are Rogers' distance coefficients. G. = *Geospiza*, and C = *Camarhynchus*. Note the similarity of the clustering here and the grouping of species in Fig. 2, where *Cactospiza pallidus* and *Platyspiza crassirostris* were placed by Lack (1947) in the genus *Camarhynchus*.

revealed to be particularly distinctive within the ground finch group; but ground finches are all so similar it is doubtful whether the exact sequence of differentiation of the six species can be reliably estimated by this technique at present. Nor can it be estimated by examination of chromosome structure and numbers, at least not with current techniques, because there is a large degree of variation within species in both chromosome features (Jo, 1983). Again, the DNA–DNA hybridization technique has the best potential of resolving these problems at the species level.

How many species evolved, how many became extinct and over what period of time?

Geological data suggest that the islands are no more than five million years old (Bailey, 1976; Hey, 1977; Cox, 1983). This sets the maximum span of time over which the full adaptive radiation of Darwin's Finches occurred. The biochemical differences between the species can be converted to times since their evolutionary separation by applying Nei's method of dating. This allows an estimation to be made of the actual time course of the radiation. Yang & Patton (1981) provide these estimates, and discuss the uncertainties of the assumptions upon which the method rests (see also Thorpe, 1982). They calculate that *Certhidea olivacea* split off from the ancestral stock about 570 000 years ago. This first differentiation may have occurred earlier, as early as 1.5–2.0 million years ago, if more conservative assumptions about the equivalence of electrophoretic distance and time are correct (Yang & Patton, 1981; see also Sarich, 1977; Thorpe, 1982). Nevertheless, the overall conclusion is that Darwin's finches have occupied the islands for a relatively short period of time, and have differentiated rapidly, for example in relation to the Hawaiian honeycreepers (estimated to be 15–20 million years by Sibley & Ahlquist, 1982).

The temporal pattern of speciation is illustrated in Fig. 4. Figure 4A shows that the unstandardized rate of speciation increased to a maximum in the interval 50 000–100 000 years B.P. (before present), and that no further speciation occurred in the most recent interval. But Fig. 4A is based on extant forms only, and a true pattern of speciation can only be obtained by incorporating extinctions, if they have occurred. This point is illustrated in Fig. 4B: modern material is used to estimate the C curve, but the total number of species evolved (T curve) could be seriously underestimated if extinctions have been numerous.

Fossils are required to provide estimates of the curves in Fig. 4B, and to identify our current position on the time axis. Fossils have seemed impossible to obtain from volcanic islands such as the Galapagos, but recently Steadman (1982b, pers. comm.) has assembled and identified a remarkable collection of bones from cracked lava tubes on Islas Floreana (Table 2) and Santa Cruz (pers. comm.). The dated fossils are no older than 2400 years B.P.: undated ones

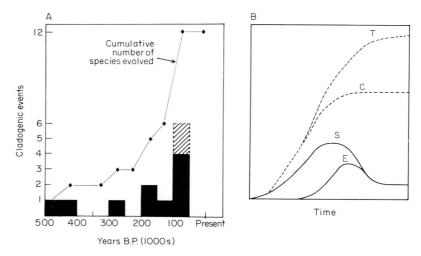

Figure 4 A, B. The temporal pattern of speciation of Darwin's finches. A, The number of speciation events occurring in each 50 000 year period, based on data in Yang & Patton (1981). The hachured part of the histogram refers to two species, *Camarhynchus psittacula* and *Cactospiza heliobates*, not studied by Yang & Patton (1981): the time of their formation has been estimated on the basis of their morphological similarity with congeners of known times of origin. The accumulation of species is shown by a continuous line. Note the absence of recent speciation: if anything, the recent period without speciation has been underestimated (see Yang & Patton, 1981), and perhaps the whole time span has been underestimated (see text). B, Hypothetical curves of speciation (S) and extinction (E) and the accumulation of species in total (T) and at any one time (C). It is assumed that there is a fixed maximum number of species sustainable in the archipelago and that this maximum is approached in dampened fashion (C) as a result of a rise and fall in the speciation rate (S) and, with a lag, a rise and fall in the extinction rate (E). The difference between T and C is solely attributable to E. The C curve has approximately the same form as the cumulative species curve in the upper diagram: the difference is that the C curve is the product of speciation and extinction whereas the equivalent curve in the upper diagram is estimated from extant forms only. Rates of speciation and extinction represented by solid line, cumulative and total number of species shown by broken line.

Table 2. Numbers of individual represented by fossils (from Steadman, 1982b, and pers. comm.) and by male specimens (skins) in Museum collections (from Lack, 1947). *Geospiza negulosa*, as used by Steadman (1982b), is synonymous with *G. difficilis* in Table 1

	I. Floreana	
	Fossils	Skins
Geospiza magnirostris	229	5
Geospiza fortis	12	181
Geospiza fuliginosa	18	86
Geospiza nebulosa	6	4
Geospiza scandens	2	102
Platyspiza crassirostris	4	24
Camarhynchus psittacula	0	3
Camarhynchus pauper	3	80
Camarhynchus parvulus	1	86
Certhidea olivacea	7	25

are probably much older. They are believed to be derived largely from disintegrated barn owl pellets (Steadman, 1982b).

Work currently in progress suggests the possibility that two *Geospiza* species have become extinct on I. Santa Cruz in the last 2400 years (D. W. Steadman, pers. comm.). The final taxonomic judgement as to whether some bones from I. Santa Cruz belong to extinct or extant species has not been rendered. If they belong to extinct species, they will provide the first evidence that Darwin's finches differentiated further than is shown by the 14 living species, and they will provide the first step in the construction of extinction and total species curves, illustrated in hypothetical form in Fig. 4B.

Fossils also have the potential of providing a documentation of evolutionary change. The fossils from I. Santa Cruz and I. Floreana have not yet been studied in sufficient detail for an analysis of evolution.

A third value of fossils is in providing estimates of changes in community membership through time. In Table 2 I have listed the minimum number of individuals represented by fossils on I. Floreana, without regard to time. For comparison I have also listed the number of specimens in museum collections from the same island as a measure of the current relative abundance of those species. The measure is only approximate, but it is correlated with the number of birds trapped in mist nets in a standard census period (Fig. 5). Two results of the comparison are worth mentioning. First, all species that are present on Floreana now, or were present when Darwin visited it, but subsequently became extinct, are represented in the fossil collections, with the trivial exception of *Camarhynchus psittacula*. The exception is not surprising because the species is rare, and occurs in the highlands, whereas the source of the fossils is in the lowlands. Second, there is a mismatch in relative abundances in fossil and modern collections. Not all differences can be attributed to collecting biases. *Geospiza magnirostris* dominates the fossil record, perhaps because the fossils were derived from owl pellets and perhaps because the barn owls preyed selectively on this finch species as its relative, the short-eared Owl, is known to do (Grant, P.R. & Grant, B.R., 1980a). *Geospiza magnirostris* specimens are rare in modern collections

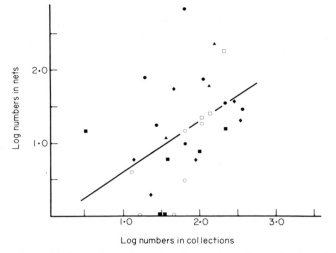

Figure 5. The relationship between numbers of specimens of *Geospiza* species from different islands in Museum collections and numbers of birds of the same species and from the same islands trapped in mist nets in a standard way in the dry season (data from Grant *et al.*, 1975; Abbott *et al.*, 1977; Smith *et al.*, 1978; Schluter, 1982b; Schluter & Grant 1982; and unpubl.). The correlation is statistically significant ($P < 0.05$). Islands include Santa Cruz but not Floreana (see Table 2). Symbols: ● *G. fuliginosa*, ▲ *G. difficilis*, ◆ *G. fortis*, ■ *G. scandens*, □ *G. conirostris*, ○ *G. magnirostris*.

from I. Floreana because it went extinct after Darwin's visit, probably for reasons associated with human settlement of the island (Sulloway, 1982a, 1982b). Other striking examples of a lack of correspondence in relative abundances are provided by *G. fortis* and *G. scandens*; they are rare as fossils and abundant in modern collections and finch communities (Abbott, Abbott & Grant, 1977; Smith *et al.*, 1978). Their scarcity as fossils cannot be attributed to their avoidance by owls (Grant, P. R. *et al.*, 1975).

Why have no more species evolved?

Lack (1945, 1947) pointed out that large islands have more finch species than small ones. In several subsequent studies a statistical relationship has been established between number of species and island area (Hamilton & Rubinoff, 1963, 1964, 1967; Harris, 1973; Power, 1975; Abbott *et al.*, 1977; Connor & Simberloff, 1978). Finch species diversity is also correlated with food resource diversity. The reasons for the relationships are usually discussed in the context of the equilibrium theory of island biogeography, that is in terms of immigration and extinction (MacArthur & Wilson, 1963, 1967). For a group of species that have evolved from a common ancestor in an archipelago, like Darwin's finches, the question that is usually not addressed is why the relationship between species number and island takes its particular form and no other. Why, for example, are there three species of finches on I. Española, eight on I. Floreana and ten on I. Isabela, and not 6, 16 and 20 on them respectively?

The two types of answers are (a) there has not been sufficient time for the evolution of 20 or more species, and (b) the archipelago is 'saturated' with species, and the upper limit to the number of species on each island and in the archipelago as a whole, set by ecological constraints, has been reached already.

Evidence for the first hypothesis comes from an attempt to predict the maximum number of co-existing ground finch species from the theory of limiting similarity (Grant, P. R., 1983b). A knowledge of the range of sizes of mainland species of finches, and the spacing rule of island species on a bill size axis, leads to the prediction that a maximum of seven species of ground finch species should co-exist. In fact only five co-exist (see Fig. 6). Ecological explanations for the absence of two species were sought and not found, therefore an hypothesis of insufficient time was tentatively accepted (Grant, P. R., 1983b). However the ecological data were crude, and the analysis was more revealing in identifying areas of ignorance than in providing support for a particular hypothesis.

Other and stronger evidence supports the alternative hypothesis of saturation. First, as discussed above, it is likely that there has been sufficient time for more than 14 species to evolve, as possibly two or more species have become extinct. Second, the potential for evolutionary change exists in present species: there is enough additive genetic variance underlying phenotypic variation in mensural traits for selection to act on in such a way as to result in the formation of a new species. Third, ecological data suggest that islands have as many finch species as they can sustain over long periods of time.

To amplify the second point, consider the absence of a micro *Geospiza* species pinpointed by the analysis in Fig. 6. For this missing species to evolve, selection should effect a 15% reduction in the size of *G. fuliginosa*. Heritabilities of quantitative morphological traits (beak size and body size) are probably large in this species, because they are known to be large in the related species *G. fortis* and *G. conirostris* (Boag & Grant, 1978; Boag, 1983; Grant, P. R. & Price, 1981; Grant, 1981a, in press a). Furthermore a selective shift of about 6% in morphological traits has been witnessed in the space of one year in the population of *G. fortis* on I. Daphne Major (Boag & Grant, 1981, in press). As finches depleted the non-renewed food supply of mainly small seeds during the drought of 1977, only those with large bills were capable of cracking the relatively large seeds that remained in moderate abundance, and they survived at a higher frequency than did those with small bills. Genetic variation governing bill size and body size was not detectably reduced during this episode of directional selection: heritabilities of all traits remained large (Boag, 1983).

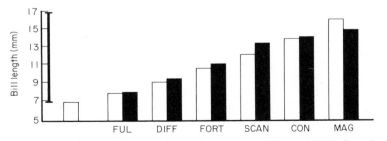

Figure 6. Bill lengths of ground finch species on the Galapagos (Grant, 1983b) Open bars show lengths predicted from limiting similarity theory (see text) and from the possible range of lengths indicated by the vertical line. Solid bars show the observed bill lengths of the six species in the genus *Geospiza: fuliginosa* (FUL), *difficilis* (DIFF), *fortis* (FORT), *scandens* (SCAN), *conirostris* (CON) and *magnirostris* (MAG). Note the absence of the smallest possible finch species. For further details see Grant (1983b).

Therefore the potential for further evolutionary change was present, and in fact realized (Price, in press a & b; Price & Grant, in press; Price, Grant & Boag, in press).

Given high heritabilities and strong selection, it is not difficult to envisage a selective shift in bill size and body size of 15% over a period of, say, 50–100 years. The fact that a micro- *Geospiza* species does not occur anywhere in the archipelago therefore suggests that the appropriate regime of selection does not exist. In other words, the feeding niche for such a species does not exist. Small seeds are consumed by *G. fuliginosa*, and the population sizes of this species on various islands are well predicted by the abundance of small seeds on those islands (Schluter & Grant, in press a). Searching for vacant niches is not a rewarding exercise, and the wealth of data on the feeding ecology of the ground finches obtained in the last decade (Grant, P. R. *et al.*, 1976; Abbot *et al.*, 1977; Smith *et al.*, 1978, Grant, B. R. & Grant, P. R., 1979, 1981, 1982; Grant, P. R. & Grant, B. R., 1980a, 1980b; Grant, P. R., 1981b; Grant, P. R. & Schluter, in press, Schluter & Grant, 1982, in press a & b; Schluter, 1982a, 1982b; Boag & Grant, in press; Millington & Grant, 1983) does not suggest that one will be found.

The third class of evidence for saturation is ecological. There are non-random distributional features of species in the genus *Geospiza* that can be interpreted as the product of interspecific competition. The outcome of competition between two, closely related, ecologically similar species has been the extinction of one of them. The result of competitive exclusion has been referred to as differential colonization (Grant, P. R., 1969; Grant, P. R. & Abbott, 1980) or size assortment (Case & Siddell, 1983). It has been an important process in the adjustment of species number to island size.

The evidence for differential colonization is as follows. First, combinations of *Geospiza* species are non-random; some combinations occur more frequently and others less frequently than expected by chance (Abbott *et al.*, 1977; Simberloff & Connor, 1981; Grant, P. R., 1981c; Grant, P. R. & Schluter, in press, Simberloff, 1983). Second, those that are under-represented are combinations of species that are morphologically and ecologically similar to each other (Grant, P. R. & Schluter, in press). Third, each of the six *Geospiza* species occurs with its most similar congener on fewer islands than average (Grant, P. R. & Schluter, in press). Fourth, there is a statistically significant under-representation of species/genus on islands compared with what would be expected from a random distribution of species in all genera (Strong, Szyska & Simberloff, 1979; Grant, P. R. & Abbott, 1980; Grant, P. R., 1981c). The trends for *Camarhynchus* species are weaker (Power, 1975; Strong, Szyska & Simberloff, 1979; Hendrickson, 1981; Alatalo, 1982; Case & Siddell, 1983). Interpretation of these results in terms of competition between closely related species is dependent upon the strength of the assumptions of the analyses and the statistical adequacy of the tests. These topics have been discussed extensively by Abbott *et al.*, 1979; Simberloff, 1978, in press; Connor & Simberloff, 1978; Strong *et al.*, 1979; Grant, P. R. & Abbott, 1980; Grant, P. R., 1981c, 1983b; Hendrickson, 1981; Strong & Simberloff, 1981; Simberloff & Connor, 1981; Simberloff & Boecklin, 1981; Alatalo, 1982; Grant, B. R. & Grant, P. R., 1982; Schluter & Grant, 1982; Schoener, 1982, in press; Case & Siddell, 1983; Colwell & Winkler, in press; Grant & Schluter, in press.

The evidence of differential colonization supports the hypothesis that a particular island is saturated with species or close to that state. It does not throw light on the question of whether the archipelago is saturated with species or not. To take the extreme case, why are there not many more species of Darwin's finches, each restricted to one island? Part of the answer lies in the recurrent patterns of resource frequency distributions among islands (Abbott et al., 1977; Schluter & Grant, in press a). Another part of the answer is that species evolved sequentially and not simultaneously. A third part is that each speciation event was followed by dispersal of the new species to other islands, and colonization of those that had the food resources which the new species was able to exploit. Thus each speciation event followed by dispersal brought the number of species in the archipelago closer to the maximum sustainable (Fig. 4). This argument is qualitative, and needs to be put into quantitatively testable form (see also Hamilton & Rubinoff, 1964, 1967).

In conclusion, lack of ecological opportunity, rather than lack of evolutionary time, is the best current explanation for why there are no more than 13 species of Darwin's finches on the Galapagos.

How is the adaptive differentiation explained?

This is the last and largest question. How can we account for the evolution of 13 species from a single ancestral species, and how can we account for their morphological, behavioural, ecological and distributional properties? These were the major questions dealt with by Lack (1945, 1947), and to a lesser extent by Bowman (1961).

Lack (1945, 1947) proposed a model of repeated speciation events (see also Grant, P. R., 1981c, for elaboration) that involved two phases: an allopatric phase in which populations of the same species on different islands underwent a small amount of evolutionary divergence, and a secondary sympatric phase in which selection reinforced and amplified the initial divergence of the original and derived populations, thereby minimizing both competition for food and interbreeding (Fig. 7). In Lack's language, speciation involved the development of ecological isolation and reproductive isolation between two populations.

Both the ecological and reproductive relationships between closely related species in the ground finch group have been the focus of much recent research. Other aspects of the adaptive differentiation of the finches have not been studied in as much detail.

Ecological isolation

Adaptive morphological differentiation of populations of the same species begins in allopatry. This is suggested by the marked differences in the food supplies on different islands (Bowman, 1961; Abbott et al., 1977; Smith et al., 1978; Boag & Grant, in press; Schluter & Grant, in press a & b), something which Lack (1947) initially overlooked, later explicitly acknowledged (Lack, 1969) but largely ignored (Lack, 1971). Evolution in allopatry is more directly demonstrated by documenting interisland variation in morphology, and showing that such variation is correlated with ecological variation (Grant, B. R. & Grant P. R., 1982; Schluter & Grant, 1982, in press a & b; Boag & Grant, in press).

Lack (1947) argued that competition for food occurred between original and derived populations in the sympatric phase of the speciation process. Bowman

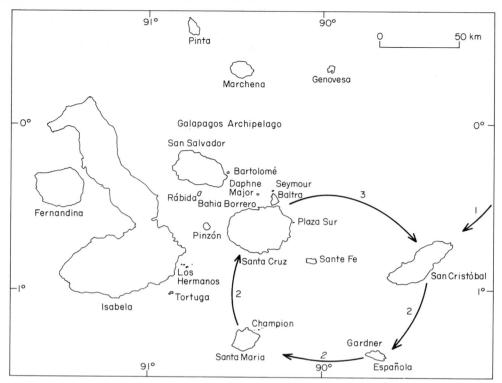

Figure 7. A representation of the allopatric model of speciation (after Grant, 1981c). Immigrants from the mainland colonized an island in step 1. Then dispersal to other islands took place, perhaps repeatedly, in step 2. In step 3 members of one of the derived populations colonized the original island and became established as a new species, interbreeding little if at all with members of the original population. Thus two species were formed from one, and in step 4 (not shown) the cycle of events was repeated several times with the eventual formation of at least 13 species, possibly more (see Fig. 4). The choice of islands to illustrate the model is arbitrary.

(1961) put forward the alternative view that competition did not occur because large ecological differences acquired in the allopatric phase enabled the populations to co-exist without interaction in the sympatric phase. The conflict of views and its resolution have been discussed in detail (Grant, P. R., 1981c, 1983b; Grant, B. R. & Grant, P. R., 1982). It is only summarized here.

Lack's evidence for competition was non-quantitative and, necessarily, inferential. For example he pointed to the apparently regular spacing of sympatric species of ground finches on a bill size axis on large islands, and the intermediate position of solitary species on this axis on small islands. Such regular spacing, and the inferred corollary of regular spacing along an axis of food size (see Abbott *et al.*, 1977 for evidence), is often used to draw conclusions about the over-dispersion of ecological niches as a result of competitive interaction (Schoener, 1974). However, given the small number of species involved in each analysis, it is not easy to detect non-random spacing along an axis. Efforts to apply statistics to data (Simberloff & Boecklin, 1981) have been criticized for their unrealistic assumptions (Schoener, in press).

Co-existing species of the genus *Geospiza* always differ in at least one bill dimension by at least 15% (Grant, P. R., 1981c, 1983b). This is consistent with an hypothesis of competition (Grant, P. R., 1981c), but a statistical null hypothesis has not been constructed and tested with this set of data. Instead, patterns of morphological variation among sympatric species have been quite accurately predicted from models based on food supply that incorporate the effects of competition. Models that do not include competitive effects do not make such accurate predictions (Schluter & Grant, in press a).

Nevertheless there are statistically non-random features of the morphology and distribution of the six ground finch species. Eleven of the 13 pairs of *Geospiza* species that co-exist differ morphologically more in sympatry than in allopatry (Fig. 8). Such a high proportion is not expected by chance (Grant, P. R., 1981c, 1983b). Morphological differences between *G. fortis* and *G. fuliginosa*, the two species that co-exist most frequently, are significantly greater in sympatry than would be expected if all their populations had been randomly combined in pairs (Grant, P. R. & Schluter, in press). These two analyses provide evidence of character displacement, that is enhanced morphological, and presumed ecological, differences in sympatry as a result of evolutionary changes of one or both species. The interplay between food factors and competition, and its evolutionary effect upon the finches, is discussed more fully in Schluter & Grant (in press a).

The overall conclusion from these tests and from those discussed on p. 124, is that interspecific competition has occurred in the past, and has left its mark on

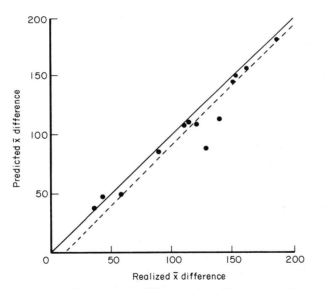

Figure 8. Prediction of multivariate beak differences, in arbitrary units, between co-occurring *Geospiza* congeners from a model of random combination of species. The predicted average differences between co-existing species are the differences between species means for all island populations, whether they co-exist or not, and realized differences are averages of the actual differences between co-existing populations of those species: thus sympatric populations (realized) are compared with sympatric and allopatric populations (predicted). If predictions were correct in all cases, all points would lie along the solid line; the broken line is a least squares best fit to the points. Most beak differences are greater than predicted (see text). *Geospiza conirostris* and *G. scandens* are predicted to occur with a multivariate difference of 67 units, and for *G. conirostris* and *G. fortis* the predicted difference is 83 units, but neither of these species pairs actually occurs on an island. These results were first presented in a symposium in 1977 (Grant, 1983b).

the morphological and distributional features of the finches. Lack's (1945, 1947) hypothesis of competitive interaction in the sympatric phase of the speciation processes is thus supported by the results of these quantitative tests.

Reproductive isolation

Lack's (1945, 1947) hypothesis postulates character displacement (divergence Grant, P. R., 1972) in courtship signals at the time of secondary contact between original and derived populations. At the start of the process, certain properties of individuals functioned as cues, signalling information about the group (population) to which those individuals belonged, with a degree of ambiguity in proportion to the lack of group-distinctiveness of the cues. At the end of the process of divergence under selection, the ambiguity had diminished or disappeared. Presumably the response systems also diverged under selection, with the end result being complete reproductive isolation of the two groups, now species. At this point the cues are species-specific.

Testing this hypothesis requires identification of courtship signals that convey species-specific information to the receiver. All species of Darwin's finches engage in similar postures and acts in courtship (Orr, 1945; Lack, 1947), and if there are reliable differences among species they are quantitative and hitherto undetected. Within some groups of finch species, such as the ground finch species, plumages are the same. Lack (1945, 1947) suggested that the most reliable cues conveying information about species identity are the ones we use, namely morphological cues and in some instances specifically bill shape. His experiments with stuffed speciments of *G. fortis* and *G. fuliginosa* supported this suggestion. However the large amount of inter-individual variation in response necessitates a repeat and extension of these experiments, which Ratcliffe (1981) has performed.

The role of morphological cues in species recognition was confirmed by the results of experiments conducted on several islands with *Geospiza* species. Discrimination by responding birds was tested in experiments each with two stuffed specimens. Males and females (of different species) responded more aggressively towards male conspecific specimens than to sympatric heterospecific ones (Ratcliffe & Grant, 1983a). Males preferentially courted conspecific females during experiments (and in natural encounters). Therefore species recognition by males responding to morphological cues is an important component of premating isolation in the genus. Results of an incomplete series of experiments suggested that recognition depends on combined morphological stimuli from head and body, rather than on stimuli from head alone (such as bill shape).

The reinforcement hypothesis of Lack can be tested with stuffed specimens. The hypothesis makes the prediction that males from sympatric populations should show significantly stronger sexual preferences for conspecific females than would males from allopatric population given the same choice situation. The prediction was tested and upheld by the results of experiments carried out with *Geospiza* species on different islands (Ratcliffe & Grant, 1983b).

Some of the species were more different in morphology when sympatric than when allopatric. But on I. Pinta the two species tested (*G. fuliginosa* and *G. difficilis*) were no more different from each other than were their tested allopatric populations, yet preference for conspecific females was enhanced in the sympatric populations. This result shows that the discrimination behaviour itself is different among island populations of the same species even when differences in the signals are the same among those populations. The question then arises as to whether enhanced discrimination in sympatry is the result of the experience of seeing heterospecifics, and possibly courting them without reward, or whether it represents an evolutionary change of behaviour. This question is unanswered by the experiments. But an evolutionary change is suggested by the enhanced ability to discriminate shown by *G. fuliginosa* living in the lowlands of I. Pinta where *G. difficilis* is virtually absent (Schluter, 1982a, 1982b; Schluter & Grant, 1982).

The reinforcement hypothesis also leads to the expectation of reproductive 'confusion' and hence interbreeding between a resident species and immigrants from another island of the same species, or of a similar and relatively undifferentiated species. A long-term field study of *G. fortis* on I. Daphne Major has yielded evidence of repeated but infrequent hybridization. Of all populations of *G. fortis* this one most resembles an allopatric species, *G. fuliginosa*. Small numbers of *G. fuliginosa* immigrate to I. Daphne where they rarely hybridize with *G. fortis* but do not establish themselves as a breeding population (Boag, 1981; Grant, P. R. & Price, 1981). In choice tests with stuffed specimens, *G. fortis* on this island show no discrimination between conspecifics and *G. fuliginosa*. Therefore, in agreement with the hypothesis, reproductive confusion exists, apparently as a consequence of morphological similarity.

Females rarely respond to stuffed specimens of males. Males, but not females, sing. Therefore song may also convey species-specific information, and female choice of males may be based partly or wholly on male song. Lack (1947) believed that song was too variable to be used by females as a reliable indicator of species identity. Bowman (1979, 1983) and Ratcliffe (1981) have used sonagraphic analysis to document song variation among and within species. These studies show that on most islands species usually sing different songs, although quantitative analyses establish a lack of discreteness in the structural features of the songs of different species on one or two islands (Ratcliffe, 1981). Moreover, song is less reliable as a species recognition cue than are some morphological cues since it is culturally acquired and mis-imprinting occasionally occurs (Bowman, 1979, 1983; Ratcliffe, 1981). This casts doubt on Bowman's (1983) untested hypothesis that female finches identify conspecific males *chiefly* by their song during courtship.

Species recognition by song has been tested experimentally with *Geospiza* species (Ratcliffe & Grant, in press). The results show that males of different species can discriminate between conspecific and heterospecific song in the absence of morphological cues. When morphological and song cues were presented simultaneously to test birds in a short series of experiments, the results were ambiguous. The general conclusion, however, is that both morphological and song cues can be, and probably are, used by birds in choosing a mate. And in choosing a mate, birds discriminate between members of their own and members of a different species more strongly when the heterospecifics are resident on the same island than when they are immigrants from another island.

It is a fundamental assumption of Lack's reinforcement hypothesis that between-group matings yield fewer offspring per capita that contribute to the

5

next generation than do within-group matings. The significance of visual discrimination in a sexual context during the experiments may be interpreted (Ratcliffe & Grant, 1983a) as a reflection of an individual's preference for a potential mate with high fitness prospects (i.e. a conspecific) rather than low fitness prospects (i.e. a heterospecific). Since hybrids are generally intermediate in morphology between parental phenotypes (Boag, 1981) it is reasonable to assume that such individuals have reduced chances of surviving the Galapagos dry season when finch diets diverge interspecifically as food becomes scarce and probably limiting to population sizes (Smith *et al.*, 1978; Grant, P. R. & Grant, B. R., 1980a, 1980b; Boag & Grant, 1981, in press; Schluter, 1982b). However reduced fitness of hybrids and their parents has not yet been demonstrated (Boag, 1981; Grant & Price, 1981).

Hybridization also occurs between resident *Geospiza* species (Boag, 1981; Grant, P. R. & Price, 1981; Grant, B. R. & Grant, P. R., 1982; Boag & Grant, in press), although at a frequency of 1% or less (see also Lack, 1945, 1947; Bowman, 1961; Harris, 1973). This means that reproductive isolation between several pairs of species is not perfect. The unanswered question for future research is the basis, or bases, of mate selection. If heterospecifics are usually avoided as potential mates why are they not always avoided, and within the range of acceptable conspecific mates what is it that governs which particular individuals will be chosen? Mis-imprinting has been implicated as the cause of the rare pairings between resident species (Boag, 1981; Ratcliffe, 1981; Grant, B. R. & Grant, P. R., 1982). Within *Geospiza* species there does not appear to be a pronounced tendency for mate choice to be based on either morphological or song cues (Grant, B. R. & Grant, P. R., 1979, 1983; Boag, 1983; Grant, B. R. in press; Millington, Price & Ratcliffe, in prep.), although two exceptions to this have been observed. In the first instance there was a weak positive association between *G. fortis* mates in mensural traits on I. Daphne Major in 1976, when the sex ratio was approximately 1 : 1 (Boag & Grant, 1978; Boag, 1981, 1983). The association has not been observed in any year since 1977, when males have consistently outnumbered females. However, in the second instance females have paired preferentially with large males during times of an unequal sex ratio, suggesting that mate choice was based either on male size or on the correlated traits of degree of blackness in the plumage and territory size (Price, in press a; see also Millington & Grant, 1983).

Lack (1945) initially attributed all differences in bill shape between morphologically similar species of finches to selection-reinforcing reproductive isolation. He later reinterpreted bill shape differences primarily in terms of selection for ecological isolation (Lack, 1947). Two recently established facts suggest that Lack's final emphasis was right. First, selection favoured large-billed *G. fortis* on I. Daphne Major in 1977, because they had a feeding advantage over the remainder of the population during a drought when breeding did not occur (Boag & Grant, 1981, in press). This study of selection in action complements inferences made about selection from numerous studies of the relationship between beak size and diets (e.g. Grant, P. R. *et al.*, 1976; Abbott *et al.*, 1977; Smith *et al.*, 1978; Grant, P. R., 1981b; Grant, B. R. & Grant, P. R., 1981, 1982; Schluter & Grant, in press a & b; Boag & Grant, in press; Grant & Schluter, 1983). Second, discrimination experiments on I. Pinta (Ratcliffe & Grant, 1983b) show that reproductive isolation may be brought about by the enhancement of the ability to discriminate and not necessarily by the divergence of morphological features. The principally ecological determination of morphological differences between species (Schluter & Grant, in press a) renders most of the theoretical objections to the hypothesis of reinforcement of reproductive isolation (Templeton, 1981) inapplicable here.

To summarize, recent observational and experimental results support Lack's reasoning about natural selection operating upon incipient species in sympatry. This does not mean that selection operated in the postulated way at all secondary contacts. In some cases substantial reproductive and ecological differences between species may have evolved in allopatry. But the evidence shows that divergence in ecological and reproductive traits occurred often enough to leave its mark on the properties of modern finch communities.

Some current and future developments

Evolutionary research on Darwin's finches in the last ten years has largely attempted to provide explanations for facts known to David Lack 35 years ago. Much of the research has involved fieldwork, including long-term population studies. These have yielded new facts. For example, it is now known that the interspecific pattern of egg size in relation to body size is quite different in Darwin's Finches from the pattern among continental finches living in a climatically similar habitat (Fig. 9). Clearly the difference represents an evolutionary shift on the islands that needs to be explained (Grant, P. R., 1982, 1983c). Another example is a colour polymorphism in the bills of young finches (Grant, P. R., *et al.*, 1979). It may be, as in domestic fowl, under simple genetic control. This has yet to be established. But if it is under simple genetic control it could be a useful tool for studying genetic processes within populations. The frequency of the colour morphs is known to vary between species, between populations of the same species (Grant, P. R. *et al.*, 1979), and between segments of a single population in an interesting way (Grant, B. R. & Grant, P. R., 1979, 1983).

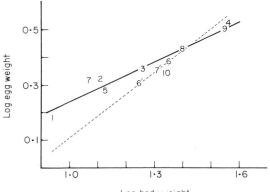

Figure 9. Relationships between log mean egg weight and mean body weight for Darwin's finches on the Galapagos and for a group of continental finches in a climatically similar part of Peru. The relationship for 12 numbered populations of 10 Darwin's finch species (slope 0.55) is shown by a solid line (see Grant, 1982 for population identities). The relationship for the six mainland species (slope 0.81) is shown by a broken line (Grant, 1983c).

Studies of life history traits such as clutch size and development patterns (Downhower, 1978; Grant, P. R. & Grant, B. R., 1980a; Boag, 1981; Grant, P. R., 1981a; Grant, B. R. & Grant, P. R., 1983; Grant, P. R. in press a), and of natural selection acting on morphological traits with which the life history traits are correlated (Price & Grant, in press), are beginning to reveal the previously unsuspected dynamic nature of Darwin's finch populations. They are showing not only large fluctuations in numbers as a result of climatic fluctuations (Grant, P. R. & Grant, B. R., 1980b; Grant, P. R. & Boag, 1980; Grant, P. R., in press, b), but pervasive natural selection (Grant, P. R. *et al.*, 1976; Boag & Grant, 1981; Grant, P. R. & Price, 1981; Price, in press b; Price, Grant & Boag, in press) and sexual selection (Price, in press a) operating on populations structured by age, size and song type (Grant, B. R. & Grant, P. R., 1979, 1983; Price, in press a). These studies have the potential of explaining why Darwin's Finches are sexually dimorphic in size (Downhower, 1976; Price, in press b; why some species are sexually dichromatic and others are not; why males generally have a more conspicuous plumage than females and why rate of acquisition of fully adult plumage varies among males in a population (Price, in press a); and why some populations are so variable in bill size and body size (Ford, Parkin & Ewing, 1973; Grant, P. R. *et al.*, 1976; Grant, B. R. & Grant, P. R., 1979, 1983; Grant, P. R. & Price, 1981). Studies of genetic and demographic processes can deepen our understanding of both adaptation and speciation.

THIS IS THE HOUSE THAT LACK BUILT

Lack's (1945), 1947) monograph on Darwin's finches may be likened to an edifice. It was built on the same site, with the same materials and around the same pillars and struts as Darwin's much simpler construction erected a century earlier. New materials were supplied by Swarth (1931, 1934) and others, and incorporated with the aid of a blueprint sketched by Stresemann (1936). It was extremely well-built and modern for its time, being widely adopted as the archetype of a new style. How well has it fared since then?

It is still standing. It has not been replaced but has been transformed. It suffered storm damage in the early 1960s and late 1970s. This has been repaired. Rotten planking has been replaced. Bricks had been put in back to front, and in the wrong place. These errors have been corrected. Structurally it is the same building, but it stands on a firmer foundation now. It has been thoroughly overhauled, modernized, elaborated and extended. Architectural details of the revisions and extensions have been published by L. M. Ratcliffe and P. T. Boag as a foreword to Lack (1983). The house of evolution has evolved.

ACKNOWLEDGEMENTS

I thank the Royal Society of London for a travel grant that enabled me to participate in the symposium on Evolution in the Galapagos Islands. Further support came from National Science Foundation (U.S.A.) grant DEB 79-21119. The organizer of the symposium, R. J. Berry, helped in numerous ways which I greatly appreciate. I thank I. Abbott, B. R. Grant, D. Schluter and D. W. Steadman for advice on the manuscript.

REFERENCES

ABBOTT, I. & ABBOTT, L. K., 1978. Multivariate study of morphological variation in Galápagos and Ecuadorean mockingbirds. *Condor, 80:* 302–308.
ABBOTT, I., ABBOTT, L. K. & GRANT, P. R., 1977. Comparative ecology of Galápagos Ground Finches (*Geospiza Gould*): evaluation of the importance of floristic diversity and interspecific competition. *Ecological Monographs, 47:* 151–184.
ALATALO, R., 1982. Bird species distributions in the Galápagos and other archipelagos: competition or chance? *Ecology, 63:* 881–887.
BAILEY, K., 1976. Potassium–argon ages from the Galapagos Islands. *Science, 192:* 465–467.
BARROWCLOUGH, G. F., 1983. Biochemical studies of microevolutionary processes. In A. H. Brush & G. A. Clark (Eds), *Perspectives in Ornithology.* Cambridge: University Press.
BOAG, P. T., 1981. *Morphological variation in the Darwin's Finches (Geospizinae) of Daphne Major Island, Galápagos.* Unpubl. Ph.D. thesis, McGill Univ., Montreal.
BOAG, P. T., 1983. The heritability of external morphology in Darwin's Ground Finches (*Geospiza*) on Isla Daphne Major, Galápagos. *Evolution, 37:* 877–894.
BOAG, P. T. & GRANT, P. R., 1978. Heritability of external morphology in Darwin's Finches. *Nature, 274:* 793–794.
BOAG, P. T. & GRANT, P. R., 1981. Intense natural selection in a population of Darwin's Finches (Geospizinae) in the Galápagos. *Science, 214:* 82–85.
BOAG, P. T. & GRANT, P. R. (in press). The classical case of character release: Darwin's Finches (*Geospiza*) on Isla Daphne Major, Galápagos. *Biological Journal of the Linnean Society.*
BOWMAN, R. I. 1961. Morphological differentiation and adaptation in the Galápagos finches. *University of California Publications in Zoology, 58:* 1–302.
BOWMAN, R. I. 1979. Adaptive morphology of song dialects in Darwin's Finches. *Journal für Ornithologie, 120:* 353–389.
BOWMAN, R. I. 1983. The evolution of song in Darwin's Finches. In A. E. Leviton & R. I. Bowman, (Eds), *Patterns of evolution in Galápagos organisms:* 237–537. Special Publication, American Association for the Advancement of Science, Pacific Division, San Francisco.
BOWMAN, R. I. & CARTER, A., 1971. Egg-pecking behavior in Galápagos mockingbirds. *Living Bird, 9:* 243–270.
BROWN, J. L., 1974. Alternative routes to sociality in jays—with a theory for the evolution of altruism and communal breeding. *American Zoologist, 14:* 63–80.
BROWN, J. L. 1978. Avian communal breeding systems. *Annual Reviews of Ecology and Systematics, 9:* 123–156.
CASE, T. J. & SIDELL, R., 1983. Pattern and chance in the structure of model and natural communities. *Evolution, 37:* 832–849.
COLWELL, R. K. & WINKLER, D., in press. A null model for null models in evolutionary ecology. In D. R. Strong, D. S. Simberloff, A. B. Thistle & L. G. Abele (Eds), *Ecological Communities: Conceptual Inssues and the Evidence.* Princeton: University Press.
CONNOR, E. G. & SIMBERLOFF, D. S., 1978. Species number and compositional similarity of the Galápagos flora and avifauna. *Ecological Monographs, 48:* 219–248.
COX, A., 1983. Ages of the Galápagos Islands. In R. I. Bowman, M. Berson & A. E. Leviton (Eds), *Patterns of evolution in Galápagos organisms:* 11–23. Special Publication, American Association for the Advancement of Science, Pacific Division, San Francisco.
CUTLER, B. D., 1970. *Anatomical studies on the syrinx of Darwin's finches.* Unpubl. M.A. thesis, San Francisco State Univ., San Francisco.
De VRIES, Tj. 1975. The breeding biology of the Galápagos Hawk, *Buteo galapagoensis.* Le Gerfaut, 65: 29–58.
De VRIES, Tj. 1976. Prey selection and hunting methods of the Galápagos Hawk, *Buteo galapagoensis.* Le Gerfaut, 66: 3–42.
DOWNHOWER, J. F., 1976. Darwin's Finches and the evolution of sexual dimorphism in body size. *Nature, 263:* 558–563.
DOWNHOWER, J. F., 1978. Observations on the nesting of the small ground finch *Geospiza fuliginosa* and the large cactus ground finch *G. conirostris* on Espanola, Galápagos. *Ibis, 120:* 340–346.
EMLEN, S., 1978. The evolution of cooperative breeding in birds. In J. R. Krebs & N. B. Davies (Eds), *Behavioral Ecology: an Evolutionary Approach.* Massachusetts: Sinauer.
EMLEN, S., 1982a. The evolution of helping. I. An ecological constraints model. *American Naturalist, 119:* 29–39.
EMLEN, S., 1982b. The evolution of helping. II. The role of behavioral conflict. *American Naturalist, 119:* 40–53.
FAABORG, J., DE VRIES, Tj., PATTERSON, C. B. & GRIFFIN, C. R., 1980. Preliminary observations on the occurrence and evolution of polyandry in the Galápagos Hawk (*Buteo galapagoensis*). *Auk, 97:* 581–590.
FAABORG, J. & PATTERSON, C. B., 1981. The characteristics and occurrence of cooperative polyandry. *Ibis, 123:* 477–484.

FORD, H. A., EWING, A. W. & PARKIN, D. T., 1974. Blood proteins in Darwin's Finches. *Comparative Biochemistry and Physiology, 47B:* 369–375.

FORD, H. A., PARKIN, D. T. & EWING, A. W., 1973. Divergence and evolution in Darwin's Finches. *Biological Journal of the Linnean Society 5:* 289–295.

FRANKLIN, A. B., CLARK, D. A. & CLARK, D. B., 1979. Ecology and behavior of the Galápagos Rail. *Wilson Bulletin, 91:* 202–221.

GRANT, B. R. (in press). The significance of song variation in a population of Darwin's Finches. *Behaviour.*

GRANT, B. R. & GRANT, P. R. 1979. Darwin's finches: population variation and sympatric speciation. *Proceedings of the National Academy of Sciences, USA, 76:* 2359–2363.

GRANT, B. R. & GRANT, P. R., 1981. Exploitation of *Opuntia* cactus by birds on the Galápagos. *Oecologia, 49:* 179–187.

GRANT, B. R. & GRANT, P. R., 1982. Niche shifts and competition in Darwin's Finches: *Geospiza conirostris* and congeners. *Evolution, 36:* 637–657.

GRANT, B. R. & GRANT, P. R., 1983. Fission and fusion in a population of Darwin's Finches: an example of the value of studying individuals in ecology. *Oikos, 41:* 530–547.

GRANT, P. R., 1969. Colonization of islands by ecologically dissimilar species of birds. *Canadian Journal of Zoology, 47:* 41–43.

GRANT, P. R., 1972. Convergent and divergent character displacement. *Biological Journal of the Linnean Society, 4:* 39–68.

GRANT, P. R., 1981a. Patterns of growth in Darwin's Finches. *Proceedings of the Royal Society of London B, 212:* 403–432.

GRANT, P. R., 1981b. The feeding of Darwin's Finches on *Tribulus cistoides* (L.) seeds. *Animal Behaviour, 29:* 785–793.

GRANT, P. R., 1981c. Speciation and the adaptive radiation of Darwin's Finches. *American Scientist, 69:* 653–663.

GRANT, P. R., 1982. Variation in the size and shape of Darwin's Finch eggs. *Auk, 99:* 15–23.

GRANT, P. R., 1983a. The endemic landbirds. In R. L. Perry (Ed.), *The Galápagos.* Key Environment series, vol. 1. Oxford: Pergamon.

GRANT, P. R., 1983b. The role of interspecific competition in the adaptive radiation of Darwin's Finches. In R. I. Bowman, M. Berson & A. E. Leviton (Eds), *Patterns of Evolution in Galápagos Organisms:* 187–199. American Association for the Advancement of Science, Pacific Division, San Francisco.

GRANT, P. R., 1983c. The relative size of Darwin's Finch eggs. *Auk, 100:* 228–230.

GRANT, P. R. (in press a). Inheritance of size and shape in a population of Darwin's Finches, *Geospiza conirostris. Proceedings of the Royal Society of London, B:*

GRANT, P. R. (in press b). Climatic fluctuations on the Galápagos islands and their influence on Darwin's Finches. *Ornithological Monographs:*

GRANT, P. R. & ABBOTT, I., 1980. Interspecific competition, island biogeography, and null hypotheses. *Evolution, 34:* 332–341.

GRANT, P. R. & BOAG, P. T., 1980. Rainfall on the Galápagos and the demography of Darwin's Finches. *Auk, 97:* 227–244.

GRANT, P. R., BOAG, P. T. & SCHLUTER, D., 1979. A bill color polymorphism in young Darwin's Finches. *Auk, 96:* 800–802.

GRANT, P. R. & GRANT, B. R., 1980a. The breeding and feeding characteristics of Darwin's Finches on Isla Genovesa, Galápagos. *Ecological Monographs, 50:* 381–410.

GRANT, P. R. & GRANT, B. R., 1980b. Annual variation in finch numbers, foraging and food supply on Isla Daphne Major, Galápagos. *Oecologia, 46:* 55–62.

GRANT, P. R., GRANT, B. R., SMITH, J. N. M., ABBOTT, I. J. & ABBOTT, L. K., 1976. Darwin's Finches: population variation and natural selection. *Proceedings of the National Academy of Sciences USA, 73:* 257–261.

GRANT, P. R. & GRANT, K. T., 1979. The breeding and feeding of the Galápagos dove, *Zenaida galapagoensis. Condor, 81:* 397–403.

GRANT, P. R. & GRANT, N., 1979. Breeding and feeding of the Galápagos mockingbird, *Nesomimus parvulus. Auk, 96:* 723–736.

GRANT, P. R. & PRICE, T. D., 1981. Population variation in continuously varying traits as an ecological genetics problem. *American Zoologist, 21:* 795–811.

GRANT, P. R. & SCHLUTER, D., in press. Interspecific competition inferred from patterns of guild structure. In D. R. Strong, D. S. Simberloff, A. B. Thistle, & L. G. Abele (Eds), *Ecological Communities: Conceptual Issues and the Evidence.* Princeton: University Press.

GRANT, P. R., SMITH, J. N. M., GRANT, B. R., ABBOTT, I. J. & ABBOTT, L. K., 1975. Finch numbers, owl predation and plant dispersal on Isla Daphne Major, Galápagos. *Oecologia, 19:* 239–257.

HAMILTON, T. H. & RUBINOFF, I., 1963. Isolation, endemism, and multiplication of species in the Darwin's finches. *Evolution, 17:* 388–403.

HAMILTON, T. H. & RUBINOFF, I. 1964. On models predicting abundance of species and endemics for the Darwin finches in the Galápagos archipelago. *Evolution, 18:* 339–342.

HAMILTON, T. H. & RUBINOFF, I., 1967. On predicting insular variation in endemism and sympatry for the Darwin's finches in the Galápagos archipelago. *American Naturalist, 101:* 161–172.

HARRIS, M. P., 1972. *Coereba flaveola* and the Geospizinae. *Bulletin of the British Ornithologists Club, 92:* 164–168.

HARRIS, M. P., 1973. The Galápagos avifauna. *Condor, 75:* 265–278.

HARRIS, M. P., 1974. *A Field Guide to Birds of Galápagos.* London: Collins.

HATCH, J., 1966. Collective territories in Galápagos Mockingbirds, with notes on other behavior. *Wilson Bulletin, 78:* 198–206.

HENDRICKSON, J. A., Jr., 1981. Community wide character displacement reexamined. *Evolution, 35:* 794–809.

HEY, R., 1977. Tectonic evolution of the Cocos-Nazca spreading center. *Geological Society of America Bulletin, 88:* 1404–1420.

JO, N. 1983. Karyotypic analysis of Darwin's finches. In R. I. Bowman, M. Berson & A. E. Leviton (Eds), *Patterns of evolution in Galápagos organisms:* 201–217. American Association for the Advancement of Science, Pacific Division, San Francisco.

KINNAIRD, M. F. & GRANT, P. R., 1982. Cooperative breeding by the Galápagos Mockingbird, *Nesomimus parvulus. Behavioral Ecology and Sociobiology, 10:* 65–73.

LACK, D., 1945. The Galápagos finches (*Geospizinae*): a study in variation. *Occasional Papers of the California Academy of Sciences, 21:* 1–159.

LACK, D., 1947. *Darwin's Finches.* Cambridge: University Press.

LACK, D., 1969. Subspecies and sympatry in Darwin's Finches. *Evolution, 23:* 252–263.

LACK, D., 1971. *Ecological Isolation in Birds.* London: Methuen.

LACK, D. 1983. *Darwin's Finches.* Cambridge: University Press. Reprinting of Lack, D. 1947, with foreword by L. M. Ratcliffe & P. T. Boag.

MACARTHUR, R. H. & WILSON, E. O., 1963. An equilibrium theory of insular biogeography. *Evolution, 17:* 373–387.

MACARTHUR, R. H. & WILSON, E. O., 1967. *The Theory of Island Biogeography.* Princeton: University Press.

MADER, W. J., 1979. Breeding behavior of a polyandrous trio of Harris's Hawks in southern Arizona. *Auk, 96:* 776–788.

MARCHANT, S. 1960. The breeding of some S. W. Ecuadorian birds. *Ibis, 102:* 349–381, 584–599.

MILLINGTON, S. J. & GRANT, P. R., 1983. Feeding ecology and territoriality of the Cactus Finch *Geospiza scandens* on Isla Daphne Major, Galápagos. *Oecologia, 58:* 76–83.

MILLINGTON, S., PRICE, T. D. & RATCLIFFE, L. M., in press. Song inheritance and mating patterns in two species of Darwin's Finches.

ORR, R. T., 1945. A study of captive Galápagos finches of the genus *Geospiza. Condor, 47:* 177–201.

POLANS, N., 1983. Enzyme polymorphisms in Galápagos finches. In R. I. Bowman, M. Berson & A. E. Leviton (Eds), *Patterns of evolution in Galápagos organisms:* 219–235. American Association for the Advancement of Science, Pacific Division, San Francisco.

POWER, D., 1975. Similarity among avifaunas of the Galápagos Islands. *Ecology, 56:* 616–626.

PRICE, T. D., in press a. Sexual selection and body size, plumage and territory variables in a population of Darwin's finches. *Evolution.*

PRICE, T. D., in press b. The evolution of sexual size dimorphism in a population of Darwin's Finches. *American Naturalist.*

PRICE, T. D. & GRANT, P. R., in press. Life history traits and natural selection for small body size in a population of Darwin's Finches. *Evolution.*

PRICE, T. D. in press. Genetic changes in the morphological differentiation of Darwin's Ground Finches. In K. Wörhmann & V. Loshchke (Eds), *Population Biology and Evolution.* Berlin: Springer-Verlag.

PRICE, T. D., MILLINGTON, S. & GRANT, P. R., 1983. Helping at the nest as misdirected parental care. *Auk, 100:* 192–194.

RATCLIFFE, L. M., 1981. *Species recognition in Darwin's Ground Finches* (Geospiza, Gould). Unpubl. Ph.D. thesis, McGill Univ., Montreal.

RATCLIFFE, L. M. & GRANT, P. R., 1983a. Species recognition in Darwin's Finches (*Geospiza,* Gould). I. Discrimination by morphological cues. *Animal Behaviour, 31:* 1139–1153.

RATCLIFFE, L. M. & GRANT, P. R., 1983b. Species recognition in Darwin's Finches (*Geospiza,* Gould). II. Geographic variation in mate preference. *Animal Behaviour, 31:* 1154–1165.

RATCLIFFE, L. M. & GRANT, P. R. in press. Species recognition in Darwin's Finches (*Geospiza,* Gould). III. Male responses to playback of different song types, dialects and heterospecific songs. *Animal Behaviour 32.*

ROTHSCHILD, W. & HARTERT, E., 1899. A review of the ornithology of the Galápagos islands. With notes on the Webster-Harris expedition. *Novitates Zoologicae, 6:* 85–205.

SARICH, V. M., 1977. Rates, sample sizes and the neutrality hypothesis for electrophoresis in evolutionary studies. *Nature, 265:* 24–28.

SCHLUTER, D., 1982a. Seed and patch selection by Galápagos ground finches: relation to foraging efficiency and food supply. *Ecology, 63:* 1106–1120.

SCHLUTER, D., 1982b. Distributions of Galápagos ground finches along an altitudinal gradient: the importance of food supply. *Ecology, 63:* 1504–1517.

SCHLUTER, D. & GRANT, P. R., 1982. The distribution of *Geospiza difficilis* in relation to *G. fuliginosa* in the Galápagos islands: tests of three hypotheses. *Evolution, 36:* 1213–1226.

SCHLUTER, D. & GRANT, P. R. (in press a). Determinants of morphological patterns in communities of Darwin's Finches. *American Naturalist.*

SCHLUTER, D. & GRANT, P. R. (in press b). Ecological correlates of morphological evolution in a Darwin's finch species, *Geospiza difficilis.*

SCHOENER, T. W., 1974. Resource partitioning in ecological communities. *Science, 185:* 27–39.

SCHOENER, T. W., 1982. The controversy over interspecific competition. *American Scientist, 70:* 586–595.

SCHOENER, T. W., 1983. Size differences among sympatric, bird-eating hawks: a worldwide survey. In D. R. Strong, D. S. Simberloff, A. B. Thistle & L. G. Abele (Eds), *Ecological communities: conceptual issues and the evidence.* Princeton: University Press.

SIBLEY, C. G. & AHLQUIST, J. E., 1982. The relationships of the Hawaiian honeycreepers (Drepaninae) as indicated by DNA–DNA hybridization. *Auk, 99:* 130–140.

SIBLEY, C. G. & AHLQUIST, J. E., in press. The phylogeny and classification of the passerine birds, based on comparisons of the genetic material, DNA. *Proceedings of the 18th International Ornithological Congress (Moscow, 1982).*

SIMBERLOFF, D., 1978. Using island biogeographic distributions to determine if colonization is stochastic. *American Naturalist, 112:* 713–726.

SIMBERLOFF, D., 1983. Morphological and taxonomic similarity and combinations of coexisting birds in two archipelagos. In D. R. Strong, D. S. Simberloff, A. B. Thistle & L. G. Abele (Eds), *Ecological communities: conceptual issues and the evidence.* Princeton: University Press.

SIMBERLOFF, D. & BOECKLEN, W., 1981. Santa Rosalia reconsidered: size ratios and competition. *Evolution, 35:* 1206–1228.

SIMBERLOFF, D. & CONNOR, E. F., 1981. Missing species combinations. *American Naturalist, 118:* 215–239.

SMITH, J. N. M., GRANT, P. R., GRANT, B. R., ABBOTT, I. J. & ABBOTT, L. K., 1978. Seasonal variation in feeding habits of Darwin's Ground Finches. *Ecology, 59:* 1137–1150.

STEADMAN, D. W., 1982a. The origin of Darwin's Finches. *Transactions of the San Diego Society of Natural History, 19:* 279–296.

STEADMAN, D. W., 1982b. *Fossil birds, reptiles, and mammals from Isla Floreana, Galápagos archipelago.* Unpubl. Ph.D. thesis, Univ. Arizona, Tucson.

STRESEMANN, E., 1936. Zur Frage der Artbildung in der Gatung *Geospiza. Orgaan der Club Van Nederl. Vogel., 9:* 13–21.

STRONG, D. R., Jr. & SIMBERLOFF, D., 1981. Straining at gnats and swallowing ratios: character displacement. *Evolution, 35:* 810–812.

STRONG, D. R. Jr., SZYSKA, L. & SIMBERLOFF, D., 1979. Tests of community-wide character displacement against null hypotheses. *Evolution, 35:* 897–913.

SULLOWAY, F. J., 1982a. The *Beagle* Collections of Darwin's Finches (Geospizinae). *Bulletin of the British Museum (Natural History), Zoology series, 43:* 49–94.

SULLOWAY, F. J., 1982b. Darwin and his Finches: the evolution of a legend. *Journal of the History of Biology, 15:* 1–53.

SWARTH, H. S., 1931. The avifauna of the Galápagos Islands. *Occasional Papers of the California Academy of Sciences, 18:* 1–299.

SWARTH, H. S., 1934. The bird fauna of the Galápagos Islands in relation to species formation. *Biological Reviews, 9:* 213–234.

TEMPLETON, A. R., 1981. Mechanisms of speciation—a population genetic approach. *Annual Reviews of Ecology and Systematics, 12:* 23–48.

THORPE, J. P., 1982. The molecular clock hypothesis: biochemical evolution, genetic differentiation and systematics. *Annual Reviews of Ecology and Systematics, 13:* 139–168.

YANG, S. H. & PATTON, J. L., 1981. Genic variability and differentiation in Galápagos finches. *Auk, 98:* 230–242.

Biological Journal of the Linnean Society (1984), *21*: 137–155.

Evolution and adaptations of Galapagos sea-birds

D. W. SNOW

Sub-department of Ornithology, British Museum (Natural History), Tring, Hertfordshire HP23 6AP

AND

J. B. NELSON

Department of Zoology, Tillydrone Avenue, Aberdeen AB9 2TN

With two endemic genera, 26% endemism at the specific, and 58% endemism at the subspecific level, the Galapagos sea-bird fauna is more highly endemic than that of any other archipelago. Of the four most distinct, hence probably oldest, endemics, three are probably of north Pacific origin and the fourth may be. The next most distinct group is of Humboldt Current origin, and the remainder, which are very little differentiated, are of Pacific or Caribbean origin. Special adaptations to Galapagos conditions include the loss of flight by a cormorant and the evolution of nocturnal habits by a gull, the latter probably as a result of kleptoparasitism or nest-predation by frigatebirds. As a group the Galapagos sea-birds show varied adaptive responses to the relatively aseasonal and highly unpredictable marine environment. Opportunistic and non-annual breeding regimes predominate. Unlike the Galapagos land-birds, the study of Galapagos sea-birds has thrown light not on speciation processes but on the consequences of natural selection acting on breeding ecology and associated behaviour.

KEY WORDS:—Sea-birds – evolution – Galapagos islands.

CONTENTS

The highly endemic Galapagos land-birds, especially Darwin's finches, have attracted so much attention from biologists that the evolutionary interest of the sea-birds has been rather neglected. Certainly one cannot claim that the sea-birds should receive equal attention, as they have not undergone an evolutionary radiation within the archipelago, for a reason that is too obvious to

137

0024–4066/84/010137 + 19 $03.00/0

need stating, and so do not provide opportunities for the study of speciation. The sea-birds, however, have had to adapt themselves to some unusual environmental conditions, and some of the ways in which they have done so are unique.

We shall be concerned with two kinds of adaptation which are more or less distinct and need separate treatment: adaptations to the peculiar features of the Galapagos environment, physical and biotic, which in several cases have resulted in unusual structural and behavioural specializations; and adaptations to the seasonality, or lack of seasonality, in the marine environment of the Galapagos, which in several cases have resulted in unique breeding strategies. Though the study of Galapagos sea-birds has not advanced our understanding of the process of species formation, it has contributed significantly to an understanding of the nature of natural selection acting upon breeding ecology and behaviour.

Table 1. Breeding sea-birds of the Galapagos

Species	Status	Geographical origin	Feeding zone	Remarks
Spheniscus mendiculus	SP	Humboldt Current	Inshore	Probably closest to *S. humboldti* (Murphy, 1936 : 468)
Diomedea irrorata	SP	N Pacific	Pelagic	Related to N Pacific spp. *albatrus*, *nigripes* and *immutabilis* (Jouanin & Mougin, 1979)
Pterodroma phaeopygia	SU?	Pacific	Pelagic	Sp. breeds only Galapagos and Hawaii; related spp. in Caribbean, S Pacific, and Indian Ocean
Puffinus lherminieri	SU	Tropical Pacific	Mainly coastal	Sp. widespread, probably reached Galapagos from west
Oceanites gracilis	SU	Humboldt Current	Inshore	Sp. breeds only Galapagos and Humboldt Current; genus is of southern origin
Oceanodroma castro	—	N Pacific?	Pelagic	Sp. widespread in Atlantic and subtropical N Pacific
Oceanodroma tethys	SU	Humboldt Current	Mainly coastal	Sp. breeds only Galapagos and Humboldt Current
Phaethon aethereus	—	Caribbean–Atlantic	Pelagic	Sp. probably a recent colonist of E Pacific
Pelecanus occidentalis	SU	Caribbean–Atlantic	Inshore	
Sula nebouxii	SU	W coast of America	Mainly coastal	Sp. characteristic of warm waters N. of Humboldt Current
Sula dactylatra	—	Tropical Pacific	Pelagic	Sp. widespread, probably reached Galapagos from west
Sula sula	—	Tropical Pacific	Pelagic	Sp. widespread, probably reached Galapagos from west
Nannopterum harrisi	GEN	N Pacific?	Inshore	Probably derived from northern *P. carbo* complex
Fregata minor	—	Tropical Pacific	Pelagic	Sp. widespread, probably reached Galapagos from west
Fregata magnificens	SU?	Caribbean–Atlantic	Mainly coastal	
Creagrus furcatus	GEN	?	Pelagic	
Larus fuliginosus	SP	North America?	Inshore	May be a derivative of *L. atricilla*
Sterna fuscata	—	Tropical Pacific	Pelagic	Sp. widespread, probably reached Galapagos from west
Anous stolidus	SU	Tropical Pacific	Inshore	Sp. widespread, probably reached Galapagos from west

GEN, endemic genus; SP, endemic species; SU, endemic subspecies; SU?, doubtfully distinct endemic subspecies.

ENDEMISM OF GALAPAGOS SEA-BIRDS

Although the proportion of endemic genera, species and subspecies is considerably lower than for the land-birds, the sea-bird fauna of the Galapagos is much more highly endemic than that of any other archipelago in the world. Of the 19 breeding species of sea-birds (Table 1), 2 are placed in endemic genera, 3 in endemic species (i.e. a total of 5 endemic species), 6 more in endemic subspecies whose validity is generally agreed, and 2 more in endemic subspecies of doubtful validity. Endemism at the species level is thus 26%, and at the subspecific level at least 58%. The contrast with Hawaii is striking. Hawaii is also of volcanic origin but is older, more isolated geographically, and has an even more highly endemic land-bird fauna. Its sea-bird fauna consists of 22 breeding species of which only three are endemic subspecies (one of doubtful validity). No other tropical or subtropical archipelago or single island supports a sea-bird fauna approaching that of the Galapagos in degree of endemism; in particular, no other has an endemic genus.

ORIGIN OF THE SEA-BIRDS

In most cases the geographical origin of the species can be inferred with some confidence. Three species have close relatives living in the Humboldt Current area of the western coast of South America: *Spheniscus mendiculus*, *Oceanodroma tethys* and *Oceanites gracilis*. The first is an endemic species, the other two endemic subspecies. There can be little doubt that the ancestors of these species reached the Galapagos by dispersal from the south-east. Another species, the booby *Sula nebouxii* (endemic Galapagos subspecies), must also have reached the Galapagos from the American coast, where it is a characteristic species of the warmer waters to the north of the Humboldt Current.

Three species have distributions which are mainly Caribbean and Atlantic, with limited extensions into the tropical eastern Pacific: *Phaethon aethereus*, *Pelecanus occidentalis*, *Fregata magnificens*. The Galapagos populations of the last two of these constitute slightly marked endemic subspecies, while the Galapagos population of *P.aethereus* is undifferentiated. The slight degree of differentiation of these three species in the eastern Pacific suggests that they are comparatively recent arrivals or, in the case of *F. magnificens* which regularly crosses the Isthmus of Panama, that the Caribbean and Pacific populations are still in effective contact.

The largest group consists of species with wide distributions in the tropical and subtropical Pacific, and in most cases also in other tropical oceans: *Puffinus lherminieri*, *Oceanodroma castro*, *Sula sula*, *S. dactylatra*, *Fregata minor*, *Sterna fuscata* and *Anous stolidus*. Only two of them (*P. lherminieri*, *A. stolidus*) are represented by endemic Galapagos subspecies. *Pterodroma phaeopygia* may perhaps be placed in this group, but its world range is much narrower, with breeding populations only in the Galapagos and Hawaii. It is probably significant that in this group the degree of differentiation of the Galapagos populations is related to feeding habits: the two that are subspecifically distinct are inshore feeders, and the others primarily pelagic.

The four species that remain are the most distinct of the Galapagos endemics: *Diomedea irrorata*, *Nannopterum harrisi*, *Creagrus furcatus* and *Larus fuliginosus*. Their relationship to other members of their families may eventually be clarified by new techniques such as DNA-hybridization, but for the present caution is necessary. Murphy (1936) considered that *D. irrorata* has affinities with the north Pacific albatrosses, its closest relative perhaps being *D. albatrus*. This is supported by Jouanin & Mougin (1979). Murphy considered that *N. harrisi* is a derivative of the northern *Phalacrocorax carbo* complex of cormorant species, not of the southern group that is so numerous in southern South America; and this is supported by the arrangement adopted by Dorst & Mougin (1979) although it seems that no detailed investigation has been made. The affinities of *Creagrus* are uncertain. Though superficially resembling Sabine's gull, *Xema sabini*, in plumage pattern, its displays are very different from those of Sabine's gull (Snow & Snow, 1968) and the plumage resemblance may be convergent. Murphy (1936) thought that *Larus fuliginosus* is closely related to the grey gull, *L. modestus*, of the Humboldt Current area, but it now seems much more likely that it is a derivative of a hooded gull such as *L. atricilla* of North America. In spite of its generally dark grey plumage, it still retains the trace of a well-defined sooty brown hood of the same extent as in *L. atricilla*, as well as a whitish eyering, and its displays are similar (Moynihan, 1962).

On the basis of their degree of divergence from other members of their groups, these four species may well have been among the earliest sea-birds to colonize the Galapagos Islands. It is, therefore, interesting that three of them are probably of northern origin and the fourth may also be. If further research confirms this, it will suggest that when the islands first became suitable for colonization by sea-birds oceanographic conditions were such that they were within the range of northern rather than of southern species.

SPECIAL ADAPTATIONS TO THE GALAPAGOS ENVIRONMENT

Intuitively one would suppose that, if they could be objectively assessed, marine environments would be found to be more uniform for sea-birds than terrestrial environments are for land-birds. The surface of the sea is much the same the world over, the food available for sea-birds near the surface is much the same in different oceans at equivalent latitudes, and cliffs and other coastal features vary much less than the country inland. One would thus not expect Galapagos sea-birds to be very different from those of other tropical islands, and it is all the more remarkable that a few of them have become adapted in unique ways to the Galapagos environment. We shall consider only the five endemic species. The endemic subspecies differ only slightly from conspecific populations elsewhere, and the differences, although presumably adaptive, cannot on present knowledge be related to the peculiar conditions of the Galapagos.

The Galapagos penguin, *Spheniscus mendiculus*, is the smallest of the four species of *Spheniscus* and has relatively the longest bill. These differences in size and proportions are in accordance with Bergmann's and Allen's rules (which, however, are generally valid for comparisons within, not between, species), and the adaptive basis may therefore be related to temperature regulation. Boersma (1977) discussed the Galapagos penguin's methods of adjusting its behaviour to two very different environments, the hot and dry Galapagos shores and the comparatively cold sea, but did not deal with the adaptive significance of size and proportions.

It is a striking fact that, although the Galapagos penguin's ancestors must have arrived in the archipelago from the east, via the south Equatorial Current, all the eastern and central islands of the archipelago are apparently now unsuitable for it. Like the flightless cormorant, it is confined to the extreme west of the archipelago (west coast of Isabela and coasts of Fernandina), where the upwelling of the Cromwell Current (Pacific Equatorial Undercurrent) produces the coldest sea surface temperatures in the Galapagos and, in consequence, the richest marine fauna. Even here, it is able to breed only when sea temperatures fall below 24°C (Boersma, 1977). Fluctuations in sea-water temperature in this area are unpredictable, with the result that the Galapagos penguin has had to evolve an opportunistic reproductive strategy, an aspect of its biology that is dealt with more fully below.

The waved albatross, *Diomedea irrorata*, is in many respects a typical member of its genus (Harris, 1973). Morphologically, its outstanding feature is its relatively very large bill; but this does not seem to be associated with special feeding behaviour, its main food, like that of other albatrosses, being fishes and squids. It is rather short-winged for its weight, and when at sea probably spends more time flying with wing-beats and less time gliding than do the other albatrosses, a difference that may be related to the less windy seas over which it ranges. The great majority of pelagic records are from the sea between the Galapagos and Ecuador and south down the Peruvian coast to about 10° S. No thorough study has been made of the flight of the Galapagos albatross in relation to its marine environment.

The flightless cormorant, *Nannopterum harrisi*, is a very large cormorant and the only one that is flightless. With a geographical range very similar to that of the Galapagos penguin, it is a bottom-feeder that seems to be dependent on the rich bottom fauna of the shallow waters in the area of upwelling of the Cromwell Current. It is probably as an adaptation for extracting bottom-living fishes and octopuses from recesses in the rocky sea-bed that the cormorant's bill is unusually large and strong. Sexual dimorphism in size is greater in the flightless cormorant than in any other member of the family, the female weighing only about 69% of the weight of the male, compared with 77–89% in other cormorants (Snow, 1966). It is possible that, as a consequence, the diets of male and female differ, and that the species thus exploits a wider range of prey than would otherwise be possible, but this aspect of its biology has not been studied.

The main evolutionary interest of this species is its flightlessness, which is clearly linked with its very large size. Snow (1966) has suggested that the following factors were important in the loss of flying ability: (1) increase in size, enabling it to exploit a wide range of bottom-living organisms and also reducing competition with the smaller Galapagos penguin (the only other bottom-feeding bird in its area); (2) restricted extent of suitable feeding gounds, which are within swimming distance of the shore; (3) absence of terrestrial predators at the nesting grounds (a general precondition for loss of flight in sea-birds); and (4) presence of frigatebirds, which persistently rob slow-flying sea-birds of their food, but cannot tackle swimming and diving birds.

The first of these factors presumably progressively reduced the cormorant's efficiency as a flying bird, while the last three would all have tended to minimize the disadvantages of flightlessness. Olson (1973) has argued that, whenever selection ceases to operate positively in favour of flying ability, resources can be diverted from the formation and maintenance of flight structures and it is to be expected that flightlessness will result.

Loss of flight, combined with general lack of agility, restricts the cormorant's range on a local scale within its already very small geographical range. It needs sheltered, gently sloping shores for landing, and as it does not move easily over the rocky shores its nesting sites must be close to its landing places. A small total population (700–800 pairs; Harris, 1974) is a necessary consequence.

The Swallow-tailed gull, *Creagrus furcatus*, is the most peculiar of the endemic Galapagos sea-birds. As already mentioned, its relationship to other gulls is uncertain, and this uncertainty is without doubt a consequence of the fact that, in becoming adapted to Galapagos conditions, it had to evolve specializations of both morphology and behaviour. Two conditions seem to have been of special importance: the unsuitability of the Galapagos shores as foraging ground for a typical gull, and the presence of frigatebirds.

The tidal range in the Galapagos is small, there is little sand, and no rivers flowing into the sea. Most of the coastline is barren and rocky. Probably, therefore, the swallow-tailed gull's ancestors found poor feeding conditions along the coast and were forced to adopt offshore feeding, most likely by plunge-diving, as their main foraging method. This would at once have exposed them to attack from frigatebirds in their flights back to land. Hailman (1964) was the first to point out what is now known as the swallow-tailed gull's unique and most striking feature, that it is entirely nocturnal. He suggested that its nocturnal habits had evolved in response to predation by frigatebirds on its nests. Certainly frigatebirds are nest-predators whenever they have the opportunity, and their presence must have been one of the main factors determining the gulls' selection of cliff nest-sites. Nests in exposed sites, where there are no holes or recesses in which the chick can hide, are almost invariably unsuccessful. We have argued, however, (Snow & Snow, 1968) that the main factor in the evolution of nocturnal habits may have been the need for immunity against frigatebirds when flying with food. The two possibilities are not, of course, mutually exclusive; both would tend to produce the same result.

Various morphological characters are associated with nocturnal behaviour and pelagic feeding (Hailman, 1964). The eyes are unusually large. The bill is blackish with a pale grey tip (an unique colour scheme in a gull), and there is a patch of white feathers at the base of the bill, the rest of the head being dark grey. The two pale spots, at the tip and base of the bill, must help to indicate the bill's orientation in the dark and probably serve to guide the begging chick. Adaptations for pelagic feeding, well away from the breeding colonies, include the combination of long wings and short legs (tern-like proportions) and perhaps the conspicuous white triangles on the upper surface of the wings which may have an intraspecific signal function.

The lava gull, *Larus fuliginosus*, is a much less strikingly modified bird than the swallow-tailed gull, but it is of great interest as it serves to re-inforce the arguments used in interpreting the evolution of the swallow-tailed gull.

The lava gull, apart from its dark plumage, is a typical hooded gull and, as mentioned above, may be a derivative of the laughing gull, *L. atricilla*. Presumably its ancestors colonized the Galapagos more recently than the ancestors of the swallow-tailed gull. One may suppose that competition with an already established gull prevented it from adopting the same strategy and becoming an off-shore feeder, so that it was forced to exploit the rather poor food resources of the Galapagos shores. In consequence it has become an accomplished scavenger, covering large distances in its foraging. Pairs are sparsely distributed, occupying long, linear territories round the coasts of the islands, and the total population of the species is perhaps lower than that of any other gull (Snow & Snow, 1969).

Hailman (1963) suggested that the lava gull's dark grey colour (even the bill is nearly all black) has a cryptic function. He noted that frigatebirds and lava gulls both feed on refuse along the shoreline and that when both are present the frigatebirds consistently rout the gulls and obtain the food. The gulls' dark colour should therefore improve their chance of successfully foraging along the shores without being detected by frigatebirds. Our observations (Snow & Snow, 1969) led us to believe that the cryptic colour functions primarily in relation to the nest. Lava gulls nest solitarily on open ground. They are exceedingly wary at the nest, and nests are hard to find. A white gull sitting on a nest would be very conspicuous, especially to patrolling frigatebirds. The evolution of more or less uniform dark grey colouring, closely matching the lava, may thus have been an essential adaptation for the ancestral lava gull when it began to breed along the Galapagos shoreline. Our interpretation and Hailman's are, of course, not mutually exclusive, and they agree in emphasizing the importance of frigatebirds.

ADAPTATIONS OF BREEDING ECOLOGY AND BEHAVIOUR

Among the strongest influences on the evolution of Galapagos sea-birds are the relatively aseasonal climate and perhaps above all the fact that in the surrounding Pacific there are complex and largely unpredictable fluctuations in the availability of the prey, predominantly flying fish and squid, upon which the more pelagic (far-foraging) sea-birds subsist.

Most biologists believe that natural selection operates upon the individual rather than upon the 'group', if the latter be defined to exclude kin. Each act of the individual affects, however slightly and indirectly, its chances of reproducing. Therefore we may look at each species' breeding ecology and behaviour (behaviour is the executive arm of ecology) as a composite response to selection pressures, persisting only because it has successfully met the challenge of establishing a breeding site, forming a pair and rearing young. Our comparison, predictably, will show that no two species have solved these problems in the same way and yet that certain response-themes are common. The diversities and the similarities throw light on the nature of the adaptive response. Ideally, this paper should examine everything concerned specifically with breeding, in all species of Galapagos sea-birds and for several island-populations of each. In fact, all we can do is select certain aspects of breeding and compare them across a few species.

Age of first breeding

Long-deferred breeding occurs widely among sea-birds and does not correlate with biomass, brood-size or the complexity of social behaviour. The feature common to all sea-birds that delay first breeding until they are four years old or more is pelagic feeding. Since this is, in the tropics, associated with uniparity, slow growth, and high mortality among young, due to temporary but extremely severe food shortages (e.g. Nelson, 1969), it may be assumed to be a necessary period of preparation for breeding in an area of high ecological constraint. It has been interpreted as an 'intrinsic' population-regulator (Wynne-Edwards, 1965), that is, breeding restraint to avoid raising numbers to the point at which there would be direct competition for, and over-exploitation of, food. This function, it is held, accounts for the flexibility in the deferred-breeding period. Deferred breeding is only one of a battery of supposed intrinsic regulators; others are: non-breeding years; variable clutch-size; infanticide and fratricide. Wynne-Edwards suggested that sea-birds are a particularly good example of such regulation, but all these features are far better interpreted as devices to maximize individual production. This accords with Lack's basic position (Lack, 1954, 1966) that birds rear as many young as they can adequately feed.

In more than 150 species of land-birds (Emlen, 1982) breeding is deferred for a variable period but the individuals concerned help their parents, and maybe other relations, to rear subsequent broods. This co-operative breeding is particularly prevalent where there are harsh and fluctuating environmental conditions which make independent breeding for inexperienced individuals particularly difficult. In the Galapagos it occurs in the mockingbird *Nesomimus parvulus*. However, nobody has commented upon the fact that co-operative breeding has never been recorded for a sea-bird. Yet the pelagic feeders of the Galapagos would seem prime candidates for such behaviour. Thus most waved albatrosses do not breed until they are 5–7 years old (Harris, 1979a). We suggest that co-operative breeding has been unable to evolve in this group because they cannot fulfil the necessary pre-condition, which is that offspring remain with their parents. The newly-independent tropical pelagic sea-bird, such as the frigate or tropicbird or waved albatross, depends upon a nomadic period to become fully competent in its very demanding feeding habitat, a period when it can take advantage of patchily-occurring food without having to return to base. A strong indication that this is vital comes from Fisher's (1975) work on the Laysan albatross of the Hawaiian group (Midway Island). As in virtually all sea-birds, mortality in the first year is higher than in subsequent years, and after (at the most) the second year there is no age-dependent difference in annual mortality rate. However, there is a marked peak in mortality in the first year in which the albatrosses return to the breeding colony as prebreeders. Apparently there is absolutely no reason to believe that these deaths are due to fighting, differential predation or anything of this sort. Circumstantial evidence indicates that they die because they cannot feed themselves adequately within the more limited foraging area that they must use if they are to spend the necessary amount of time establishing themselves in the breeding colony. A newly-independent pelagic feeder, largely inexperienced and facing its highest incidence of mortality, would presumably be unable to remain colony-based, with its parents, and therefore unable to become a helper in co-operative breeding. Since co-operative breeding is obviously a viable method of maximizing inclusive fitness via kin-selection, we must identify the strong selection pressures which preclude it in pelagic sea-birds.

Frequency and timing of breeding

The frequency of breeding, together with brood-size and breeding success, determines the lifetime-productivity of an individual, which is the crucial parameter. Once it has started to breed, a sea-bird may breed at regular, fixed intervals of more or less than a year, or at irregular intervals of more or less than a year, the latter including, in some species, non-breeding (rest) years. In strong contrast to sea-birds of temperate latitudes who experience predictable seasons with similarly predictable photoperiodic cues and food, tropical sea-birds live in relatively aseasonal climates with less-marked environmental timers and with (especially in the Galapagos) markedly unpredictable food. So, whereas temperate sea-birds have clearly-defined annual cycles, the limits of which are set at one end by natural selection penalizing late fledgelings (e.g. Potts, 1969; Brooke, 1978; Nelson, 1978) and at the other by inclement weather and constraints of food and foraging time, may tropical sea-birds breed opportunistically. Of the 17 Galapagos sea-bird species whose breeding periodicity is known, two (*Diomedea irrorata* and *Pterodroma phaeopygia*) have a rigidly fixed annual cycle; five have a much looser annual cycle with considerable variation in timing (within each species) on different islands (*Oceanodroma tethys, O. castro, Phaethon aethereus, Sula sula, S. dactylatra*); eight breed more often than once a year (*Puffinus lherminieri, Pelecanus occidentalis, Sula nebouxii, Larus fuliginosus, Creagrus furcatus, Anous stolidus, Nannopterum harrisi* and *Spheniscus mendiculus*); and two breed less often than once a year, almost certainly only once every two years (*Fregata minor* and *F. magnificens*). On occasions, *Sula sula* also breeds less frequently than once a year but more than once in two years (Snow, 1965a,b; Snow & Snow, 1967; Harris, 1969b,c, 1973; Nelson, 1970, 1976, 1978). Thus the great majority of Galapagos sea-birds breed non-annually and, over their lifetimes, are likely to lay in most months of the year.

Whether the breeding cycle is relatively short and breeding more frequent than once a year, or long and less frequent, is related perhaps chiefly to the species' foraging regime. Because they are fed more often, the chicks of committed inshore feeders, notably *Nannopterum harrisi* and *Spheniscus mendiculus* but also *Sula nebouxii, Pelecanus occidentalis* and *Larus fuliginosus*, predictably grow faster than those of pelagic feeders (*F. minor, F. magnificens, Phaethon aethereus, Sula sula, S. dactylatra* and *Diomedea irrorata*) and their breeding cycles are therefore shorter. Also, the energy-budgets of these two categories differ in that extreme inshore feeders expend little energy travelling to and from their feeding areas or indeed in locating patches of ocean where food is present, whereas extreme offshore feeders probably spend much. Each breeding attempt is therefore less of an investment for inshore feeders. For both reasons, inshore feeders can breed more often (and also can lay larger clutches). Nevertheless the biennial breeding cycle of the frigatebirds is an extreme case. It is a consequence of the exceptionally prolonged period for which the free-flying young continue to be fed by their parents. This prolongation is not confined to the Galapagos population (although there is evidence that Galapagos juvenile frigates fare worse than, for example, those on Aldabra in the more seasonal tropics of the Indian Ocean). It is a characteristic of the genus, and a nine-month period of post-fledging care has been recorded for *F. andrewsi* of Christmas Island (Indian Ocean) (Nelson, 1976) and *F. aquila* of Ascension Island (Atlantic Ocean) (Stonehouse & Stonehouse, 1963). It is probably correlated with the specialized feeding techniques of frigatebirds (Nelson, *loc.cit.*).

At the other extreme, *Nannopterum harrisi* may make several breeding attempts in one year (a male bred seven times in 24 months and a female laid eight clutches within 36 months (Harris, 1979b), but most of these attempts were unsuccessful). Of 11 cases in which the juvenile (i.e. fledged chick) from one breeding attempt survived at least three months, the mean interval between the egg which produced that juvenile and the next clutch was 13.3 months for males and 9.3 for females (Harris, *loc.cit.*). When eggs were lost the mean intervals were 6.0 months (males) and 6.7 months (females). Interestingly, lost clutches were not replaced (replacements are eggs which are produced merely in the time required for the development of new ova without intervening regression of the ovary, usually a matter of days or at most two to three weeks). Apparently *Nannopterum* requires four to six months between breeding attempts.

Spheniscus mendiculus breeds, at most, twice a year although this is never achieved by the entire population, some individuals of which are markedly out of phase with the others (Boersma, 1977). Although there may be laying in any month, it occurs in irregularly-spaced waves which may be related to upswings in food, and there may be periods of several months when no breeding takes place. Individuals may miss a laying 'wave' because of moult, during which the penguin cannot maintain its body temperature in water and therefore cannot feed. It lives on its fat reserves and usually cannot initiate breeding until it has rebuilt these.

In the Galapagos, frequent and opportunistic breeding is clearly an effective strategy for maximizing lifetime productivity. Since periods of plenty and scarcity are largely unpredictable, a fixed breeding periodicity could not be timed to coincide with plenty. In addition, it would prevent birds from exploiting temporary upswings if these occurred outside the fixed breeding period. Opportunistic breeders can utilize such upswings, at least for the energetically costly processes of site-establishment, pair-formation and egg-laying. Of course, breeding attempts started thus often run into famines and crash (e.g. Nelson, 1969) but this could not be avoided by a fixed periodicity of breeding.

Among inshore feeders with limited feeding grounds such as *Nannopterum* and *Spheniscus*, staggered breeding may be advantageous in spreading the birds' impact on their food resources. This could be particularly important in the cormorant, which may specialize on octopuses (Snow, 1966). Reef-living octopuses probably breed continuously and may produce two generations a year, passing from the egg to the breeding adult stage in as little as four to six months (P. R. Boyle, *pers. comm.*). Such octopuses, weighing 300–400 g, are very rich in protein, though low in lipids and carbohydrate, and would be a nutritious component in the cormorant's diet, especially important in view of its limited foraging range of only 100–300 m from the shore. It is now clear that in the tropics, cephalopods are a crucially important food for many sea-birds, and their astonishingly rapid growth and population turnover must be a highly significant factor in the marine ecosystem.

The biennial cycle of *Fregata minor* referred to above means that an individual cannot maintain its breeding site from one breeding cycle to the next, because it may be taken over in the interim. Therefore a prolonged pair-bond is also impracticable (Nelson, 1968, 1976). Whilst the Galapagos form of *F. minor* is not different in these respects from those of other localities, the case illustrates the far-reaching effect of feeding regime (here both tropical and highly specialized) upon the evolution of behaviour and breeding ecology. Briefly, *F. minor* shows very little aggressive (territorial) or pair-bonding behaviour. Instead, males display communally and those which attract a female quickly form the pair-bond on the display site and build there. It has recently been discovered on Aldabra (Reville, 1980) that *F. minor's* habit of clustering, which gives its colonies a characteristic spatial pattern, is not shared by its congener, *F. ariel*, even though the two species nest in the same area of mangroves. In a detailed analysis of spatial distribution and the build-up of breeding groups, Reville showed that whereas *F. minor* 'clumped' at 15–23 sites per clump, *F. ariel* tended to be regularly spread. This is because the latter fill patches of suitable habitat simultaneously, whereas *F. minor* fills them in succession, 'spares' (unsuccessful males) going from one display nucleus to another. One consequence of this is greater synchrony and much greater hatching success in *F. minor* (54.5% over two years compared with 20.1% in *F. ariel*). Apparently, 'spare' males (non-breeding males, or would-be breeders) are less likely to interfere in synchronized breeding clumps, and such males are the main cause of egg-loss. The next link in the chain is the discovery that in *F. minor* the sex-ratio was skewed in favour of males whereas in *F. ariel* it was 1 : 1. Thus, in male *F. minor* there was greater competition to nest, which could explain why more of them joined the display-nuclei and tried to mate before all the females were paired.

The size of clutch, egg and brood

Sixty-eight per cent of Galapagos sea-bird species are uniparous. The exceptions are *Pelecanus occidentalis* (2–3 young), *Sula nebouxii* (2–3), *Sula dactylatra* (2), *Spheniscus mendiculus* (2), *Nannopterum harrisi* (3) and *Larus fuliginosus* (2). Uniparity is a derived condition (Wynne-Edwards, 1955, 1965) and, mainly since Ashmole's (1963) stimulating discussion, has been recognized as part of the adaptive syndrome of tropical, pelagic sea-birds (e.g. Lack, 1967; Nelson, 1970). In at least 10 Galapagos species there is evidence for the dramatic effects of the sudden famines that are such a feature of this complex region of the eastern tropical Pacific in which several major currents and upwellings swirl around the archipelago. Such species include *Puffinus lherminieri*, *Phaethon aethereus*, *Sula sula*, *S. nebouxii*, *S. dactylatra* and *Fregata minor*—mostly single-egg, offshore feeders whose colonies may on occasions be littered with dead chicks and abandoned eggs. The earlier stages (site-establishment, nest-building and pair-formation) may also be temporarily abandoned. Whilst comparable events are not unknown elsewhere, it is in the Galapagos Islands that they are best-documented and perhaps commonest, and the underlying oceanographic causes are fundamental determinants of Galapagos sea-birds' adaptive strategies.

The species which lay clutches of more than one egg are mainly or entirely inshore feeders. Boersma (1977) does not mention starvation as a major cause of mortality among young *Spheniscus mendiculus* although he identified the main cause of death among adults as lack of food prior to moulting. Although single chicks grew faster than broods of two, few actually starved. This does not mean that food was plentiful. Most nests that failed were deserted before hatching because adults were unable to maintain their own body weight. Moult commonly occurs just prior to or during the breeding 'season', perhaps because penguins have to deposit fat rapidly before moulting and the feeding conditions which initiate breeding are suitable, also, for laying down reserves. It means, however, that they have to recoup the considerable loss of weight which attends moult and then, in the case of the female, find the resources to produce the clutch.

Nannopterum also appears to escape mass starvation of young. Like *Spheniscus* it loses many clutches and small young, the former apparently due to infertility or early death of the embryo (Harris 1979b) prior to or after desertion. Despite its clutch of three eggs rarely more than one juvenile per brood survives (Snow, 1966; Harris, *loc.cit.*).

Sula nebouxii is the only sulid in the Galapagos, and one of the only two in the world, that both lays clutches of more than one and sometimes rears more than one chick. It is also partly an inshore feeder. *Nannopterum* and *Spheniscus* are restricted to the western-most Galapagos Islands (Isabela and Fernandina)—a distribution which, as already noted, coincides with the cold and exceptionally productive upwelling of the Cromwell current. *Sula nebouxii*, too, is always associated with upwellings or their fringes, not only in the Galapagos but over the whole of its markedly discontinuous distribution. However, unlike *Nannopterum* and *Spheniscus* it (perhaps particularly the female) also feeds well offshore. Especially on Daphne but also on Española, it has been known to suffer catastrophic chick mortality (Nelson, 1978).

Of the six polyparous species all except *Sula dactylatra* sometimes rear broods of two. However, in all of them the average brood-size is much less than two (except perhaps for *Larus fuliginosus*, for which there are few data). Expressed as young reared per breeding cycle, approximate figures culled from several sources are: *Spheniscus mendiculus* (0.2 young), *Pelecanus occidentalis* (1.0), *Sula nebouxii* (0.45–0.8), *S. dactylatra* (0.63), *Nannopterum harrisi* (0.40). Nevertheless, at a crude, pooled average of 0.55 young per breeding cycle (excluding *Sula dactylatra*, which always reduces its brood to 1–see below), these species produce on average about twice as many young as the uniparous, pelagic species (*Diomedea irrorata* (0.16–0.20), *Puffinus lherminieri* (0.26), *Pterodroma phaeopygia* (0.06) (an artificially low figure since most losses are due to introduced rats), *Oceanodroma castro* (0.3), *O. tethys* (0.23), *Phaethon aethereus* (0.32–0.55), *Sula sula* (0.08), *Fregata minor* (0.15), *F. magnificens* (0.5), *Creagrus furcatus* (0.33), *Anous stolidus* (0.40). In addition, polyparous species breed more often.

Egg-size, like clutch-size, varies adaptively both within and between species. Large eggs are laid by nidifugous species or, among nidicolous species, by the most pelagic ones. *Phaethon aethereus* and *Fregata minor*, perhaps the most pelagic of the Galapagos pelecaniforms, lay eggs which, as a proportion of female weight, are three or four times heavier than those of the inshore-feeding *Nannopterum* and *Pelecanus* of the same order, whilst the eggs of the small, pelagic procellariids such as *Oceanodroma* spp. are some ten times heavier. A relatively large yolk, much of which remains in the allantois of the newly hatched chick and is available as food, is, however, not the only way of reducing the risk of breeding failure due to hatching young during a temporary famine. Whereas *Sula sula* lays a large egg (5.1% of the female's weight), *S. dactylatra* lays two smaller eggs (3.6%) some five days apart. This lengthens the period for which there is a viable chick, just as effectively as providing five days' more food in the yolk. But, being a pelagic feeder, *S. dactylatra* is best suited by a single chick and so the brood (if two) is reduced to one by fratricide. Its two-egg clutch has been misinterpreted as a device for rearing two young in good times (Lack, 1966), and fratricide just as erroneously construed as a brood-size regulator matching productivity to prevailing food conditions (Wynne-Edwards, 1962). In fact, it is simply a means of maximizing productivity by increasing the chances of survival of one (and only one) chick. If food is adequate when the second one hatches it is killed, automatically, by its sibling (Dorward, 1962; Nelson, 1978). If it were allowed to live longer it would endanger the longer-term survival of them both, for should both become strong they would resolutely compete for food, if and when it became scarce, and thus give two weak fledglings, both of which would probably succumb. It is interesting that on the same island, for example Española, *Sula nebouxii* practises mere sibling-competition whereas in *S. dactylatra* fratricide is the rule. In the first-named, therefore, brood reduction is not inevitable whereas in the second it is. Presumably this relates to the lower probability of the inshore *nebouxii* facing severe famine, compared with its more pelagic congener. Similarly in several pelicans, all of which are largely inshore or fresh-water feeders, brood reduction even within the same species grades from mere competition to outright fratricide, although the ecological correlates have not been specifically identified.

Finally, the suddenness with which famines in the Galapagos come and go should dispel any notion that they are related to the numbers of sea-birds. They affect pelagic species most because the inshore feeders already have a richer source and less extensive foraging trips, though even this is not enough to shield them entirely. But inshore feeders are more likely to experience direct competition from their own and other species, especially if they specialize on particular kinds of prey. Hence the clutch-size, growth, productivity and patterns of mortality are different in inshore and pelagic Galapagos sea-birds.

Composition of the breeding cycle and attachment to area, site and mate

The recent and rapid growth of interest in optimality studies and evolutionarily stable strategies encourages the search for the probable costs and benefits of alternative ways of doing things. Breeding cycles differ not only in overall length but in composition—the time and energy devoted to each of the components. As in other aspects of breeding, the extreme range of conditions around the Galapagos has resulted in extreme behavioural adaptations.

A breeding cycle comprises: the behavioural events that begin with a return to the colony or, in cases of continuous occupancy, with renewed response to territorial and sexual stimuli; egg-laying and incubation; intensive chick-care; rearing of young to fledging; post-fledging care of young; and events that follow the departure of the independent offspring and precede the departure of the adult or, if still at the colony, their lapse into the post-breeding phase of moult and recovery. Breeding is thus much more than the period during which the adults are engaged with eggs or unfledged young. We shall take the three main stages in sequence.

Pre-laying activities

Unfortunately there is still no way of assessing the investment of energy except for egg-production, incubation and as food provided for the chick. The mere duration of other components is of limited value and, even for this, information is often not available.

The time spent in the pre-laying phase by experienced breeders (first-time breeding is another matter) is divided between establishing or re-establishing the site, forming or re-forming the pair, and (as the case may be) building the functional or symbolic nest. Two aspects are significant: first, the time and manner in which each species typically structures these preliminaries and second, the degree of flexibility each species shows. Both of these are closely related to their ecology. *Fregata minor* spends no time establishing a breeding site. It is no exaggeration to say that the male never prospects for, settles on and defends a breeding site as such. He settles on a *display* site, usually close to other displaying males, and if he is eventually joined by a female who remains for two or three hours (many leave during this period) they form a pair bond, by mutual display. This takes up to three days. They spend the first day together on the site, displaying and vocalizing frequently, take their first flights on the second day, the male begins collecting nest material by the third day, and this becomes a major activity on the fourth (Reville, 1980). Only then, when building is about to begin, does the male aggressively defend his nest-site by lunging at and occasionally grappling briefly with intruders. Even at this definitive stage there is no special site-ownership display. This route to the ownership and defence of a site contrasts with that taken by most temperate, seasonal breeders in which the male establishes a site, displays on it (both in site-defence and as advertisement for a female) and then forms a pair on it. Such males are tied to a particular site before pair-formation whereas the frigate is not. As for flexibility, the frigatebird theoretically needs only the short time required for forming a pair, time to build its flimsy nest and for the female to develop her single egg. In practice, though, a male may have to display for three or four weeks before he secures a mate (Reville, 1980). The speed with which the pre-laying preliminaries can be completed may be advantageous in permitting this energy-costly phase to be conducted during a perhaps briefly favourable period. And the ephemeral attachment to site and mate presumably means that frigatebirds have not been selected to invest time and energy in site- and pair-bonding displays, in strong contrast to many other sea-birds. Opportunistic breeding is facilitated by the frigatebird's system. Although the same principles apply to *Nannopterum harrisi* there are interesting differences. *Nannopterum*, too, is an opportunistic breeder in which, like *Fregata*, the pair-bond is forged (often) before the site attachment is firmed. Snow (1966) shows that courtship begins on the water and that the site is chosen later, by the pair. But there is much more courtship after pair-formation than in *Fregata* and much more nest-building, and whereas *Fregata* has no greeting interaction at re-union on the nest, *Nannopterum* has a ritualized presentation of nest-material and often performs part of the courtship display. *Nannopterum* retains great flexibility in this early stage and Harris (1979b) records that the interval between display on the water and egg-laying varied from 10 days to six weeks. Nevertheless, it has more highly developed site-attachment behaviour and pair interactions, and although both site and mate are usually changed in successive cycles, they also sometimes remain unchanged (Harris, *loc.cit.*).

The often short duration of pre-laying activities in opportunistic tropical sea-birds is not matched in temperate, seasonal breeders. *Phalacrocorax aristotelis,* for example, though possessing many displays homologous with those of *Nannopterum,* spends weeks or months in the pre-laying phase and retains site and mate over several breeding cycles (Snow, 1960). Likewise, whereas *Sula sula* normally requires about three weeks to complete site-establishment, pair-formation and nest-building, and can do it in less, *Sula bassana* in the North Atlantic normally requires a complete season prior to the one in which it first breeds, and thereafter two to four months each year. Again, the site and mate are normally held for life, food and breeding are highly seasonal and the date of return to the colony is predictable. All these factors are lacking in the Galapagos and this component of breeding is therefore structured differently.

Fregata never nests twice on the same site or with the same mate except by chance, while *Nannopterum* is highly flexible but sometimes does re-use a site and maintain its pair-bond for two or even three cycles. *Phaethon aethereus*, a hole-nester, illustrates the extraordinary variability which is so marked in Galapagos birds. Snow (1965a) showed that the tropicbirds on South Plaza had an annual and seasonal cycle, no birds breeding out-of-season, whilst on nearby Daphne eggs were laid in all months and there was little synchrony. On Plaza Harris (1969a) found that *Phaethon* displayed at the nest hole for at least two months prior to laying, and Snow implies that a proportion, possibly high, of nest holes are re-used by the same pair in successive years. On Daphne, fierce competition for holes resulted in rapid turn-over, year-round breeding and lower success. There are no comments on the duration of display, but one might predict that it would be shorter.

Diomedea irrorata spends part or most of a season prior to the one in which it first breeds (at 3–7 years; Harris, 1979a) at the colony, displaying. Once paired, the nest-site is chosen within an area to which both partners return, the male first, in succeeding years. Thus the territory, which is vigorously defended, is (as in so many temperate, seasonal breeders) the focus for reunion, and return to it is highly seasonal (late March). The fact that the nest-site itself is not fixed within the territory (indeed the egg may be moved as much as 40 m within a few days) is irrelevant so long as the territory is fixed. The mean laying date for experienced breeders is about a month after they return and since the pair-bond is already formed and there is no nest to be built, this period represents consolidation of the bond and the time required for the female to form the large egg. Harris (*loc.cit.*) comments that after her return the female "spends little time at the colony", which underlines the strength of the bond formed during the year or two preceding breeding and the role of the fixed territory in facilitating re-union.

Diomedea's annual, seasonal breeding cycle is inflexible, but this is readily understandable in view of the species' feeding behaviour. An incubating adult averages spells of three weeks continuous absence from the colony (though these shorten towards hatching) during which, unlike *Nannopterum* and *Spheniscus*, it can range extremely widely. It is almost three times as heavy as *Fregata* and can go longer without food. When it has young to feed it can virtually avoid all risk of returning empty-handed, not only because of its foraging range but also because it feeds its young mainly on oil produced by the proventriculus. If *Fregata*, which also takes long foraging trips, encounters food only at intervals, it must 'decide' when to return with a bolus which cannot long avoid being digested. *Diomedea*, by contrast, can manufacture and store oil and return with a guaranteed load. This mechanism must greatly buffer the effects of patchy and scarce food, particularly if (as seems likely) *Diomedea* feeds partly at night, when squids are thought to be more accessible. It is therefore not surprising that starvation among young *Diomedea* is virtually unknown. Almost all breeding failure results from loss of eggs, or very small young in the intensive guard period (Harris, *loc.cit.*). A greater contrast with *Fregata* and its starvation-prone offspring could hardly be imagined.

Egg-laying and incubation

Significant features of the incubation period, apart from its total length (which is largely a function of egg-size, prolonged incubation being correlated with large eggs and slow development, as a response to a low rate of acquisition of food), are the length of incubation stints (related to foraging mode) and the nature of nest-relief. The term "prolonged incubation", as defined by Whittow (1980), implies an extra adaptation, the embryo taking longer to develop than an egg of that size would be expected to do, judged by the regression line relating incubation time to egg weight.

The energetics of incubation are complex, especially in pelagic feeders which require lengthy foraging stints thereby condemning the partner to continuous incubation and loss of weight. *Fregata minor* incubating for up to 17 days at a time lost up to 20% of its weight (Nelson, 1968). This proportion approaches the safe limit for a sea-bird without large initial fat reserves and may require special physiological adaptations, although these have not been proposed. The eggs of Procellariids, both in burrows (e.g. *Pterodroma phaeopygia*) and in the open (e.g. *Diomedea irrorata*), are especially resistant to temperature extremes and can withstand at least six days without attention (Harris, 1969b). Why adult *Fregata* should be able to withstand 17 days continuous incubation whereas *Sula sula* deserted after about 6 days (nearly 70% of the eggs laid on Tower Island in the first half of 1964 were deserted; Nelson, 1969) is unknown. The details of incubation behaviour in relation to feeding requirements would provide a suitable subject for an optimality study, since there are many theoretical options each with their 'trade-offs'. One partner could do all the incubating whilst the other provided the food. This would be practicable only in an inshore forager, and no Galapagos sea-bird does so. An inshore feeder may make several short excursions per day (*Nannopterum*) or fewer, longer ones (*Spheniscus*). Eggs may be left unattended when the ambient temperature is within a certain range, thus releasing both adults for foraging. This presupposes lack of predators and would be a predictable strategy for *Pterodroma phaeopygia,* which may be why so many eggs are lost to introduced predators (Harris, 1970). Short absences would not usually be appropriate because absences of more than a few minutes incur long recovery time in relation to the exposure that produces the drop in temperature of the developing embryo (Drent, 1973). As the egg cools, the rate declines until it is hardly perceptible, and so absences should be either very short, or long.

From the functional viewpoint, it seems most probable that long incubation is merely the necessary prelude to the slow growth of the chick after hatching. It is the slow growth and resistance to starvation, so notable in pelagic sea-birds, that is adaptive—not the long incubation period. Of what conceivable advantage is long incubation *per se?* Similarly, the ability of the embryo to survive in the egg whilst unattended is adaptive, but not a long incubation period.

In our opinion and contrary to the assumptions of many, the long foraging trips of pelagic feeders, reflected in long incubation stints with their attendant risk of desertion, do not stem from competition for a limited resource. This assumption crops up repeatedly in the Galapagos sea-bird literature and beyond (e.g. Whittow, 1980). The 'competition' is purely hypothetical although the 'limited resource' is real enough. Inshore competition is much more likely, and to that extent Diamond's (1978) suggestion that one compensation for the greater travelling distance of pelagic feeders may be reduced competition for food, seems plausible. But to conceive of direct competition between frigatebirds during 17 days of oceanic foraging, covering hundreds or thousands of miles, strains credulity. The same applies in essence to other pelagic feeders.

Chick care

The three significant phases are: intensive guard stage; the fledging period (time between hatching and either acquiring flight or taking to the water); and the post-fledging period until independence.

The intensive guard stage bears no simple relationship to the fledging period. *Creagrus* chicks may seek the shelter of a crevice or overhang soon after hatching, whereas the equally downy *Diomedea* chick is brooded continuously for "several weeks" (Harris, 1973). *Phaethon* chicks (the only down-covered pelecaniforms) are brooded in their holes usually for at least 50 days. Altricial young require continuous brooding until they can regulate their own body temperature (four to five weeks for pelecaniforms) but this may be far too short to guarantee their safety from predators or other intruders. The cost/benefit equation differs for each species in relation to food. In the Galapagos, it must be an advantage to have a short guard stage. *Nannopterum* chicks grow thick down in less than two weeks, which is a shorter period than in other cormorants. *Fregata minor* leaves its chick, on its exposed platform, at about five weeks of age. Without benefit of shade, and unable to seek any, the frigate chick faces severe heat-regulation problems and it may be that the loose and flimsy structure of the nest allows air-currents to remove heat from the axillary and vent regions. The function of the black 'cape' consisting of precociously developed scapulars is unknown; it may screen out UV or help the chick to regain heat after a night-time's exposure. *Fregata* loses a significant proportion of young in the immediate post-guard stage. The gain in adult foraging time must justify this cost, corroborating, again, the difficulty frigatebirds experience in gathering food. Long incubation periods precede slow development of the chick and so correlate with long fledging periods. But, as with the guard stage, so with post-fledging care, there is no simple relationship with fledging period.

The post-fledging period faces sea-birds with various options, and the choice is clearly dictated by the feeding ecology of the species. It is a critical stage, for by then parental investment in their offspring has been substantial and to jeopardize the chick's survival for want of a final effort would seem a poor strategy. The first option is to feed the free-flying offspring for a period consonant with its requirements, which means until it has become reasonably proficient. This, however, must fit in with the species' feeding habits. *Fregata* feeds its fledged juvenile for nine months or more but *Phaethon*, an equally pelagic feeder, does not feed its fledged young at all. But the juvenile frigatebird can practise scavenging, fishing and kleptoparasitism near to the colony and also remain in a position to receive food from its parents. The young tropicbird is neither a scavenger nor a kleptoparasite. Furthermore, and unlike the frigatebird, it can roost on the sea. Its energy budget is different. Instead of post-fledging feeding, its parents provide it with enough food in the nest for it to lay down a fat reserve, which presumably helps to tide it over the immediate post-fledging period. Similarly, young *Diomedea irrorata* take 167 days to fledge but once fledged, are never fed thereafter. The young, like the tropicbird, could not adequately learn to feed if restricted to local waters and once away, foraging far, return to the colony would be totally impracticable if only because there would be little chance of coinciding with a parental visit. No sea-bird in the world manages to do both; either they go, or they stay.

Here, it is appropriate to comment on parental strategy with regard to investment in the chick, and the point is relevant, also, to deferred breeding, non-breeding years, abandonment of young in times of food shortage and many other aspects of breeding. Basically, an adult has two options—to accept stress (limited or extensive) and risk shortening its reproductive life, or to ensure that it is not stressed, whatever the cost to any single breeding attempt. The options, in terms of a lifetime's production of young, do not apply in the same way to inshore and pelagic feeders. A frigatebird, laying one egg every two years, with a low success rate, and with a long deferred-breeding period, would gain less over its lifetime from incurring stress and reducing life expectancy, even if that stress increased its success, than would an inshore feeder with a larger clutch, higher success and shorter deferred-breeding period. The prediction is that the latter category—in the Galapagos *Nannopterum* would be a good test case—should show greater evidence of stress among adults, both directly (by weight loss) and indirectly by lower probability of deserting eggs and young, than the pelagic feeders. The adult pelagic feeders are less expendable than the inshore feeders.

REFERENCES

ASHMOLE, N. P., 1963. The regulation of numbers of tropical oceanic birds. *Ibis, 103:* 458–473.
BOERSMA, P. D., 1977. An ecological and behavioral study of the Galapagos penguin. *The Living Bird, 15:* 43–93.
BROOKE, M. de L., 1978. Some factors affecting the laying date, incubation and breeding success of the Manx Shearwater, *Puffinus puffinus. Journal of Animal Ecology, 47:* 477–495.
DIAMOND, A. W., 1978. Feeding strategies and population size in tropical sea-birds. *The American Naturalist, 112:* 215–223.
DORST, J. & MOUGIN, J. L., 1979. Order Pelecaniformes. In Peters' *Check-list of Birds of the World,* vol. 1 (2nd edition): 155–193. Cambridge, Mass.: Museum of Comparative Zoology.
DORWARD, D. F., 1962. Comparative biology of the White Booby and the Brown Booby *Sula* spp. at Ascension. *Ibis, 103:* 174–220.
DRENT, R., 1973. The natural history of incubation. In D. S. Farner (Ed.), *Breeding Biology of Birds:* 262–311. Washington: National Academy of Sciences.
EMLEN, S. T., 1982. The evolution of helping. I. An ecological constraints model. *The American Naturalist, 119:* 29–39.
FISHER, H. I., 1975. The relationship between deferred breeding and mortality in the Laysan Albatross. *Auk, 92:* 433–441.
HAILMAN, J. P., 1963. Why is the Galápagos Lava Gull the color of lava? *Condor, 65:* 528.
HAILMAN, J. P., 1964. The Galápagos Swallow-tailed Gull is nocturnal. *Wilson Bulletin, 76:* 347–354.
HARRIS, M. P., 1969a. Factors influencing the breeding cycle of the red-billed tropicbird in the Galapagos Islands. *Ardea, 57:* 149–157.
HARRIS, M. P., 1969b. The biology of Storm petrels in the Galápagos Islands. *Proceedings of the California Academy of Sciences, 37:* 95–166.
HARRIS, M. P., 1969c. Breeding seasons of sea-birds in the Galápagos Islands. *Journal of Zoology, London, 159:* 145–165.
HARRIS, M. P., 1970. The biology of an endangered species, the Dark-rumped Petrel (*Pterodroma phaeopygia*) in the Galápagos Islands. *Condor, 72:* 76–84.

HARRIS, M. P., 1973. The biology of the Waved Albatross *Diomedea irrorata* of Hood Island, Galapagos. *Ibis, 115:* 483–510.

HARRIS, M. P., 1974. A complete census of the flightless cormorant (*Nannopterum harrisi*). *Biological Conservation, 6:* 188–191.

HARRIS, M. P., 1979a. Survival and ages of first breeding of Galápagos sea-birds. *Bird-Banding, 50:* 56–61.

HARRIS, M. P., 1979b. Population dynamics of the Flightless Cormorant *Nannopterum harrisi*. *Ibis, 121:* 135–146.

JOUANIN, C. & MOUGIN, J.-L., 1979. Order Procellariiformes. In Peters' *Check-list of Birds of the World*, vol. 1 (2nd edition): 48–118. Cambridge, Mass: Museum of Comparative Zoology.

LACK, D., 1954. *The Natural Regulation of Animal Numbers*. Oxford: Clarendon Press.

LACK, D., 1966. *Population Studies of Birds*. Oxford: Clarendon Press.

LACK, D., 1967. Inter-relationships in breeding adaptations as shown by marine birds. *Proceedings of the XIV International Ornithological Congress:* 3–42.

MOYNIHAN, M., 1962. Hostile and sexual behavior patterns of south American and Pacific Laridae. *Behaviour*, Suppl. 8.

MURPHY, R.C., 1936. *Oceanic Birds of South America*. New York: Macmillan.

NELSON, J. B., 1968. *Galapagos: Islands of Birds*. London: Longman Green.

NELSON, J. B., 1969. The breeding ecology of the red-footed booby in the Galapagos. *Journal of Animal Ecology, 38:* 181–198.

NELSON, J. B., 1970. The relation between behaviour and ecology in the Sulidae with reference to other seabirds. *Oceanography and Marine Biology, 8:* 501–574.

NELSON, J. B., 1976. The breeding biology of frigatebirds—a comparative review. *The Living Bird, 14:* 113–155.

NELSON, J. B., 1978. *The Sulidae: Gannets and Boobies*. Oxford: Oxford University Press.

OLSON, S. L., 1973. Evolution of the rails of the South Atlantic islands (Aves: Rallidae). *Smithsonian Contributions to Zoology, 152:* 1–53.

POTTS, G. R., 1969. The influence of eruptive movements, age, population size and other factors on the survival of the shag (*Phalacrocorax aristotelis* (L.)). *Journal of Animal Ecology, 38:* 53–102.

REVILLE, B. J., 1980. *Spatial and temporal aspects of breeding in the frigatebirds* Fregata minor *and* F. ariel. Ph.D. Thesis, Aberdeen University, U.K.

SNOW, B. 1960. The breeding biology of the shag *Phalacrocorax aristotelis* on the island of Lundy, Bristol Channel. *Ibis, 102:* 554–575.

SNOW, B. K., 1966. Observations on the behaviour and ecology of the Flightless Cormorant *Nannopterum harrisi*. *Ibis, 108:* 265–280.

SNOW, B. K. & SNOW, D. W., 1968. Behavior of the Swallow-tailed Gull of the Galápagos. *Condor, 70:* 252–264.

SNOW, B. K. & SNOW, D. W., 1969. Observations on the Lava Gull *Larus fuliginosus*. *Ibis, 111:* 30–35.

SNOW, D. W., 1965a. The breeding of the Red-billed Tropic Bird in the Galapagos Islands. *Condor, 67:* 210–214.

SNOW, D. W., 1965b. The breeding of Audubon's Shearwater (*Puffinus lherminieri*) in the Galapagos. *Auk, 82:* 591–597.

SNOW, D. W. & SNOW, B. K., 1967. The breeding cycle of the Swallow-tailed Gull *Creagrus furcatus*. *Ibis, 109:* 14–24.

STONEHOUSE, B. & STONEHOUSE, S., 1963. The frigate bird *Fregata aquila* of Ascension Island. *Ibis, 103:* 409–422.

WHITTOW, G. C., 1980. Physiological and ecological correlates of prolonged incubation in seabirds. *American Zoologist, 20:* 427–436.

WYNNE-EDWARDS, V. C. 1955. Low reproductive rates in birds, especially sea-birds. *Acta XI Congressus Internationalis Ornithologici:* 540–547.

WYNNE-EDWARDS, V. C., 1962. *Animals Dispersion in Relation to Social Behaviour*. Edinburgh and London: Oliver and Boyd.

WYNNE-EDWARDS, V. C., 1965. Social organisation as a population regulator. *Symposia of the Zoological Society of London, 14:* 173–178.

Biological Journal of the Linnean Society (1984), 21: 157–164. With 1 figure.

The evolution of breeding strategies in the flightless cormorant (*Nannopterum harrisi*) of the Galapagos

R. TINDLE

Charles Darwin Research Station, Galapagos, and Beatson Institute, Garscube Estate, Switchback Road, Glasgow

The evolution of breeding strategies in male and female flightless cormorants is thought to have been shaped by three principal factors: Firstly, an unpredictable food supply; secondly, a prolonged breeding season; and thirdly, the inbred nature of colonies.

These factors have selected for opportunistic breeding and for females (but not males) to attempt to raise two broods per season by deserting their offspring to the further care of their mates, while they re-breed with new males. That this strategy maximizes the inclusive fitness of both sexes is examined in terms of genetic pay-off and the relative food-providing efficiency of male and female parents. Parental investment is examined in relation to female desertion. Mate selection is discussed and some theoretical aspects of breeding strategy evolution are considered.

CONTENTS

INTRODUCTION

During 1970–80 a detailed study was made of a group of approximately 100 flightless cormorants nesting at Punta Espinosa, Fernandina, Galapagos Islands. This was 5–10% of the population of this endemic species. Virtually all birds were banded. The colony was visited regularly throughout the period, and all nests systematically checked the birds noted. In 1977–80, five months were spent in residence at the colony and groups of nests were observed continuously. The population dynamics of the species is described in Harris (1979) and the detailed breeding biology in Tindle & Harris (1982). The ecology and distribution of the birds is given in Snow (1966) and Harris (1974).

0024–4066/84/010157 + 08 $03.00/0

Disassortative mating and parental investment data are given in Tindle (1983). Briefly, approximately 90% of the clutches are laid between March and October, coinciding with the prevalence of cold, highly productive waters throughout the species range. Nests are built in groups of 1–20 on bare lava a few metres above high water where there is easy access to-and-from the sea. Courtship commences in water, then moves on to land where a flimsy display nest of alga and marine detritus is constructed and consolidated. Three eggs (two, one and four in order of frequency) are laid several days apart, 10–40 days later. Both sexes incubate, and asynchronous hatching occurs at 33–37 days. The mean brood size is 1.7 young per clutch, mainly the result of egg infertility and chick predation by the Galapagos hawk. Both sexes feed the young which take to the water at 50–70 days. Very few parents fledge two offspring, due to starvation of (usually) the younger. At 70–90 days the females desert, and the fledglings are fed to between five and nine months by males alone. When conditions for breeding remain suitable, deserting females lay second clutches with new mates. Even if not re-laying immediately, the interval between successive nesting attempts is less for successful females than successful males. Whether successful or not, partners are almost invariably changed between breeding attempts. Young reach adult size within one year and both sexes commence breeding at 30 months. Annual adult survival is 85% for both sexes. The Punta Espinosa colony is extremely in-bred. Flightless cormorants are extremely sedentary; despite much banding and searching on nearby coast-lines, only two adults moved into the area, and all 240 banded adults, immatures and juveniles remained within the colony limits during the study.

The predominant factors shaping the evolution of the breeding strategy of this species are: (1) unpredictable food supplies both within and between seasons; (2) a breeding season of more than adequate duration (c. eight months) to allow a pair to raise one brood, but of insufficient duration to allow a second successive brood to be raised; (each brood requires c. five months) and (3) the in-bred nature of the colony.

THE ENVIRONMENT

The effect of the capricious interplay of warm, poorly productive, and cold, highly productive waters on the breeding of sea-birds in Galapagos has been discussed by Boersma (1978). In an 'ideal' year the cold upwelling Cromwell current system provides an abundance of prey species during March to October. More frequently, fluctuating local sea temperatures produce unpredictable food supplies during this season (see Maxwell, 1974). Occasionally, warm waters may persist well into the second half of the year; the 'El Nino' phenomenon (Wyrtki et al., 1976). During 1970–80, considerable variation occurred in reproductive effort and reproductive success of the flightless cormorant. Wherever measurements were available there was a direct correlation with sea temperature. The availability of food is probably both the ultimate and proximate factor controlling the timing of breeding in this species (Harris, 1979). Thus a breeding strategy has evolved whereby birds breed as soon as environmental conditions become favourable, and if the first breeding attempt fails, but favourable conditions ensue, they re-breed again as soon as possible. Females (but not males) attempt to raise two successive broods in a season. This is accomplished by the female deserting her mate and brood at *c*. 70 days after hatching, and re-breeding with another male. Even if the deserting female does not re-lay again that season, she is ready to breed again before her first mate (Harris, 1979). Thus under any conditions, female desertion strategy speeds up offspring production.

An immediate question is, of course, why is it the female who invariably deserts and leaves her mate 'holding the baby'? Why doesn't the male pre-empt the female at desertion? (see later).

SEXUAL DIMORPHISM AND MATE SELECTION

Male flightless cormorants are about 15% heavier than females, and stouter billed. Cause and effect of sexual dimorphism are difficult to disentangle. Sexual selection for dimorphism in body size also produces associated differences in trophic structures that result secondarily in differential niche utilization (Selander, 1972). Dimorphism in several bird species has probably evolved because of sexual differences arising from divergence of male and female feeding niches (Selander, 1966). Since male flightless cormorants do not defend territories or compete extensively for females, intrasexual selection is probably not a major selective force in determining male body size. On the other hand, intersexual selection may be important (see below).

However, what is needed is a quantitative model that considers, for each sex, the relative advantages and disadvantages of their particular size. Such a model is lacking not only in flightless cormorants, but in all species. This is because animal body weight has such wide ranging effects on morphology, physiology, behaviour and ecology that any model with sufficiently few parameters to be manageable and make quantitative predictions is bound to be an oversimplification (Reiss, 1982).

I found evidence of a disassortative mating among 22 pairs of breeding birds. The heavier a male cormorant, the lighter and more slender billed his mate; $Y = 5.8082 - 0.6526 \times r = 0.43$ $P = 0.05$ (Tindle, 1982). This suggests a mate selection mechanism is operative in which heavier males are preferentially choosing smaller females as mates and *vice versa*. In a monogamous species in which parental care is shared between partners (as in the flightless cormorant), the ability to select an efficient food provider and a co-operative mate presumably has high selective value (Wilson, 1975). In flightless cormorants, females should be selecting males as reliable providers capable of latterly raising a juvenile single handed, and males selecting females on the ability to pack in as much chick-feeding as possible prior to their desertion. I have no evidence that heavy males are better food gatherers than lighter males.

However, I have studied in detail the foraging habits of males versus females. Both sexes devoted an equal amount of time to foraging, (*c*. 4 h per day) foraged in the same places, (within 300 m of the shore adjacent to the breeding colony) and caught prey at the same rate (1 prey item per 10–12 min of foraging) (Tindle, 1983). Males however, were able to remain submerged for longer in deeper water, presumably because of their greater bulk and strength. Males can also take larger prey than females. Since flightless cormorants take both mid-water and bottom-dwelling species (Harris, 1979), in times of food shortage males may be better able to exploit deeper water than females. 'Optimal' pairs may be those in which there is a maximal discrepancy in body size, and thus a minimum of food-niche utilization overlap. This would be an obvious advantage in minimizing intra-family competition should food be scarce during raising of the young. However, as yet, evidence on the relative breeding success of pairs composed of large males and small females versus pairs with a smaller size difference between the partners, is inconclusive.

Other factors may influence mate choice. Following Darwin's hypothesis that secondary sexual characters attractive to females will co-evolve with female preference (Darwin, 1871; Harvey & Arnold, 1982) larger male flightless cormorants may be preferred as mates simply because they *are* large. In a widow bird (*Euplectes progne*), female choice selects for extreme tail length regardless of the detrimental effects that excessive tail length may have on the male's fitness (Andersson, 1982). Furthermore, O'Donald (1979) has suggested that females have low thresholds of response to particular male phenotypes that they prefer to mate with, but they will mate with other males if stimulated to a higher level of threshold i.e. an inferior male must court more assiduously.

It has been suggested that in general females should choose older males as mates, with a demonstrated capacity for survival, the female thereby maximizing the likelihood that her offspring will have a long reproductive life (Maynard-Smith, 1977). Evidence on the breeding success of flightless cormorant females with older mates compared to females with younger mates is inconclusive. Since the flightless cormorant colony is extremely inbred it is also likely that incest-avoiding mechanisms affect mate choice as they do in Florida scrub jays (Emlen, 1978) and mice (Yamasaki *et al.*, 1976). On theoretical grounds, there should be selection for methods for assessing degrees of relatedness (Bertram, 1976).

Mate changing is exceptional among sea-birds. There is probably a selective advantage in pairs staying together. In Kittiwakes, established pairs are more reproductively successful than new pairs (Coulson, 1976). In flightless cormorants, mate changing is the rule. It maximizes gene flow in the in-bred colony. For the individual, it ensures that it's genes reside in varied genetic milieux in it's offspring (and thereafter), i.e. genes do not become locked up in a less fit genome which might result were the bird to repeatedly breed with the same related mate.

WHY DO FEMALE FLIGHTLESS CORMORANTS DESERT THEIR OFFSPRING? WHY DON'T MALES DESERT?

The duration of the season of food abundance is more than adequate for a pair to raise one brood, but not long enough for it to raise two consecutive broods. Since the partners have some degree of relatedness to each other, what is the best strategy for each to maximize his or her reproductive fitness under such conditions? If the male is selected to 'tolerate' his mate's desertion prior to completing the raising of her first brood, so that she can re-breed with another partner, he will have a genetic interest in the second brood (by virtue of shared genes among himself, his mate, and her new mate). Meanwhile he ensures the success of his first investment by continuing to care for the first brood until it is independent. Should the second brood succeed, the male's genetic pay-off would be less than two but greater than one. But the deserter's (i.e. female's) genetic pay-off would be two. So why doesn't he pre-empt his mate at desertion and re-breed himself? (see Maynard-Smith, 1977). The answer appears to lie in the relative food-gathering efficiency of males and females. Prior to desertion, females (but not males) are stressed to catch enough food for themselves and their young (Tindle, 1983). When the offspring were aged 40–70 days, female parents spent significantly more time foraging than their mates (females 5.3 h per day, males 4.1 h per day, $F_{1,139} = 28.19$ $P < 0.001$). Thus there is considerable doubt that a female could raise a juvenile single handed. Males, we know, can. For a male to desert and re-breed might not therefore be a high pay-off strategy since his investment in the first brood would probably come to nought.

For males to be naturally selected for faithfulness to the brood in face of their mate's desertion, they must be extremely sure of paternity. It is therefore surprising to find that males are extremely susceptible to cuckoldry prior to egg-laying. Females spend about 65% of the day unguarded by their mates during the period immediately prior to egg-laying (Tindle, 1983) allowing ample opportunity for copulations by philander males. However, that a high rate of infertility occurs in eggs (Snow, 1966), might suggest that the chances of fertilization from an occasional promiscuous copulation would be low. Nonetheless, deserted males should attempt to covertly fertilize their deserting females thereby cheating the female's new mate into raising the deserted male's offspring. Should the second brood succeed, his genetic pay-off would increase from greater than one, to two. This would seem easy to do as the new pair frequently make their nest near or even alongside that of the deserted male.

If females are to accommodate two consecutive broods within a limited season it seems paradoxical that so much time is spent in prolonged courtship in this species. However, on grounds of fidelity there are good reasons for males to reject 'fast' females (Dawkins & Krebs, 1978). Furthermore, if a function of courtship is to assess potential mates as brood-rearing co-operators, there will be a selective advantage in taking time to 'get it right'.

SEX RATIO

It might be expected that a consistently male biased sex-ratio would prevail in the colony so that deserting females could find suitable second males. In fact the ratio was skewed to females during 1971–75 (mean M : F ratio 0.80), and to males during 1976–79 (mean M : F ratio 1.17). The change was due predominantly to differential male/female mortality during 1975–76. Deserting females had ample choice of new mates from among males whose first breeding attempts had failed, and those who had not previously attempted to breed that season. In only one of 11 years (1978) when virtually all birds bred and most pairs were successful, was the availability of second mates a limiting factor.

Had suitable second mates been limiting, this would presumably have favoured earlier and earlier desertion by the females, in order to secure a new male. However since availability of second mates is not usually limiting and since females are breeding faster than males, parent flightless cormorants might be expected to depart from the Fisher model of investing equally in offspring of both sexes and be selected to invest preferentially in females, (Fisher, 1958;

Hamilton, 1967; Trivers & Wilard, 1973). In so doing, parents would maximize their inclusive fitness. In times of food shortage this strategy would have an added advantage in that female offspring, being smaller, are presumably cheaper to raise than male offspring. How an equilibrium sex ratio is maintained in flightless cormorants remains to be determined.

PARENTAL INVESTMENT

Trivers (1972) predicts that at any given moment, the parent that has invested least is most likely to desert as he or she will have less to lose should the offspring fail to survive in the hands of the remaining partner. This view has been criticized on the grounds that desertion decisions should be based on future expected pay-offs in inclusive fitness, not on past investments (Dawkins & Carlisle, 1976). There is evidence from redwinged blackbirds, that a parent's investment is determined by the expected future benefits minus the expected costs of the present reproductive attempt, rather than by past cumulative past investment (Robertson & Biermann, 1979). In this context it is pertinent to enquire about parental investment in the flightless cormorant.

Table 1 includes all parameters of investment except courtship and the gametic investment (see footnote*). It can be seen that the pattern of investment is dissimilar between the sexes. Notably, males invest significantly more time than their mates in fetching alga nest-building material. Females spend about 50% longer foraging for the young and they feed the brood

Table 1. Time investment budgets of male and female flightless cormorants in reproduction*†

	Investment, measured in hours	
	Male	Female
Gathering of nesting material:		
Before first egg laid	9.3³	0.9 X² = 7.03 P < 0.01
From first egg to offspring age 70 days	18.3	15.7
Total daylight hours spent at nest-site:**		
Before first egg laid	77.0	71.4
During incubation	211.0	211.0
During rearing to offspring age 70 days	359.0	325.0
Time spent foraging to feed offspring to 70–90 days old‡	12.3	18.3
(Sub-total at time of female desertion)	686.9	642.3 X² = 1.49 P > 0.5
Offspring 90–120 days§	5.3	0

   * Forty nests were placed under constant observation, spread over all phases of the breading cycle. Observers quantified the contribution of each parent to nest construction, to incubation and brooding, and to feeding of offspring. Details in Tindle (1983).
   † No attempt was made to quantify time devoted to courtship. Males and females invest more or less equal amounts of time. Gametic investment is omitted; eggs are more costly than sperm. The flightless cormorant's egg is small relative to female body weight.
   ** Both partners roost at nest during darkness 18.30–06.00 hours.
   ‡ Calculated as cumulative feeds given to the offspring multiplied by the time required by the adult to catch an item of prey.
   § Systematic observations ceased when the young reached 120 days. The male sporadically feeds the juvenile to 5–9 months. Thus post-mate desertion investment in offspring will exceed 5.3 h.

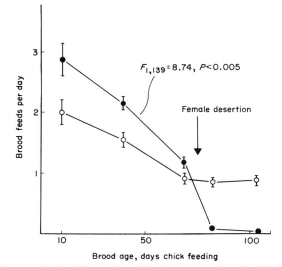

Figure 1. Feeds given to the brood by male and female parents (±s.e.). ○ Male; ● female.

significantly more frequently than their partners (Fig. 1). At 70 days of age, just prior to the female parent's desertion, the brood has received about 30% more feeds from it's mother than it's father. Although the overall investment in time given to reproductive activities does not differ significantly between males and females (Table 1), in real terms females probably invest more (prior to desertion), since foraging is likely to be bioenergetically more expensive than alga fetching. The female's strategy is thus a compromise between packing in as much chick feeding as possible, to maximize chances of success of the first brood, and deserting as early as possible in order to raise a second brood while the season lasts, of twelve females who successfully raised first broods in 1977, earlier deserters were more successful in raising second broods than later deserters.

REFERENCES

ANDERSSON, M., 1982. Female choice selects for extreme tail length in a widow bird. *Nature, 299:* 818–820.

BERTRAM, B. C. R., 1976. Kin selection in Lions and in Evolution. In P. P. G. Bateson & R. A. Hinde (Eds), *Growing Points in Ethology:* Cambridge: University Press.

BOERSMA, P. D., 1978. Breeding patterns of Galapagos penguins as an indicator of oceanographic conditions. *Science, 200:* 1487–1493.

COULSON, J. C., 1976. The influence of the pair bond and age on the breeding of the Kittiwake gull, *Rissa tridactyla. Journal of Animal Ecology, 35:* 269–274.

DARWIN, C. 1871. *The Descent of Man and Selection in Relation of Sex.* London: Murray.

DAWKINS, R. & CARLISLE, T. R., 1976. Parental investment and mate desertion: a fallacy. *Nature, 262:* 131–133.

DAWKINS, R. & KREBS, J. R., 1978. Animal signals; information or manipulation? In J. R. Krebs & N. G. Davies (Eds), *Behavioural Ecology:* Oxford: Blackwell Scientific Publications.

EMLEN, S. T., 1978. Co-operative Breeding. In J. R. Kregs & N. B. Davies (Eds), *Behavioural Ecology:* 245–281. Oxford: Blackwell Scientific Publications.

FISHER, R. A. 1958. *The Genetic Theory of Natural Selection* (2nd edition) New York: Dover Publications, Inc.

HAMILTON, W. D. 1967. Extra-ordinary sex-ratios. *Science 156:* 477–488.

HARRIS, M.P. 1974. A complete census of the Flightless Cormorant (*Nannopterum harrisi*). *Biological Conservation, 6:* 188–191.

HARRIS, M. P. 1979. Population dynamics of the Flightless Cormorant (*Nannopterum harrisi*). *Ibis, 121/2:* 135–146.

HARVEY, P. H. & ARNOLD, S. J. 1982. Female mate choice and runaway sexual selection. *Nature, 297:* 533–534.

MAYNARD-SMITH, J. 1977. Parental investment, a prospective analysis. *Animal Behaviour, 25:* 1–9.

MAXWELL, D. C. 1974. *Marine Primary Productivity of the Galapagos Archipelago.* Ph.d Thesis. Ohio State University, USA.

O'DONALD, P. 1979. Theoretical aspects of Sexual selection: variation in threshold of female mating response. *Theoretical Population Biology, 15/2:* 191–204.

REISS, M. 1982. Males bigger, females biggest. *New Scientist,* 226–229. October 28.

ROBERTSON, R. J. & BIERMANN, G. C. 1979. Parental Investment Strategies determined by expected benefits. *Zeitschritt Tierpsychologie, 50:* 124–128.

SELANDER, R. K. 1966. Sexual dimorphism and differential niche utilisation in birds. *Condor, 68:* 113–151.

SELANDER, R. K. 1972. Sexual selection and Dimorphism in Birds. In B. Campbell (Ed.) *Sexual Selection and the Descent of Man. 1871–1971:* 180–230. Chicago: Aldine-Atherton

SNOW, B. K. 1966. Observations on the behaviour and ecology of the flightless cormorant (*Nannopterum harrisi*). *Ibis, 108:* 265–280.

TINDLE, R. W. 1983 (in prep.)

TINDLE, R. W. & HARRIS, M. P. 1982. Breeding strategy and population dynamics of the Flightless Cormorant (*Nannopterum harrisi*). *Journal of Field Ornithology,* (accepted for publication).

TRIVERS, R. L. 1972. Parental Investment and Selection. In B. Campbell (Ed.) *Sexual Selection and the Descent of Man 1871–1971:* 136–179. Chicago: Aldine

TRIVERS, R. L. & WILLARD, D. E. 1973. Natural selection of ability to vary the sex ratio of offspring. *Science, 179:* 90–92.

WILSON, E. O. 1975. *Sociobiology.* 697 pp. Cambridge: Belknap Press.

WYRTKI, K., STROUP, E., PATZERT, W., WILLIAMS, R. and QUINN, W. 1976. Predicting and observing El Niño. *Science, 191:* 343–344.

YAMASAKI, K., BOYSE, E. A., MIKE, V., THALER, H. T., MATHIESON, B. J., ABBOTT, J., BOYSE, J., ZAYAS, Z. A. & THOMAS, A. L. 1976. Control of mating preferences in the major histocompatibility complex. *Journal of Experimental Medicine 144:* 1324–1335.

Biological Journal of the Linnean Society (1984), *21:* 165–176. With 6 figures.

Evolutionary divergence of giant tortoises in Galapagos

THOMAS H. FRITTS

U.S. Fish and Wildlife Service, Denver Wildlife Research Center, Museum of Southwestern Biology, University of New Mexico, Albuquerque, New Mexico 87131, U.S.A.

The giant tortoises in the Galapagos Archipelago diverge considerably in size, and in shape and other carapace characteristics. The saddleback morphotype is known only from insular faunas lacking large terrestrial predators (i.e. Galapagos and Mauritius) and in Galapagos is associated with xeric habitats where vertical feeding range and vertical reach in agonistic encounters are adaptive. The large domed morphotype is associated with relatively cool, mesic habitats where intraspecific competition for food and other resources may be less intense than in xeric habitats. Other external characteristics that differ between tortoise populations are also correlated with ecological variation. Tortoises have radiated into a mosaic of ecological conditions in the Galapagos but critical data are lacking on the role of genetic and environmental controls on phenotypic variation. Morphological divergence in tortoises is potentially a better indicator of present ecological conditions than of evolutionary relationships.

KEY WORDS:—*Geochelone* – turtles – Galapagos – evolution.

CONTENTS

INTRODUCTION

Perhaps more than any other vertebrate group in Galapagos, the giant tortoises are poorly understood. Research has been hampered by a variety of factors: (1) the extreme reduction of populations by man prior to the formation of the Galapagos National Park; (2) the difficulty of locating and handling individuals which in most cases are quite dispersed and isolated in interior regions of the islands and (3) the confusion around the taxonomic relationships of the different populations. Presently, research on evolutionary aspects of tortoises in Galapagos is more feasible than ever before. MacFarland, Villa & Toro (1974a) provided an overview of population numbers and mapped principal distributions of extant populations. Additional information gathered as a result of work by Craig MacFarland and by personnel of the Charles Darwin

0024–4066/84/010165 + 12 $03.00/0

Research Station (CDRS) and the Galapagos National Park are currently available. These data facilitate location and study of tortoises and their habitats. Taxonomic confusion persists, although most recent authors follow Mertens & Wermuth (1955) in recognizing a single species *Geochelone elephantopus* (Harlan) with subspecific names frequently applied to the individual populations. Unfortunately, this taxonomic usage by Mertens & Wermuth (1955) was apparently an arbitrary one not substantiated by revisionary studies. The most exhaustive reviews of taxonomic relationships of tortoises in Galapagos were those of Van Denburgh (1914) and Rothschild (1915). Van Denburgh and Rothschild recognized 13 species (and two unnamed populations known from inadequate material) presumed to be from 11 islands on the basis of large museum collections available to them. Several years ago, I initiated morphological studies of museum specimens and living tortoises in Galapagos with the objectives of clarifying the taxonomy of tortoises, understanding the morphological variation within tortoises in relation to ontogenetic, sexual, geographic and ecological differences and documenting the evolutionary relationships of the tortoises in Galapagos. The taxonomic history and relationships of this group are extremely complex and will be discussed in a paper still in preparation. The present discussion is based in part on data presented by Fritts (1983) and focuses on morphological variation related to biological and environmental factors and preliminary evolutionary conclusions.

MORPHOLOGICAL DIVERGENCE

The existence of considerable morphological divergence between tortoise populations in Galapagos has been known since Darwin's visit to Galapagos (Darwin, 1889). Some populations are composed of tortoises with highly domed carapaces and an anterior opening of the carapace that can be protected by the retracted forelimbs. This domed carapace shape resembles that of tortoises in many parts of the world. Tortoises in other populations have diverged markedly from this common carapace shape; they have a large anterior opening of the carapace and a carapace shape resembling a Spanish saddle. Tortoises with this carapace shape are frequently termed saddlebacks and are known only from Galapagos (Van Denburgh, 1914; Fritts, 1983) and a single island in the Indian Ocean area (Arnold, 1979). The latter area was inhabited by a distantly related tortoise species now extinct. Thus the Galapagos saddlebacks are critical to understanding this unusual morphotype.

Using either morphometric data or qualitative characteristics, considerable divergence related to interpopulational differences is discernible in Galapagos tortoises (Fritts, 1983). Domed tortoises attain the largest sizes. Saddleback tortoises are markedly smaller in size and have proportionately longer necks and forelimbs (Fritts, 1983; Van Denburgh, 1914). That some of these differences have a genetic basis is suggested by the early appearance of differences even when young are raised in identical environments.

Based on my observations and measurements of young tortoises reared at the CDRS from both domed and saddlebacked populations, the elevation of the anterior margin of the carapace that results in the saddleback condition occurs in tortoises less than five years old. Under identical conditions, young of saddleback populations start to develop the saddle carapace whereas young of domed populations do not. This strongly suggests that this difference in carapace shape is genetically controlled and not merely an environmental effect. This genetic basis is important to evaluating response to selective pressures.

Some evidence suggests that interbreeding of individuals from divergent populations results in lower hatchability and survival (MacFarland *et al.*, 1974a). Eggs deposited at the CDRS in a corral containing animals from various islands had a lower hatch success rate and the resulting young had a higher incidence of abnormalities than did the eggs and young resulting from captive individuals housed only with tortoises from the same island (C. G. MacFarland, pers. comm.). Conservation programmes at the CDRS have attempted to avoid lowered reproductive success and genetic contamination by excluding tortoises of unknown origin and segregating tortoises from different islands into separate enclosures when held at the research station (MacFarland, Villa & Toro, 1974b).

Tortoises from Galapagos have had exceptionally low reproductive success in captive zoo herds (MacFarland *et al.*, 1974a). The reproductive success at the San Diego Zoo (Shaw, 1967) where tortoises from several islands were mixed, was markedly lower than that at the Honolulu Zoo (Throp, 1972) where only tortoises collected on southern Isabela by the Townsend expedition were housed together. Whether this difference reflects problems due to genetic incompatibility or environmental problems cannot be determined.

SIGNIFICANCE OF VARIATION IN CARAPACE SHAPE

Distortion diagrams illustrate a basic difference between the domed and saddleback carapace shape (Fig. 1). Basically the saddleback has a vertical distortion of the anterior carapace through allometric growth in the frontal (vertical) plane. The anterior carapace has increased in height relative to the remainder of the carapace and the attachment of the base of the neck to the carapace is higher relative to the rest of the body. This vertical distortion of the anterior carapace results in a radically different carapace shape and a higher point of attachment for the base of the neck. It also is related to a relatively longer neck and longer forelimbs than in domed individuals.

In both field and captive situations, differences in behaviour can be seen

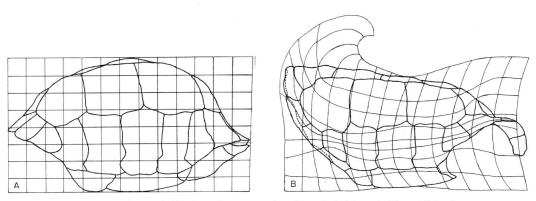

Figure 1. Relative shape of (A) domed carapace (southern Isabela) and (B) saddleback carapace (Pinzon) using distortion diagrams superimposed on carapace outlines.

Table 1. Sequential elements of agonistic behaviour and submission in tortoises from Galapagos

Agonistic encounters	Submission
1. Vertical extension of neck and forelimbs	1. Withdrawal of head and neck
2. Gape	2. Lower body
3. Display of neck coloration	3. Expulsion of air (hiss)
4. Bite	4. Turn away
	5. Retreat
Dominance re-enforced by biting forelimbs and hindlimbs	

between saddleback and domed tortoises. Saddlebacks are less tolerant of other tortoises in close proximity, are more attentive of movements by other animals (including tortoises and man) and are more frequently involved in agonistic behaviours.

Agonistic behaviour in *Geochelone* in the Galapagos appears to be a ritualistic competition for height (Fig. 2). The competing tortoises attempt to achieve dominance by raising the head as high as possible and, if necessary, by gaping and biting each other on the head or upper jaw. Normally, the tortoise reaching higher than the other is dominant as judged by submissive behaviour in the other tortoise. The sequence of related behaviours is outlined in Table 1. I have observed these encounters in the field, in captive tortoises and in a series of experimental agonistic encounters between a saddleback tortoise from Isla Española and a domed tortoise of unknown origin (probably from Isabela).

Figure 2. Agonistic behaviour between tortoises of saddleback (left) and domed (right) morphotypes at the Charles Darwin Research Station.

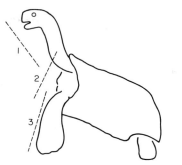

Figure 3. Schematic representation of the components of vertical reach: neck length (1), height of anterior carapace (2) and forelimb length (3).

Under experimental conditions, the saddleback was able to extend the head higher and was dominant even though the saddleback male was considerably smaller (860 mm in curved carapace width) than the domed male (1198 mm). Apparently vertical reach is more important than overall size in determining dominance in agonistic situations. The greater vertical reach resulted from a combination of factors: a longer neck, an elevated anterior opening of the carapace and longer forelimbs (Fig. 3).

A relative index to the vertical reach of tortoises of differing morphotypes can be computed using measurements given by Van Denburgh (1914): height of anterior carapace, neck length and forelimb length. When the sum of these three measurements is plotted against a measure of overall size (curved carapace width) it is apparent that tortoises of the saddleback populations have a relatively long vertical reach for their size (Fig. 4). If all tortoises were similar in

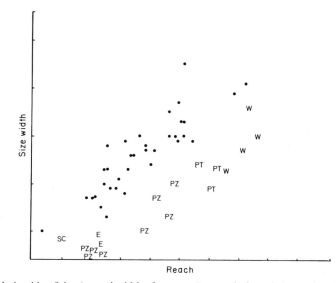

Figure 4. Relationship of size (curved width of carapace) to vertical reach (sum of neck length, height of anterior carapace and forelimb length). Data from Van Denburgh (1914). Dots represent tortoises of domed morphotypes; letters represent tortoises of saddleback populations: Española (E), Pinzon (PZ), Pinta (PT), San Cristobal (SC) and Volcan Wolf (W).

shape and body proportions, the largest tortoises would consistently be those with the longest vertical reach. However, the variation in carapace shape and correlated proportions of the body results in an exaggeration of the vertical reach in populations with the saddleback morphotype. Tortoises from saddleback populations tend to be smaller in overall size (Fritts, 1983; Van Denburgh, 1914). Body size is correlated with elevation of the island or volcano inhabited (Fritts, 1983). Large tortoises tend to occur on islands with maximal elevations and mesic habitats related to cloud capture at high elevations. Small tortoises tend to occur on islands with low elevations and markedly drier and warmer habitats.

I have proposed two hypotheses to explain size variation between populations (Fritts, 1983).

(1) Small tortoises have lower energy and water budgets and can use the limited shade of the sparse vegetation better than can larger tortoises; thus, small tortoises occupy xeric habitats.

(2) Large tortoises can exploit the abundant food, water and shade of moist, high habitats and can use their large size to resist cooling during overcast or foggy periods; thus, large tortoises occupy the mesic habitats at high elevations.

The saddleback morphology appears to be associated with the islands with the smallest areas and the least ecological diversity (i.e. those having predominantly dry conditions). In contrast, domed tortoises are present on larger islands with maximal ecological diversity and more mesic habitats. Xeric habitats with limited food, water and shade and subject to climatic variation would produce intense intraspecific competition for these resources and this environmental variation could result in selective pressures that:

(1) favoured increase in vertical feeding range without increased body size in extremely dry situations where food and water are limited, and

(2) favoured increase in vertical reach to facilitate agonistic behaviour important in the defence of food, water and shade in dry conditions.

If, under xeric conditions, selection favoured reduction in body size without changes in carapace shape and body proportions, the smaller tortoise with lower energy and water requirements would be less dominant in agonistic encounters and less able to defend food, water or shade due to a shorter vertical reach than larger individuals. However, selection for increased vertical reach together with decreased body size could increase fitness in xeric conditions. Because saddleback tortoises occur in lower quality habitats subject to extremely xeric conditions, the emphasis of agonistic and asocial behaviour potentially tends to disperse individuals, reducing intraspecific competition and allowing defence of resource when competition does occur.

The saddleback morphology appears to be related to a simple ontogenetic shift of a morphotype that occurs in several domed populations (southern Isabela and Santiago = El Salvador). Carapace shape and the anterior opening of the carapace are relatively constant through ontogeny; however, the largest individuals (especially males) exhibit an increased opening of the anterior carapace (Fritts, 1983). This difference presumably results from accelerated growth of vertical dimensions in relation to other carapacial dimensions. This trait, visible only in the largest and possibly oldest tortoises, may allow increased vertical reach related to feeding and increased dominance. Apparently accelerated growth of the vertical plane early in the ontogeny results in the saddleback morphotype. The difference between initiating the accelerated growth late in the ontogeny or much earlier may involve only a slight genetic change in loci controlling growth even though such a shift results in radical alteration of shell morphology.

On the basis of the varying degrees of development of the saddleback morphotype, (Fritts, 1983) shifts in the ontogenetic sequence of growth patterns appear to have occurred more than once from the ancestral tortoises colonizing Galapagos. Such shifts appear to be related to various degrees of aridity and ecological complexity. Intraspecific competition for extremely limited resources resulting from aridity would be most likely on small, low islands such as Española and Santa Fe. Similar but less marked selective pressures for increased reach and small body would be expected on slightly larger and more ecological complex islands such as Floreana (= Sta Maria), Pinta and Pinzon. Large islands or volcanoes where conditions were xeric due to lack of exposure to moist winds, or due to relatively recent volcanic activity (Volcan Wolf on Isabela and Fernandina), are exceptional conditions where saddleback morphotypes are known. The variation in ecology between these seven islands is great and probably accounts for the varying degrees of development of the saddleback morphology and variation in overall size reported by Fritts (1983). As island area and maximal elevation vary, so too would the ecological stability vary in response to seasonal, annual and long-term climatic changes. The degree of departure from a domed morphotype is inversely related to the ecological stability of the island inhabited (Fritts, 1983). Climatic variation is well documented in Galapagos (Colinvaux, 1972) and rainfall varies considerably with altitude as well as between islands and from year to year. Colinvaux & Schofield (1976) documented a drier climate than present during the late Pleistocene.

COMPARISONS OF SIMILAR POPULATIONS

Some tortoises that are superficially similar in size and shape differ in other morphological characteristics. Tortoises from Española and Pinzon are similar in carapace shape but differ in the degree of striations on carapacial scutes, the colour of the scutes and the extent of flaring and prominence of marginal scutes (Table 2). Tortoises on Pinzon may grow to a slightly larger size but samples from Española are limited and may not represent the entire size range.

Similar differences can be seen in two domed taxa, *Geochelone vicina* and *G. guentheri*, recognized by Van Denburgh (1914). Van Denburgh (1914: 338) was somewhat confused by the variation observed in tortoises from southern Isabela (i.e. Sierra Negra and Cerro Azul areas). He recognized *guentheri* as a

Table 2. Comparison of saddleback tortoises from Española and Pinzon

Trait	Española	Pinzon
Scutes	Brown, striated	Black, smooth
Anterior marginals	Recurved, well developed	Less flared, blunted
Size	Small	Moderate
Yellow on face and neck	Extensive	Moderate

tortoise with a less elevated carapace, marginals reduced in size and smooth unstriated carapacial scutes. In Van Denburgh's interpretation *vicina* had a more domed carapace, more pronounced marginals and striated scutes. Tortoises collected near Cabo Rosa (a coastal inlet between Sierra Negra and Cerro Azul) were considered by Van Denburgh to most closely resemble tortoises from Caleta Tagus (further to the north on Volcan Darwin). Van Denburgh pointed out that some apparent intermediates existed in the collections and that both *vicina* and *guentheri* morphotypes were present on Sierra Negra and near coastal areas.

Many of the specimens from southern Isabela available to Van Denburgh were collected by natives and lacked precise locality and altitudinal information. However, even Beck (1903) had previously commented that as one moved up Sierra Negra the tortoises had black scutes with distinctive eroded or scarred areas, characteristics indicative of *guentheri* (*sensu* Van Denburgh, 1914).

On the basis of examining the specimens available to Van Denburgh, other museum material and living tortoises in Galapagos, I have confirmed the marked morphological differences occurring on southern Isabella (Table 3; Fig. 5). The distribution of the two morphotypes appears to be related to moisture and altitudinal gradients (Fig. 6) with *guentheri* morphotypes occurring in mesic habitats usually associated with high elevations and *vicina* morphotypes occurring in relatively drier habitats usually associated with low or intermediate elevations. The problematic population to the west of Cabo Rosa, which Van Denburgh considered to resemble *microphyes* more than *guentheri*, strongly resembles *guentheri* except in having scutes less eroded and a more elongate carapace in relation to carapace width. This population is exceptional for the *guentheri* morphotype in occurring at a relatively low elevation. However, this locality is conspicuously more mesic than other lowland localities, perhaps because of its exposure to moist winds from the south-east and its position between two major volcanoes (Sierra Negra and Cerro Azul). An adjacent locality further to the east receives less exposure to moist winds and has tortoises with the *vicina* morphotype.

Are the differences seen in *guentheri* genetically controlled and related to higher, cooler and wetter habitats or are they merely environmentally determined phenotypes with no genetic basis? Tortoises of the *guentheri* morphotype in highland areas of Cerro Azul were wiped out by colonists and only remnants of former populations exist today. There are several localities on

Table 3. Comparison of morphotypes *vicina* and *guentheri*
(*sensu* Van Denburgh, 1914)

Trait	*vicina*	*guentheri*
Anterior marginals	Thin edge, pointed	Thick edge, blunt
Marginal sutures	Deeply notched	Shallowly notched
Nuchal notch	Deep, wide angle	Shallow, narrow angle
Vertebral scutes	Convex, striated, normal, brown	Flat, smooth, eroded, black
Supracaudal	Indented	Continuous with adjacent flare
Plastron of females	Flat	Moderate concavity
Anals	Flat	Thickened
Carapace profile	Domed	Flatter

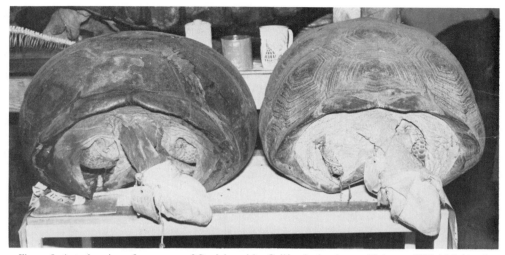

Figure 5. Anterior view of carapaces of *Geochelone vicina* California Academy of Sciences 8227 (right) and *G. guentheri* CAS 8206 (left) showing differences in doming of carapace, striation of scutes and projection of marginals.

southern Isabela where tortoises occur or previously occurred for which no information is available on which morphotype is present (Fig. 6). Additional information on the ecology and ontogeny of these populations is critical to address these questions.

The smooth, black scutes of the *guentheri* morphotype are frequently present in adult females of other populations. Since females do not grow to the large size of males, the slow growth as adults potentially results in unstriated scutes darker in colour than the striated brownish scutes of faster growing males and juveniles. The cool, moist habitats occupied by *guentheri* (perhaps the most mesic in Galapagos) may result in a slowing of growth of all adults with corresponding alteration of scutes. Individuals occupying drier and warmer habitats at lower elevations may exhibit more normal growth rates producing striation and brown coloration.

Other characteristics in which *vicina* and *guentheri* differ (size and shape of marginal scutes, degree of doming of carapace, extent of flare over hindlimbs and tail) are potentially subject to environmental influence as well. Juveniles when growing rapidly have a highly domed carapace, prominently pointed marginals and other attributes not as conspicuous in those growing slower.

In the light of these observations it is appropriate to ask whether similar differences between tortoises from Española and Pinzon are related to ecological differences and whether or not they are genetically or environmentally controlled.

Relative to tortoises from Española, tortoises on Pinzon occur in cooler, less xeric habitats. Pinzon has a maximal elevation of 458 m whereas Española reaches 206 m. Even though the Española and Pinzon tortoises differ markedly in carapace shape and size from those on southern Isabela, they differ from each other in similar aspects. Española tortoises are limited to a low, dry area and have striated, brown scutes and pronounced flared marginals. Pinzon tortoises

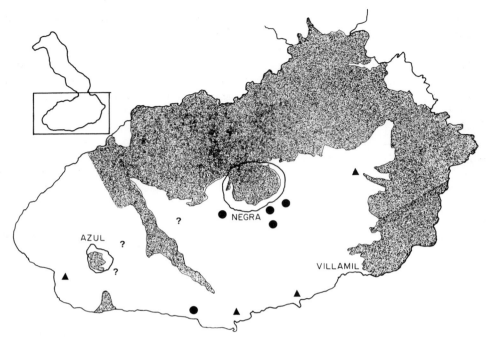

Figure 6. Distribution of morphotypes of *Geochelone vicina*—▲ and *G. guentheri*—● on southern Isabela in relation to Sierra Negra and Cerro Azul. Note single coastal population of *guentheri* and proximity of *vicina* morphotype. Stippled areas represent major lava exposures. Adapted from a map by D. Weber.

can and do occur at higher elevations where fog and drizzle produce less xeric conditions and have unstriated, black scutes and less pronounced marginals with little flare. The tortoises from southern Isabela are on the opposite extreme of environmental variation within the archipelago as tortoises from Española and Pinzon. However, morphological variation in both pairs of morphotypes is concordant with differences in altitude and moisture.

The concordance of morphological variation may be genetically fixed or environmentally labile. Clearly some morphological differences may be less useful in measuring evolutionary divergence than others. The unknown basis of some differences cautions against making phylogenetic statements without additional information. Even if many aspects of carapace morphology are genetically controlled and subject to evolutionary pressures, the morphology may be so strongly related to present ecological conditions and fitness that it is not reflective of previous evolutionary relationships.

CONCLUSIONS

Morphological divergence within tortoises from Galapagos is understandable in relation to present ecological conditions. Divergence in morphological characters correlates with ecological differences. The major morphological differences potentially result from relatively minor genetic differences controlling growth patterns rather than major genetic innovations. Such an hypothesis is concordant with the relatively slight genetic differences detected by

Marlow & Patton (1981) using electrophoretic techniques. A parallel evolutionary divergence has occurred in the geospizine finches (Bowman, 1961) without major genetic alteration detectable using electrophoretic methods (Patton, this volume). Evolution of major morphological differences in response to the unique environments on oceanic islands is well known (Carlquist, 1974). However, the rate at which such evolutionary changes occur is, in most cases, unknown. Geological dates for the oldest rocks known in the Galapagos Archipelago are relatively recent.

The divergence in carapace shape of some populations of tortoises in Galapagos involved opening the anterior carapace, exposing the head and base of neck and abandoning in part an antipredator defence employed by terrestrial turtles in many parts of the world. No native mammalian predators occur in the Galapagos Archipelago. Once free from predator pressure, some tortoise populations apparently responded to selection for increased height in relation to size. By modifying previous developmental patterns, a diversity of tortoise morphotypes developed which allowed occupation of a wide array of environmental conditions in the relatively small but ecologically complex Galapagos Archipelago.

ACKNOWLEDGEMENTS

The assistance of personnel of the Galapagos National Park and the Charles Darwin Research Station is gratefully acknowledged. Miguel Cifuentes and Craig MacFarland were generous with their time and knowledge in planning field work. Research on tortoises was supported in part by the National Science Foundation, U.S.A. (DEB 76-10003). I am grateful to Patricia Fritts, Howard Snell, Michael McCoid, Charles Crumly and Susan Palko for assistance in the collection and analysis of data. The curators of the British Museum (Natural History), Oxford University Museum, California Academy of Sciences, U.S. National Museum of Natural History, American Museum of Natural History and Harvard University Museum of Comparative Zoology provided access to specimens and facilities.

REFERENCES

ARNOLD, E. N., 1979. Indian Ocean giant tortoises: their systematics and island adaptations. *Philosophical Transaction of the Royal Society of London, B, 286:* 127–145.

BECK, R. H., 1903. In the home of the giant tortoise. *Annual Report of the New York Zoological Society, 7:* 160–174.

BOWMAN, R. I., 1961. Morphological differentiation and adaptation in the Galapagos finches. *University of California Publications in Zoology, 58:* 1–302.

CARLQUIST, S., 1974. *Island Biology.* New York: Columbia University Press.

COLINVAUX, P. A., 1972. Climate and the Galapagos Islands. *Nature, Lond., 240:* 17–20.

COLINVAUX, P. A. & SCHOFIELD, E. K., 1976. Historical ecology in the Galapagos Islands. I. A Holocene pollen record from El Junco Lake, Isla San Cristobal. *Journal of Ecology, 65:* 989–1012.

DARWIN, C., 1889. *A Naturalist's Voyage, Journal of Researches into the Natural History and Geology of the Countries Visited During the Voyage of H.M.S. Beagle Round the World* (2nd Edition). London: John Murray.

FRITTS, T. H., 1983. Morphometrics of Galapagos tortoises: evolutionary implications. In R. I. Bowman, M. Berson & A. Leviton (Eds), *Patterns of Evolution in Galapagos Organisms:* 107–122. San Francisco, California: Pacific Division of the American Association for the Advancement of Science.

MacFARLAND, C. G., VILLA, J. & TORO, B., 1974a. The Galapagos giant tortoises *(Geochelone elephantopus).* Part I: Status of the surviving populations. *Biological Conservation, 6:* 118–133.

MacFARLAND, C. G., VILLA J. & TORO, B., 1974b. The Galapagos giant tortoises *(Geochelone elephantopus).* Part II: Conservation methods. *Biological Conservation, 6:* 198–212.

MARLOW, R. W. & PATTON, J. L., 1981. Biochemical relationships of the Galapagos giant tortoises *(Geochelone elephantopus)*. Journal of Zoology, London, 195: 413–422.

MERTENS, R. & WERMUTH, H., 1955. Die rezenten schildkröten, krokodile und brückenechsen. Eine kritische liste der huete lebenden ARTE und rassen. *Zoologische Jahrbücher Abteilung für Systematik, Okologie und Geographie der Tiere, 83:* 323–440.

ROTHSCHILD, LORD., 1915. The giant land tortoises of the Galapagos Islands in the Tring Museum. *Novitates Zool. 22:* 403–417.

SHAW, C. E., 1967. Breeding the Galapagos tortoise-success story. *Oryx, 9:* 119–126.

THROP, J. L., 1972. Notes on breeding the Galapagos tortoise *Testudo elephantopus* at Honolulu Zoo. *International Zoo Yearbook, 9:* 30–31.

VAN DENBURGH, J., 1914. The gigantic land tortoises of the Galapagos Islands. *Proceedings of the California Academy of Science Series, 4, 2:* 202–374.

Biological Journal of the Linnean Society (1984) *21:* 177–184.

Seals of the Galapagos Islands

W. NIGEL BONNER

British Antarctic Survey, Natural Environment Research Council, Madingley Road, Cambridge CB3 0ET

The Galapagos archipelago has been colonized by two species of otariid, the fur seal, *Arctocephalus galapagoensis,* and the sea lion, *Zalophus californianus wollebaeki.* The former probably arrived from South America and the latter from North America, both by way of periodic incursions of colder water forming the east Pacific corridor. The terrestrial behaviour of both these otariids is affected by the high ambient temperatures in the Galapagos. Patterns of breeding behaviour of otariids which lead to intense polygyny and sexual dimorphism appear to be modified. The habit of prolonging lactation, widespread in otariids, is carried to an extreme in the Galapagos fur seal. The reason for this is unclear.

KEY WORDS:—*Archtocephalus* – fur seal – Galapagos – lactation – pinnipedia – sea lion – thermal stress – *Zalophus.*

CONTENTS

INTRODUCTION

An archipelago lying on the Equator seems an unlikely site for pinnipeds, yet two otariid species are well established in the Galapagos. Ancestral pinnipeds diversified in the Miocene, perhaps as a response to the development at that time of upwelling processes in the oceans which generated sufficient secondary production to support new populations of predators (Lipps & Mitchell, 1976). Although Miocene oceans were warmer than those of today (Gaskin, 1976), the ancestral pinnipeds encountered thermoregulatory problems on entering the sea. In consequence, adaptations such as increased body size, relative reduction of surface area and the development of insulating layers, were selected for and became characteristic of the group (Bonner, 1982). These adaptations to conserve heat rendered it less easy to dissipate excess heat and made the pinnipeds less well fitted for life ashore in a warm climate. In consequence, they are animals characteristic of cool or cold regions, with the great majority of the group's biomass being found in polar seas.

0024–4066/84/010177+08 $03.00/0

Despite this, the Galapagos have been successfully colonized by a sea lion and a fur seal.

As with most other otariids, considerable confusion exists in the literature about the identity and taxonomic status of the Galapagos seals. The present position is more-or-less clear, but the literature is scattered with names based on misinterpretation of the scanty available material, false assumptions or at times even faulty labelling.

The Galapagos fur seal

There is no doubt that the fur seals of the Galapagos are closely related to their neighbours on the South American continent, *Arctocephalus australis*. King (1954) concluded from a rather small sample of skulls that the Galapagos seals were somewhat smaller than those on the mainland, but stated there was no doubt that they were conspecific, not even meriting subspecific rank. However, she cautiously pointed out that the name *A. australis galapagoensis* was available should they prove to be separable. Repenning, Peterson & Hubbs (1971), in a careful examination of the entire genus, showed that Galapagos skulls were distinct. The Galapagos seal is characterized by delicate, high-pointed post-canine teeth with no accessory cusps, and by posteriorly-diverging dental arcades. The rostrum and nasals are short compared with *A. australis,* and the palate wider. The most obvious characteristic, however, is the much smaller skull of the Galapagos seal, which is smaller than that of any other adult fur seal, and the lack of development of sagittal and occipital crests. This lack of bony development gives the skull a very feminine look compared with other species of *Arctocephalus*. The femininity of the male skull reduces the difference between the male and the female, and this is reflected in a lesser general degree of sexual dimorphism in the Galapagos fur seal.

These differences are as great as any that occur between species of *Arctocephalus* (*A. pusillus* excepted) and Repenning and his colleagues were in no doubt the Galapagos fur seal should be afforded full specific status as *Arctocephalus galapagoensis* (Heller, 1904).

The sea lion

For a long period it was believed that the sea lion on the Galapagos was the South American sea lion, *Otaria flavescens*. In fact, this species is a very rare visitor to the islands, only one individual (a dead male) having been reported (Wellington & de Vries, 1976). Sivertsen (1954) carefully examined a series of sea lion skulls from the Galapagos and compared them with *Zalophus californianus* skulls from the west coast of North America. He concluded that Galapagos skulls were readily distinguishable by their smaller size. The maximum size of Californian skulls in his sample with all sutures closed was 309 mm, while Galapagos skulls of the same suture age did not exceed 276 mm. Additionally, the Galapagos skulls were more slender than those from North America. Sivertsen compared 10 ratios in the two groups and found significant differences in all of them. In general, the Galapagos skull is more slender (except that the relative breadth of the brain case is greater). The greatest difference was in the ratio width at preorbital processes/width of brain case at jugals, which was 0.93–1.23 in northern skulls, but only 0.73–0.88 in Galapagos skulls. Sivertsen regarded these differences as sufficient to justify the new species *Zalophus wollebaeki*. Scheffer (1958), reviewing the family, felt they were of subspecific, rather than specific, importance and relegated Sivertsen's species to the subspecies *Zalophus californianus wollebaeki* (Sivertsen, 1953), a view now generally accepted.

ORIGIN OF THE GALAPAGOS SEALS

The centre of evolution of otariids was in the north-east Pacific (Repenning, 1975), but one of the periodic incursions of colder water up to the west coast of South America established the East Pacific corridor, allowing ancestors of *Arctocephalus* to reach the Southern Hemisphere. Here speciation took place, probably in South America, where *A. australis* seems to provide something near to an ancestral form. From South America a reciprocal excursion back up the east Pacific corridor allowed *Arctocephalus* seals to reach the islands and coast of California and Mexico, to give rise to *A. townsendi*, while another movement up the corridor (or perhaps even the same one) established the stock of *A. galapagoensis* on the Galapagos.

Zalophus probably originated in the north-east Pacific, possibly in the area currently occupied by *Zalophus californianus californianus*. From here colonization of Japan took place, probably by movement along a cold northern route, while the Galapagos were reached by the Eastern Pacific corridor.

It is not possible to date these movements, or say which one came first, but they are likely to have occurred in the late Pleistocene, with the establishment of *Zalophus* probably being later than that of *Arctocephalus*. The subsequent warming of the sea around the Galapagos isolated the populations, allowing the development of recognizable taxa.

NATURAL HISTORY OF THE SEALS

The seals of the Galapagos show interesting similarities in their natural history. Both are remarkably tame animals, though this is not an unusual characteristic in island forms, and the behaviour of both is much affected, perhaps even dominated, by the vital need to avoid overheating in the equatorial sun.

The fur seal: The largest colonies of fur seals are found on Isabela and Fernandina in the west, Pinta, Marchena and Genovesa in the north and Santiago in the centre of the Archipelago. Other islands support smaller groups. Fur seals select rugged terrain with caves or overhanging ledges of lava where the seals can shelter from the sun. It is characteristic of fur seal habitat that there is always access to deep water (Trillmich, 1979),.

The breeding season in Fernandina lasts from August to November, with a peak of births in the first week of October. Breeding bulls establish large territories ashore (as much as 200 m²), which reflect the low density of the females, which seek out sheltered spots. Because of the broken nature of the ground, it is impossible for most bulls to survey the whole of their territories, and while they attempt to repel intruding males, it is sometimes possible for a rival to enter a territory and copulate with one of the bull's females (Trillmich, in press). All territories have access to the sea, so that the bulls can enter the water at midday, to cool off and drink. During territory tenure, which may last up to 51 days, the territorial bulls do not feed.

After the peak of pupping, the territorial system breaks down, and the big bulls leave to replenish their reserves. This allows smaller bulls, or even half-grown males, a chance to copulate with oestrous females (Trillmich, in press).

A female gives birth about two days after coming ashore, and remains with her pup for about a week before leaving to enter the sea to feed. This corresponds to the onset of oestrus, when the female is mated. Over 90% of matings take place on land. There follows a series of foraging excursions by the female, each lasting 1–4 days. Trillmich & Mohren (1981) have noted that the females spend most time at sea away from the pup around full moon.

Lactation lasts an astonishing 2–3 years. The milk demand of the pup increases during the first year and declines thereafter. Yearlings may spend 70% of their mothers' time ashore suckling, while unweaned 2-year-olds may suck for 40–50% of their mothers' time ashore (Trillmich, in press).

Pups moult into the yearling coat at about 6 months, and at about a year they begin to feed independently, but still rely heavily on their mothers' milk.

Trillmich (in press) has carefully studied the future reproductive potential of females suckling young. All females mate at the post-partum oestrus, but only some 15% give birth the following year if they are still feeding a pup. Of females without dependent young, some 70% give birth. The presence of an older sibling depresses the chances of survival of the new pup. If the older sibling is a yearling, the new pup almost always dies in the first month; if the older sibling is a 2-year-old, the young pup's chance of survival is about 50%. This arrangement clearly reduces the potential fecundity of the female to at best one pup every two years, and Trillmich calculates that if a female, which first pups at age 5 years, survives till age 15 years she can raise only about five offspring.

The sea lion: Galapagos sea lions are distributed generally throughout the archipelago and breed on all the major islands and on many small islets. In contrast to fur seals, they prefer gently sloping open beaches, either sandy or rocky. Trillmich (1979) noted that any such areas easily accessible at low tide and lacking an extended intertidal with rough rocks or difficult laval formation were likely to harbour sea lion colonies. Nelson (1968) drew attention to the difficulty experienced by sea lions in making their way back to the water over intertidal rocks at ebb tide.

The breeding season is extended to cover most of the cool season of mists and drizzle, the garua, which lasts from about May–June to December–January. Births may extend over 6–8 months with the peak of pupping 1–3 months earlier on Fernandina (the most western island) than on Espanola, the island furthest to the south-east (Trillmich, 1979).

The breeding bulls are intensely territorial, claiming strips of the coast which they patrol and defend vigorously against intruders. Nearly all territories border the sea. The bull's territorial activities take place mainly in the water, just offshore, where he swims up and down, vocalizing incessantly.

A female gives birth to a single pup and stays with it on the beach for about the first week. After this she regularly goes off to feed by day, returning to suckle her pup at night, at which time the males also may spend more time ashore. Oestrus occurs within the first month after parturition (Trillmich, 1979) (though Peterson & Bartholomew (1967) suggested 2 weeks for the Californian subspecies, this was an underestimate (Odell, 1975)). Bulls investigate oestrous females by sniffing at them and following them about. Copulation can occur on land, but generally takes place in the water. Bulls attempt to herd females in their territories, but are unable to prevent them moving about and entering the territories of other males. In low density colonies bulls may hold territories for up to 3 months or longer. In general, the denser the population and the nearer the peak of the breeding season, the shorter the period of territory maintenance (Trillmich, 1979).

Pups begin to enter the sea and swim when about 1–2 weeks old. They moult into the adult pelage at about 5 months and shortly after begin to fish for themselves, becoming increasingly independent of their mother's milk. Nevertheless, the mothers will continue to suckle their pups occasionally until another one is born, for up to 3 years or even longer (Trillmich, 1979).

Shark-mobbing behaviour: A feature of sea lion behaviour that deserves mention is their relation with sharks. The responses of breeding bulls towards sharks, *Carcharinus galapagoensis*, swimming into their territories have been reported as evincing 'paternal' behaviour which serves to protect the young from shark attack. This was first referred to by Eibel-Eibesfeldt (1975) and later developed by Barlow (1972). He described collective responses from neighbouring territorial bulls which converged towards an intruding shark, driving it away from the rookery area. Barlow claimed that this represented protective behaviour towards the young seals and established a paternal role for the bulls. These views were challenged by Miller (1974) who concluded that the actions described by Barlow could be interpreted in several ways and that is was unnecessary to postulate a paternal role by the seals. More recently Trillmich (unpubl.) considered this phenomenon. He concluded shark-mobbing was rare (he observed only three cases in the course of two years at the Galapagos, none of which involved territorial bulls, despite having spent nine months on sea lion colonies). He felt that the mobbing observed could best be explained in terms of self-protection, though this might on occasion have the effect of protecting a pup. Sea lions are most at risk from sharks when fishing, and particularly when breaking up their catch. Pups, living on milk, are little exposed to danger from sharks. Furthermore, the probability of a close kinship between a pup and a territorial bull is low, as territories change so often.

DISCUSSION

The seals of the Galapagos are successful animals which have come to terms with what is, for a pinniped, an unusual environment. Both have changed sufficiently for their present populations to be recognized as separate taxa. Some of their differences may be attributable to founder effect, but some are adaptive, though it may not be easy to determine which are which.

The Galapagos fur seal is conspicuously smaller than any other member of its genus, and the size difference between the sexes is less. The situation is less clear between sea lions from California and the Galapagos, but skull measurements indicate a smaller animal at the latter. Can these differences be related to the Galapagos environment? It is possible that they can.

Bartholomew (1970) constructed an evolutionary model for polygynous pinnipeds which convincingly accounts for many of their observed characteristics, particularly their large size and extreme development of sexual dimorphism. The concepts behind the model are complicated, but may be briefly summarized here. By adopting gregarious breeding habits, animals which are widely dispersed when feeding at sea can make use of special situations, such as oceanic islands, where appropriate terrain and absence of terrestrial predators allow them to breed successfully on land—despite the fact that their main adaptations are for an aquatic medium. In their breeding aggregations, males are more widely spaced than females, because of the former's testosterone-induced aggressiveness. From this it follows that many males are excluded from the breeding females, resulting in competition between males and selection for male epigamic characters. Those males which are most vigorous and aggressive can maintain their position on the beaches longest and can pass on a disproportionate number of genes to the next generation.

To the extent that the epigamic characters of the successful males are genetically determined, there will be a strong positive feed-back to reinforce selection for these. Clearly the longer a male can maintain his position on the beach among the oestrous females, the greater the opportunity he will have to pass on genes. This leads to selection for increased energy reserves (in the form of a blubber layer) to enable prolonged fasting during the breeding season. Since fasting is more possible for a large mammal than a small one, because of the former's low weight-relative metabolic rate, large size will be a clear advantage here, though this is likely to be less important than its effect on fighting potential (Trillmich & Trillmich, 1983).

The same considerations do not apply to the females and hence selection has led to the marked sexual dimorphism seen in the polygynous pinnipeds.

Such a system appears to be of considerable antiquity in the Pinnipedia, as the Allodesminae, a group that flourished in the Miocene, already showed marked dimorphism and other characters that could be associated with this pattern of behaviour (Mitchell, 1966).

Bartholomew's model is crucially dependent on the maintenance of advantage to only a few of the available males by the establishment of territories where a successful bull can exclude other males and mate with all the females in his territory. Such conditions pertain in several pinnipeds, such as the Antarctic fur seal, *Arctocephalus gazella* and the Elephant seals, *Mirounga spp*.

In *A. galapagoensis,* on the other hand, the general tendency to increased size and sexual dimorphism appears to have been reserved, since these characters are both less than in the parent stock of *A. australis*. This might be explained by the constraints applied by the climate of the Galapagos on the terrestrial breeding pattern of the seals. Because of the risk of overheating the females seek out shaded places when they come ashore. Such places are not necessarily close together, decreasing female density. The males attempt to control large territories, but are less successful in excluding other males than some other fur seals. Under these circumstances, the system proposed by Bartholomew breaks down and selection for increased size may even be replaced by selection for decreased size.

Trillmich (in press) has pointed out that the effort of moving about over the very rugged and extensive territories occupied by the fur seals may exert a selection pressure for smaller size, as the cost of climbing decreases with decreasing body weight, while the loss of heat generated by any strenuous activity is easier for a smaller animal with a relatively larger surface area.

The situation in the sea lions is less clear. Peterson & Bartholomew (1967) noted that *Zalophus* cannot be considered as one of the highly polygynous pinnipeds. They concluded that the social system was breaking down in an evolutionary sense, based on the occurrence of numerous aquatic copulations, a view supported by Odell (1975). Because of high temperatures on the rookeries during the daytime, *Zalophus* males remain in the water, where their high level of mobility and the frequent visits of the females to the sea for thermoregulatory purposes make prolonged effective territorial maintenance difficult and permit aquatic copulation, which may be an important factor contributing to relaxing the social structure. Conditions in the Galapagos are probably only a little more extreme than in California, so it is understandable that the differences between the sea lion subspecies are less than those between the Galapagos and South American fur seals. Prolonged suckling may have a similar explanation. In terms of the Bartholomew model, advantage will accrue to males who are larger and beter able to compete with others. To the extent that prolonging suckling may result in larger ultimate size, this might be expected to be selected for. It is certainly a common characteristic in the Otariidae. Of those species well enough known for observations on suckling patterns to be available, prolonged suckling has been recorded in all save the migratory species which have a pelagic phase at the end of the breeding season which would make prolongation of suckling impossible (Bonner, in press).

In a similar way, females might be expected to obtain an advantage in terms of fitness, both for themselves and their offspring, by increased size, since larger females might more easily withstand the stresses of gestation and lactation.

This could account for the condition of *Zalophus,* but will scarcely of itself serve to explain the strange reproductive pattern of *A. galapagoensis.* If selection is acting to produce smaller animals, as seems likely, the prolongation of suckling cannot act in the opposite direction. However, prolonged suckling can have other effects besides enhancing growth. It postpones the time at which the young must fend for themselves and it enhances the social bonds between one generation and the next. Both these factors may be important in an environment where predators may be significant. Many otariids inhabit waters where sharks are a potential hazard, and this is certainly the case at the Galapagos. Prolonging the period during which the mother provides nourishment for the offspring might enable the young to reach a stage of development where its swimming skills and agility would render it better able to deal with shark attack. We know little of the social behaviour of otariids at sea, but it is possible that if the young, having taken to the sea, tend to remain with their mothers, the resulting groups of adults and juveniles might be able to avoid sharks more successfully.

This hypothesis is, of course, speculative. The extent of shark predation on otariids is all but unknown. *Artocephalus australis* at the northern limit of its range in Peru, where sharks are not known to have any effect on the seals, sometimes appear to suckle their young for as long as the Galapagos fur seals (Trillmich &

Majluf, 1981). Extended suckling in otariids is clearly a field demanding further research.

ACKNOWLEDGEMENTS

Professor R. J. Berry suggested I prepare this paper to fill a gap in the programme of the symposium on 'Evolution in the Galapagos Islands' held at the Linnean Society. Dr Fritz Trillmich unstintingly shared his wide experiences of Galapagos seals with me and gave me much help. Sheila Anderson and Dr Richard Laws read the manuscript and made helpful comments. To all these people I am very grateful.

REFERENCES

BARLOW, G. W., 1972. A paternal role for bulls of the Galapagos Islands sea lion. *Evolution, 26:* 307–310.

BARTHOLOMEW, G. A., 1970. A model for the evolution of pinniped polygymy. *Evolution, 24:* 546–559.

BONNER, W. N., 1982. *Seals and Man—a Study of Interactions.* Seattle: Washington University Press.

BONNER, W. N., In press. Lactation strategies in pinnipeds—problems for a marine mammalian group. *Symposium of the Zoological Society of London.*

EIBEL-EIBESFELDT, I., 1955. Ethologische Studien am Galapagos—Seelöwen, *Zalophus wollebaeki* Sivertsen. *Zeitschrift für Tierpsychology, 12:* 286–303.

GASKIN, D. E., 1976. The evolution, zoogeography and ecology of Cetacea. *Oceanographical Marine Biology Annual Review, 14:* 247–346.

HELLER, E., 1904. Mammals of the Galapagos Archipelago, exclusive of the Cetacea. *Proceedings of the Californian Academy of Science,* Ser. 3, *3:* 233–250.

KING, J. E., 1954. The otariid seals of the Pacific coast of America. *Bulletin of the British Museum (Natural History) Zoology, 2:* 311–337.

LIPPS, J. H. & MITCHELL, E., 1976. Trophic model for the adaptive radiation and extinction of pelagic marine mammals. *Palaeobiology, 2:* 147–155.

MILLER, E. H., 1974. A paternal role in Galapagos sea lions? *Evolution, 28:* 473–476.

MITCHELL, E., 1966. The Miocene pinniped *Allodesmus. University of California Publications in Geological Sciences, 61:* 1–105.

NELSON, B., 1968. *Galapagos—Islands of Birds.* London: Longmans.

ODELL, D. K., 1975. Breeding biology of the California sea lion, *Zalophus californianus. Rapport et procès-verbaux des réunions. Conseil permanent internationale pour l'exploration de la mer, 169:* 374–378.

PETERSON, R. S. & BARTHOLOMEW, G. A., 1967. The natural history and behavior of the California sea lion. Special Publication *1. American Society of Mammalogists.*

REPENNING, C. A., 1975. Otarioid evolution. *Rapport et procès-verbaux des réunions. Conseil permanent internationale pour l'exploration de la mer, 169:* 27–33.

REPENNING, C. A., PETERSON, R. S. & HUBBS, C. L., 1971. Contributions to the systematics of the southern fur seals, with particular reference to the Juan Fernandez and Guadalupe species. In W. H. Burt (Ed.), *Antarctic Pinnipedia* (Antarctic Research Series 18): 1–34. Am. Geophys. union.

SCHEFFER, V. B., 1958. *Seals, Sea Lions and Walruses. A Review of the Pinnipedia.* Stamford: Stamford University Press.

SIVERTSEN, E., 1953. A new species of sea lion, *Zalophus wollebaeki,* from the Galapagos Islands. *Kongelige Norske videnskabernes forhandlinger B, 26:* 1–3.

SIVERTSEN, E., 1954. A survey of the eared seals (Family Otariidae) with remarks on the Antarctic seals collected by M/K 'Norvegia' in 1928–1929. *Det Kgl. Norske Videnskaps–Akad. Oslo. Sci. Res. Norweg. Antarctic Exped, 1927–1928, 36:* 1–76.

TRILLMICH, F., 1979. Galapagos sea lions and fur seals. *Noticias de Galapagos, 29:* 8–14.

TRILLMICH, F., in press. The natural history of the Galapagos fur seal (*Arctocephalus galapagoensis,* Heller 1904). In *Key Environments*—Galapagos Islands. Pergamon Press.

TRILLMICH, F. Shark-mobbing in Galapagos sea lions: self-defence or paternal behaviour? (unpubl.)

TRILLMICH, F. & MAJLUF, P. 1981. First observations on colony structure, behavior and vocal repertoire of the South American fur seal (*Arctocephalus australis* Zimmermann, 1783) in Peru. *Zeitschrift für Säugetierkunde, 46:* 310–322.

TRILLMICH, F. & MOHREN, W. 1981. Effects of the lunar cycle on the Galapagos fur seal, *Arctocephalus galapagoensis. Oecologia, 48:* 85–92.

TRILLMICH, F. & TRILLMICH, K. G. K., 1983. The mating systems of Pinnipeds and marine iguanas: Convergent evolution of polygyny. *Biological Journal of the Linnean Society, 21:* 209–216.

WELLINGTON, G. M. & DE VRIES, T. 1976. The South American sea lion, *Otaria byronia,* in the Galapagos Islands. *Journal of Mammalogy, 57:* 166–167.

Biological Journal of the Linnean Society (1984) *21:* 185–207. With 11 figures

Variation among populations of Galapagos land iguanas *(Conolophus):* contrasts of phylogeny and ecology

HOWARD L. SNELL, HEIDI M. SNELL AND
C. RICHARD TRACY

*Department of Zoology and Entomology, Colorado State University,
Fort Collins, Colorado 80523*

A phylogenetic scheme derived via multivariate analyses of adaptively neutral scale characteristics is compared to patterns of ecological adaptation in body size and shape, hatchling size, clutch size, and reproductive seasonality, in extant populations of Galapagos land iguanas (genus *Conolophus*). Three groups of land iguana populations are identified, the oldest being the population of Isla Santa Fé, followed by the populations of the central islands (Santa Cruz, Plaza Sur and Baltra), the youngest populations are those of the western islands (Fernandina and Isabela). Patterns of ecological similarity among these populations are not concordant with phylogenetic lineage. Populations most similar in ecological characteristics are often phylogenetically divergent. Adaptation to local conditions by iguana populations is apparently more important than phylogenetic constraint in explaining variation in ecological characteristics. The assumption that phylogenetically closely-related organisms are also ecologically more similar than less closely-related organisms is not supported by this evidence. Some previous studies may have been misled by using ecological characteristics to derive phylogenetic lineages, resulting in circular support of the assumption.

KEY WORDS:—Ecology – Galapagos – iguanas – morphology – phylogeny – reproduction – taxonomy – variation.

CONTENTS

INTRODUCTION

Interest in the evolutionary relationships of Galapagos organisms has been high since the significance of Darwin's descriptions of interisland variation in finches,

Present address: Department of Biology, Texas Christian University, Fort Worth, Texas 76129, U.S.A.

0024–4066/84/010185 + 23 $03.00/0

mockingbirds, and tortoises was realized (Darwin, 1839). Such studies have usually been systematic, concentrating on phylogenetic relationships within a particular taxon (Van Denburgh, 1914; Rick, 1963; Bowman, 1963; Dawson, 1966), but some recent work has concentrated on ecological relationships between populations of Galapagos land birds (Grant, this volume), and local variation in ecological parameters that relate to morphological variation in Galapagos tortoises (Fritts, 1983, and this volume).

Systematic relationships of Galapagos lizards were initially investigated as a result of the early expeditions to Galapagos (Gray, 1831; Bell, 1843; Heller, 1903; Van Denburgh, 1912; and Van Denburgh & Slevin, 1913). Lava lizards (*Tropidurus*), geckos (*Phyllodactylus*) and marine iguanas (*Amblyrhynchus*) have been treated in recent reviews (Eibl-Eibesfeldt, 1962; Lanza, 1973; Wright, 1983). Systematics of Galapagos land iguanas had been investigated only briefly (Gray, 1831; Bell, 1843; Rothschild & Hartert, 1899, cited in Heller, 1903) until short reviews by Heller (1903) and Van Denburgh & Slevin (1913) of the genus *Conolophus*. Karyotypic studies of Galapagos lizards in general (Paull, Williams, & Hall, 1976), and comparisons of components of land and marine iguana blood (Higgins, 1978) have been carried out in attempts to explain the origins of various Galapagos lizards.

Ecological studies of Galapagos lizards have usually dealt with aspects of single populations (Stebbins, Lowenstein & Cohen, 1967; Vagvolgyi & Vagvolgyi, 1978; Werner, 1978, 1983; Christian & Tracy, 1981, 1982, 1983), although some inter-island comparisons of ecological and behavioural characteristics in land iguanas, marine iguanas, and lava lizards have been made (Carpenter, 1966a, 1966b, 1969, 1970; Stebbins & Wilhoft, 1966).

The present study combines a phylogenetic analysis with inter-population comparisons of morphology and reproductive ecology. A common assumption of evolutionary theory is that closely-related species tend to be ecologically similar (Collier *et al.*, 1973; Pianka, 1974; Ricklefs, 1979). This apparent relationship between ecological and phylogenetic similarity is widely used in discussions of competition (Den Boer, 1980, for review). In this paper, we develop a phylogenetic scheme for extant populations of Galapagos land iguanas based on patterns in scutellation. We selected scutella characteristics which are heritable (heritabilities reach values of 0.75, pers. obs.), and which showed no apparent adaptive advantage. Variation in such characteristics provides an unbiased picture of phylogenetic affinities. In the light of the phylogenetic scheme that we create, we then examine variation in several ecological characteristics to determine whether or not similarities and differences in ecological characteristics parallel phylogenetic lineage.

Status *of* Conolophus *Populations*

Conolophus occurs (or has occurred in historic times) on the islands of Isabela, Fernandina, Santa Cruz, Plaza Sur, Santa Fé, Baltra and Santiago (= San Salvador) (Fig. 1). *Conolophus pallidus* was named as a new species from Santa Fé in 1903 (Heller, 1903). At the same time the subspecific status of *C. subcristatus pictus* (Rothschild & Hartert, 1899, cited in Heller, 1903) from Fernandina was not upheld (Heller, 1903). Subsequent synonymys (Van Denburgh & Slevin, 1913; Etheridge, 1982) have continued using the names *C. pallidus* for land

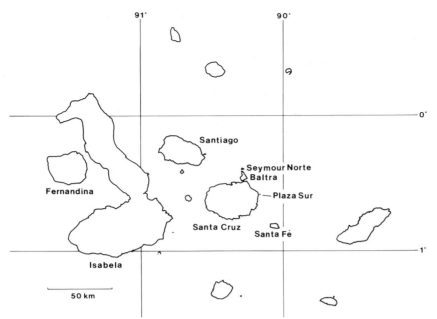

Figure 1. Map of the Galapagos Archipelago. All names are those proposed by the Ecuadorian Instituto Geográfico Militar, with the exception of Santiago whose official name is San Salvador. For detailed locations on Isabela see Fig. 2.

iguanas from Santa Fé and *C. subcristatus* for land iguanas from all other populations. The most recent synonymy (Etheridge, 1982) errs in attributing the synonymy of *C. subcristatus pictus* with *C. subscristatus* to Van Denburgh (rather than to Heller); and in omitting Plaza Sur from the range of *C. subcristatus*.

Only two populations of land iguanas, Fernandina and Plaza Sur, remain undisturbed. Others have all been variously affected by predation, competition, and/or habitat alteration from humans, feral dogs, cats, rats, pigs, goats or burros. Two populations are extinct in their natural habitat (Baltra and Santiago), although approximately 15–20 adults from Baltra remain on Seymour Norte where they were introduced in 1932 and 1933 (Banning, 1933; Honegger, 1979). There has apparently been no successful recruitment of juveniles into the adult population on Seymour Norte, but one pair of iguanas from Seymour Norte which have been placed into the Servico Parque Nacional de Galapagos (SPNG) and Charles Darwin Research Station (CDRS) breeding programme has successfully reproduced (Snell & Snell, 1981). Therefore, the only island population known to have become extinct (with no known living members) is that of Santiago. Two land iguanas found at Sullivan Bay on Santiago in late 1976, or early 1977, were thought to be native to that island (Dagmar Werner, pers. comm.), but they, and several other iguanas found on Bartolomé (a small islet at Sullivan Bay) were actually introduced from Isla Plaza Sur sometimes during the 1960s (letter from Edgar Potts in the Darwin Foundation files).

The causes of the extinction of *Conolophus* on Santiago is unknown. The presence of land iguanas on that island was described by Darwin who reported that their burrows were so dense at James Bay that it was difficult to pitch a tent (Darwin, 1839). The California Academy of Sciences expedition to Galapagos in 1905–06, seventy years after Darwin's visit, found only skeletal remains of land iguanas on Santiago (Van Denburgh & Slevin, 1913; Slevin, 1935; Slevin, in Fritts & Fritts, 1982). While Santiago has high densities of feral goats and pigs, the pigs are limited in their distribution, leaving many areas where iguanas presumably could do quite well. Even though competition for food between land iguanas and feral goats could influence the carrying capacity for iguanas, the amount of food available for iguanas in areas of the highest goat-densities appears to be no less than that in some localities on Isabela and Fernandina where iguanas survive (pers. obs.). Although they do not occur there now, feral dogs were inferred to exist on Santiago at the time of the California Academy of Sciences expedition to Galapagos (Slevin, 1935), and Heller (1903) has attributed the extinction of land iguanas on Santiago to dogs. If feral dogs were the cause of extinction of *Conolophus* on Santiago then the unanswered question becomes, what happened to the feral dogs? It is possible that the feral animals at

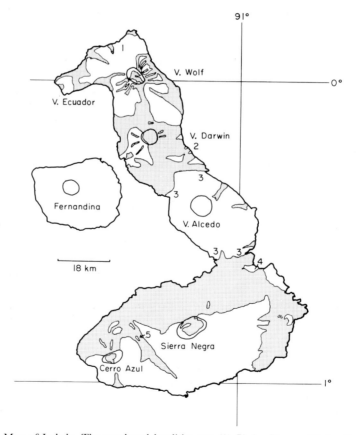

Figure 2. Map of Isabela. The numbered localities are: (1) Piedra Blanca; (2) Los Letreros; (3) Alcedo, four sites; (4) Cartago Bay; (5) Cerro Azul. The stippled areas correspond with barren lava flows, uninhabitable by land iguanas. The extent of the lava flows were determined from CDRS maps by Daniel Weber, aerial photographs in the CDRS files, and personal observation.

present on Santiago have altered the physical environment in a detrimental manner, precluding successful recruitment of a land iguana population. For example, decreased vegetation cover could lead to drier soil. The eggs of land iguanas, like those of most lizards, need to absorb water from soil, so drier soil could cause a decrease in nesting success, leading to extinction. In any event, the extinction of land iguanas on Santiago remains a mystery.

Of the disturbed populations, Santa Fé is probably the least influenced by human activity. Feral goats were eradicated from Santa Fé in 1971 (Hamann, 1978) and human predation, which used to be common, has stopped. (Van Denburgh & Slevin, 1913; Slevin, 1935; Bernardo Gutierrez, pers. comm.) Both vegetation and the iguana population appear to be recovering (Hamann, 1978; Christian, 1980). The three northern volcanoes of Isabela (Ecuador, Wolf and Darwin; Fig. 2) have populations of iguanas undisturbed except for possible predation by feral cats. Alcedo, the central volcano of Isabela, is inhabited by feral burros that may destroy iguana nests by trampling, and may compete for food used by iguanas. Iguanas appear fewer in number on Alcedo than on the other volcanoes north of Isthmus Perry.

Conolophus populations from southern Isabela are extinct except for two relics, one on the north-eastern slopes of Cerro Azul, and one at Cartago Bay. A third relic population may have been destroyed by the 1979 eruption of Volcán Chico (eastern Sierra Negra). Lava flows from that eruption extended from 10 to 20 km, possibly through an area where land iguanas had been seen by Park Wardens. Land iguanas from Santa Cruz were thought to have been exterminated by feral dogs before 1906 (Heller, 1903; Slevin, 1935), but small populations remained at Conway Bay (Snell & Snell, 1981), Cerro Colorado (William Reeder, pers. comm.) and east of Tortuga Bay (Andre DeRoy, pers. comm.). These populations were heavily attacked by feral dogs in the 1970s, and the only land iguanas from Santa Cruz known to be reproducing currently are those in the SPNG/CDRS captive and semi-captive rearing programmes.

MATERIALS AND METHODS

Fieldwork for this study was carried out from 1977 to 1982. In numerous trips during that period, the following populations were sampled: Isla Baltra (animals from Seymour Norte); Isla Plaza Sur; Isla Santa Cruz; Cerro Azul, Cartago Bay, Volcán Alcedo, Los Letreros (Volcán Darwin), and Piedra Blanca (Volcán Wolf), from Isla Isabela (Fig. 2); Isla Fernandina; and Isla Santa Fé (Fig. 1). All animals used in the analyses were alive when sampled and those not subsequently placed in the SPNG/CDRS conservation programme were released.

Meristic characteristics recorded from sampled iguanas included the following scale counts: the number of scales around the parietal scale (SP), around the mental scale (SM), around the rostral scale (SR), and around the inguinal scar (SIS); the number of supralabial (SUPRALB) and infralabial (INFRALB) scales (total of both right and left sides); the number of scales along the mid-dorsal line from (but not including) the parietal to (and including) the rostral (SLH); the number of scale rows between the suboculars and the subralabials (SO); and the number of femoral pores (FP, total of right and left rows).

Morphological measurements consisted of: snout–vent length (SVL), tail length (TL), left forelimb length (LFL), left hindlimb length (LHL), third and fourth finger lengths (3F, 4F), fourth toe length (4T), and mass. In addition, photographs and detailed notes on coloration were taken of some iguanas from each population.

Reproductive characteristics recorded included the sizes of egg clutches and hatchlings, and the seasons of oviposition and hatchling emergence (data for some populations were gathered from the literature). We studied the reproductive ecology of one population, Plaza Sur, in detail for four years from 1979 to 1982. Clutch and egg size were measured there by excavating nests within a day of oviposition. Emerging hatchlings were captured in fenced enclosures. For the land iguanas from Santa Cruz and Isabela, we collected reproductive data from the animals in the SPNG/CDRS rearing programme. As part of another study, we made comparisons between years for all three of these populations to investigate the effect of food abundance on reproductive parameters (Snell & Tracy, in prep.). Because food abundance apparently has no significant effect on the reproductive characteristics measured, we assumed that the animals in the captive programmes provide data similar to those of animals in the field.

Phylogenetic relationships of the populations sampled were assessed by a principal components analysis on the meristic data. That analysis produced factor scores for each case (individual iguana). The factor scores can be thought of as measures of position along the principal component axes, and represent derived variables that combine the effects of several different measured variables. Plots of these factor scores were then examined for groupings without an *a priori* bias for group composition. Since some insular populations appeared to group together (Fig. 3), mean factor scores for each population were entered into a cluster analysis that provided a hierarchical depiction of similarity among populations.

A common problem with the use of cluster analysis is the lack of standard tests for significance of the clustered groupings (Sneath & Sokal, 1973). To test for significant differences among the clustered groupings in our analysis, we used a stepwise discriminant function analysis in the following manner: at the first level, we tried to discriminate between all groups in the cluster analysis. If the resulting **F** matrix yielded any non-significant differences among groups ($P > 0.05$), we combined the two most similar groups (as indicated by the cluster analysis) and ran another discriminant function analysis using the new groups. This procedure was continued until we reached a point where the differences among all groups were significant ($P < 0.05$). We then concluded that those were the smallest discernible discrete groups. These analyses were all performed via BMDP statistical programs (Dixon *et al*, 1981), BMDP4M was used for the factor analysis, the cluster analysis was performed by BMDP2M, and stepwise discriminant function analysis by BMDP7M.

Ecological characteristics used in this study included morphological variation in size and shape characteristics considered to have ecological importance and which demonstrate ecotypic variation (Ballinger, 1983) and reproductive characteristics such as clutch size, hatchling size, and reproductive seasonality. Analysis of size and shape variation between populations was carried out in a manner broadly similar to that for the scutella data. The only difference was that to analyse shape in addition to size, six derived variables were added. These were: LFL/SVL, LHL/SVL, T4/SVL, and mass/SVL (allowing examination of relative LFL, LHL, T4, and mass); LFL/LHL, and T4/LHL (allowing examination of changes in relative body proportions). While statistical treatment of ratios has been criticized (Atchely *et al.*, 1976; Atchely, 1978; Atchely & Anderson, 1978), careful and intelligent use of ratios can be enlightening, especially in comparisons of shape (Corrucci, 1977; Hills, 1978; Dodson, 1978).

Statistical comparisons among populations for differences in clutch and hatchling size were analysed separately by analysis of variance and Student–Newman–Keuls tests. The analyses of variance were performed by BMDP7D (Dixon *et al.*, 1981), and the Student–Newman–Keuls tests were done by hand. The data on reproductive seasonality were analysed graphically.

RESULTS

The number of iguanas sampled from each population is shown in Table 1. Some analyses had smaller sample sizes due to data missing from a particular lizard. Results of the three separate areas of analysis are presented independently.

Phylogenetic relationships

Sexual dimorphism was not consistently present in the scale counts. In some populations, there were significant differences between mean values for males and females for a few variables (Table 2), but the differences were small and no pattern was apparent. Tests for sexual dimorphism, using the factor scores from the principal components analysis of scale counts, showed only one population to have significant differences between males and females for the first factor score (Plaza Sur; $P < 0.01$; Students t-test), and none of the populations were sexually dimorphic ($P > 0.05$) for the second factor score. Therefore, we combined data for males and females for this section of analyses.

Principal components analysis of meristic data yielded two significant principal components (eigenvalues > 1.0), which together explain 52.3% of the

Table 1. Numbers of iguanas sampled from each population. For Plaza Sur only the 320 adults were used for most analyses

Population	Number sampled
Baltra	14
Santa Cruz	24
Plaza Sur	1206 (320 adults)
Cartago Bay	24
Alcedo	7
Piedra Blanca	5
Los Letreros	20
Cerro Azul	10
Fernandina	114
Santa Fé	62
Total	1496

Table 2. Summary of comparisons between males and females within populations for mean values of scale characters. Because of small numbers of one sex or the other in some populations from Isabela, all Isabela populations were combined. Comparisons made via Student's t-test. ns—not significant at 0.05 level. See text for explanation of characters

Population	SIS	SO	SM	SR	SP	SLH	FEMPOR	SUPRALB	INFRALB
Fernandina	ns	ns	ns	ns	ns	ns	< 0.05	ns	ns
Plaza Sur	< 0.05	ns	ns	ns	ns	ns	< 0.05	ns	ns
Baltra	ns	ns	ns	ns	ns	ns	ns	ns	ns
Santa Cruz	ns	ns	ns	ns	ns	ns	ns	ns	ns
Santa Fé	ns	ns	ns	ns	ns	< 0.05	ns	< 0.05	ns
Isabela	ns	ns	ns	ns	ns	ns	<0.05	< 0.05	ns

total variance. The sources of the principal components (factors) are not readily discernible by inspection of the variables associated with them (Table 3). For example, the first factor (explaining 39.7% of the total variance) is principally associated with SUPRALB, SM, SR, and INFRALB. The second (12.6% of the total variance) is highly associated with SP, SLH, and SIS. Both principal components are associated primarily with scale characteristics of the head, but not exclusively (note that SLH loads relatively high with both factors). Such an apparently random pattern is not surprising for characteristics assumed to have little adaptive value and which probably vary more through genetic drift than in response to selection–pressure differences between the sampled populations.

Some population groups are discernible in a scatter plot of the two factor scores (Fig. 3). However, there is some confusion among the populations from

Table 3. Factor loadings (after varimax rotation) of variables on the two significant factors from the principal components analysis of the scale data. Variables with high positive loadings are highly correlated with the factor in a positive manner. Variables with high (numerically) negative loadings are highly correlated with the factor in a negative manner. Variables with intermediate factor loadings are less correlated with the factor. See text for explanation of variables

Variable	Factor 1	Factor 2
SIS	−0.195	−0.554
SO	0.629	0.256
SM	0.757	0.063
SR	0.713	0.292
SP	−0.085	0.844
SLH	0.612	0.536
FEMPOR	0.417	0.058
SUPRALB	0.824	0.176
INFRALB	0.666	−0.101

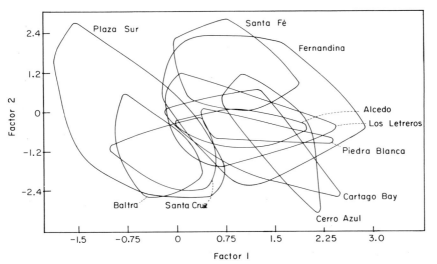

Figure 3. Scatter plot of factor scores of the two significant factors from the principal components analysis of the scale data. Lines surround the outermost points from each locality, the points were not plotted for clarity. Three groups are present: Plaza Sur, Baltra, and Santa Cruz; all Isabela populations; and Santa Fé. Fernandina cannot be distinguished from Isabela or Santa Fé via this graph.

Isabela and Fernandina. Cluster analysis and subsequent discriminant function analysis revealed seven statistically distinct populations and groups of populations derived from the ten populations sampled (Fig. 4). Distinct populations are Fernandina, Cerro Azul, Northern Isabela (made up of Los

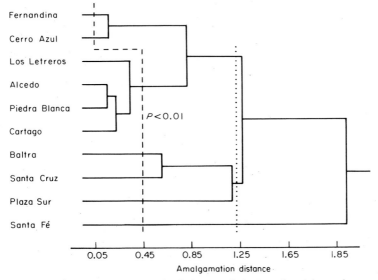

Figure 4. Phenogram of the extant *Conolophus* populations produced by a cluster analysis of the population means of the factor scores from the scale data. The dashed line crosses populations, or groups of populations, separable by stepwise discriminant function analysis (performed on the scale variables, rather than the factor scores to increase resolution) at the indicated level of significance. The dotted line crosses the three groups of populations into which the stepwise discriminant function can classify individual cases with 95% accuracy.

Letreros, Alcedo, Piedra Blanca, and Cartago Bay), Baltra, Santa Cruz, Plaza Sur, and Santa Fé. The sampled populations within Northern Isabela are not statistically distinct $(P > 0.05)$, and will be treated as a single group in subsequent discussion. It is not surprising to see that Cerro Azul is distinct from the rest of Isabela as it is certainly the most isolated population on Isabela. It is interesting that Cerro Azul and Fernandina cluster together rather than Cerro Azul and Northern Isabela. From a strictly geographical standpoint Cerro Azul and Fernandina appear more isolated from one another than either is from Northern Isabela (Fig. 2).

The pattern of divergence of the populations suggests an early separation of the population on Santa Fé. Then the central islands (Plaza Sur, Santa Cruz, and Baltra) split from the western islands (Fernandina and Isabela). Plaza Sur then diverges from the central islands, followed by the western islands forming two groups (Fernandina/Cerro Azul and Northern Isabela). Finally, Baltra and Santa Cruz diverge, and the last split is between Fernandina and Cerro Azul. The dendrogram in Fig. 4 serves as a summary of phylogenetic relatedness for later comparisons of variation in ecological characteristics.

Morphological analysis

In all populations, mean measures for males are significantly larger than for females for all variables $(P < 0.05;$ Students t-test).

Males and females were entered separately into the cluster analyses, but not the principal components analysis, since the principal components analysis simply describes the variation present in the data without concern for actual or potential group membership and cannot be confused by sexual dimorphism. The cluster analysis could be misleading if males and females were lumped together when sexual dimorphism is present, because apparent differences among populations then could be due to different proportions of males and females in the samples, or to differing degrees of sexual dimorphism. For the same reasons, ontogenetic variation was also a problem in the derived variables used to investigate shape. For example, several of the derived variables vary depending upon the age and size of the iguana (Fig. 5). To avoid the confounding effects of ontogenetic variation, we used only adult animals in this analysis. The minimum SVL of a reproducing female on Plaza Sur (out of more than 120 females observed nesting) was approximately 33 cm. Because iguanas from Plaza Sur are the smallest in the archipelago, and females are smaller than males, 33 cm was assumed to represent the minimum adult size, and no iguanas smaller than that were used. For two of the populations, Piedra Blanca and Alcedo, we had insufficient data to analyse size and shape.

The principal components analysis of the morphological data identified three significant principal components (eigenvalues > 1.0), explaining 90.4% of the total variance. These factors (principal components), in contrast with those of the phylogenetic data, are identifiable from an examination of the associated variables (Table 4). Variables associated highly with the first factor (explaining 51.4% of the total variance) are LHL, SVL, mass, mass/SVL, and T4, representing all of the measurements and one derived variable, and all reflect information on animal size. The second factor (explaining 23.7% of the total

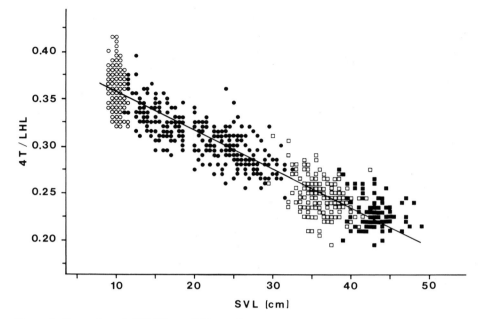

Figure 5. Regression of 4T/LHL on SVL on demonstrate ontogenetic variation in one shape variable. 4T/LHL = -0.00404 (SVL) $+0.397$; $r = 0.9468$; $N = 1206$; $P \ll 0.0001$. Open circles are hatchlings, closed circles are juveniles, open squares are adult females, and closed squares are adult males. No effort has been made to identify multiple points. All iguanas represented by this graph are from Plaza Sur.

variance) is associated primarily with LFL/SVL; LHL/SVL and T4/SVL. These are all of the linear measurements relative to body length, and thus reflect differences in shape. The third factor (explaining 15.3% of the total variance) is most highly associated with T4/LHL, and to a lower degree with T4/SVL and LFL/LHL (negatively associated). This principal component is also a shape

Table 4. Factor loadings (after varimax rotation) of variables on the three significant factors from the principal components analysis of the size and shape data. See Table 3 for explanation of factor loadings, and the text for explanation of the variables

Variable	Factor 1	Factor 2	Factor 3
SVL	0.972	-0.166	-0.078
LFL	0.939	0.246	-0.197
LHL	0.976	0.171	-0.039
T4	0.899	0.236	0.318
MASS	0.971	-0.076	-0.058
LFL/SVL	0.047	0.963	-0.263
LHL/SVL	0.161	0.897	0.119
T4/SVL	-0.087	0.711	0.691
LFL/LHL	-0.186	0.295	-0.670
T4/LHL	-0.260	0.131	0.819
MASS/SVL	0.963	-0.032	-0.052

factor. In general, high factor scores for principal component 1 indicate large body size; high factor scores for principal component 2 indicate long limbs and digits relative to body length; and high factor scores for principal component 3 indicate long digits relative to limb length, long digits relative to body length, and short forelimbs relative to hindlimbs.

Cluster analyses of measures of size and shape were performed separately for males and females, as were the associated discriminant function analyses. With females there are four groups that vary significantly in terms of size and shape (Fig. 6). These groups include animals from Baltra, Plaza Sur, Fernandina, and a group composed of the remaining populations: Cerro Azul, Cartago Bay, Santa Fé, Santa Cruz, and Los Letreros. For the males there are six groups (Fig. 7); Baltra, Fernandina/Cerro Azul, Santa Fé/Cartago Bay, Santa Cruz, Los Letreros and Plaza Sur.

The overall similarities of these groups can be seen from Figs 6 & 7. A better understanding of the comparative positions of the groups in terms of the three principal components dealing with variation in size and shape can be obtained from tri-coordinate plots of the group means (Figs 8, 9). This allows us to characterize the female populations as follows: females from Baltra are the largest of the four distinct populations, they have relatively shorter limbs than the others, but longer digits relative to limb length. Females from Fernandina are intermediate in size, they have relatively shorter limbs than all populations except the population from Baltra, and longer digits relative to limb length than all populations but the population from Baltra. The composite group ranks second in size, has the relatively longest limbs, but ranks third in digit length relative to limb length. Females from Plaza Sur are the smallest of the four populations; they have the second-longest limbs relative to body length, and they have the shortest digits relative to limb lengths.

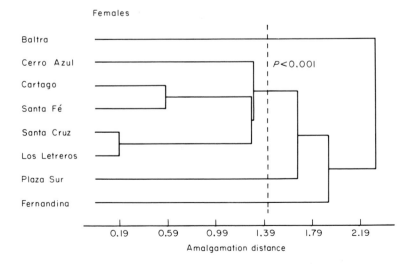

Figure 6. Dendrogram of size and shape similarity of female populations produced by a cluster analysis of the population means of the factor scores from size and shape data. The dotted line crosses populations, or groups of populations, distinguishable by stepwise discriminant analysis (performed on the factor scores) at the indicated level of significance.

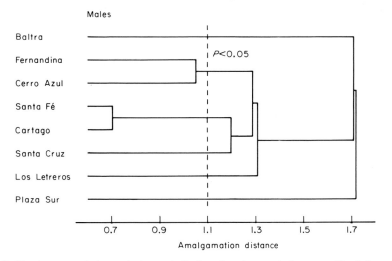

Figure 7. Dendrogram of size and shape similarity of male populations, see Fig. 6 legend for explanations.

Among males, those from Baltra are the largest, have relative limb lengths that are intermediate, and rank second in digit length relative to limb length. Males from Fernandina/Cerro Azul are intermediate in size, have relatively short limbs, and have digit lengths relative to limb lengths that are intermediate. Males from Cartago Bay/Santa Fé are intermediate in size, rank

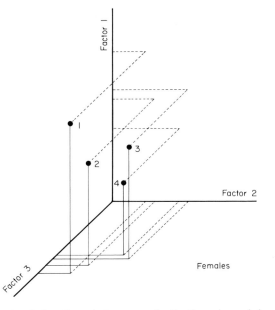

Figure 8. Three-dimensional plot of population means for the three size and shape factor scores of the four distinguishable female populations. The populations are (1) Baltra; (2) Fernandina; (3) Santa Fé, Santa Cruz, Cerro Azul, Cartago Bay, and Los Letreros; (4) Plaza Sur. All axes increase away from the origin, see text for explanation of the factors. The dotted lines allow ranking of populations on all axes.

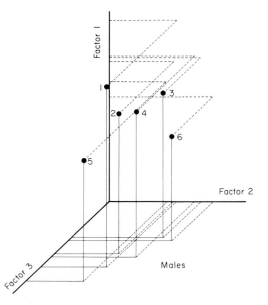

Figure 9. Three-dimensional plot of population means for the three size and shape factor scores of
the six distinguishable male populations. The populations are (1) Baltra; (2) Fernandina and Cerro
Azul; (3) Santa Fé and Cartago Bay; (4) Santa Cruz; (5) Los Letreros; (6) Plaza Sur. All axes
increase away from the origin, see text for explanation of the factors. The dotted lines allow ranking
populations on all axes.

fifth in relative limb length, and have the shortest digits relative to limb length.
The Santa Cruz males are large (second only to Baltra in terms of size) and are
intermediate in their shape characteristics. Males from Los Letreros are small
(larger only than Plaza Sur males), they have the shortest limbs relative to the
size of their bodies, and the longest digits relative to their limb lengths. Males
from Plaza Sur are the smallest, their relative limb lengths are the greatest, and
they have short digits relative to limb size.

Reproductive characteristics

We were able to determine clutch size for six populations: Plaza Sur and
Baltra (our own fieldwork); Santa Cruz and Cartago Bay (our own fieldwork
and that of SPNG Park Wardens); Santa Fé (Van Denburgh & Slevin, 1913;
Slevin, in Fritts & Fritts, 1982; Christian & Tracy, 1982); and Fernandina
(Werner, 1983). There were significant differences among the means of these
populations ($F_{5,157} = 90.10$, $P < 0.0001$), but not all populations were distinct
(Table 5). Clutch size is often considered to be a consequence of female size
(Fitch, 1970; Werner, 1983). However, even though significant, the correlation
of size of the female with size of the clutch is sometimes weak (Congdon,
Gibbons & Green, 1983). A regression of clutch size with body size of females,
across populations (Fig. 10) was not significant, even though within some
populations the regression is significant (Werner, 1983). Consequently, we
conclude that interpopulation variation in clutch size in *Conolophus* is not simply
a reflection of variation in body size.

Table 5. Means, sample sizes and results of Student–Newman–Keuls test for comparisons of clutch size among populations. ns—not significant at the 0.05 level. See text for sources of data

Population	Mean clutch size	N	Plaza Sur	Santa Fé	Baltra	Fernandina	Santa Cruz
Plaza Sur	5.9	124					
Santa Fé	9.0	5	< 0.05				
Baltra	12.0	2	< 0.05	ns			
Fernandina	13.5	15	< 0.05	< 0.05	ns		
Santa Cruz	13.7	11	< 0.05	< 0.05	ns	ns	
Cartago Bay	17.7	6	< 0.05	< 0.05	< 0.05	< 0.05	< 0.05

Hatchling sizes were more difficult to obtain and were available for only four populations in samples sufficient for analysis of variance; Plaza Sur, Baltra, Cartago Bay, and Santa Cruz. As in clutch size, there were differences among means of hatchling sizes ($F_{3,596} = 10.59$, $P < 0.0001$), but only one population is distinct (Table 6). A mean mass of 37.8 g with a range of 32.5–43.5 g ($N = 7$) has been reported for hatchlings from Fernandina (Werner, 1983). This value also appears to be distinct only from that for hatchlings from Plaza Sur. An anecdotal value is reported for Santa Fé as "about 40 g" (Christian & Tracy, 1982).

A potentially important difference in methodology exists between our determination of hatchling size from Plaza Sur and the other populations analysed. Hatchlings from Plaza Sur were sampled as they emerged from natural nests, whereas hatchlings from the other populations were sampled as they hatched from eggs incubated in the laboratory. A significant proportion of

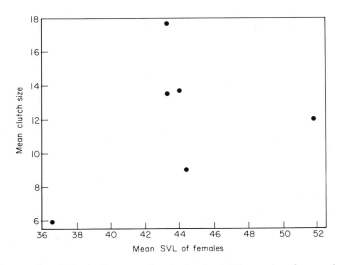

Figure 10. Scatter plot of clutch size versus female size (cm). Mean values from each population with known clutch size were used, rather than individual values, because we were unable to assign clutches to particular females in some populations. The regression is not significant ($r = 0.392$; $N = 6$; $P > 0.05$). See text for sources of data on clutch size.

Table 6. Means, sample sizes, and results of Student–Newman–Keuls test for comparisons of hatchling size among populations. ns—not significant at the 0.05 level

Population	Mean hatchling mass (g)	N	Baltra	Plaza Sur	Cartago Bay
Baltra	34.9	11			
Plaza Sur	35.6	421	ns		
Cartago Bay	37.6	125	ns	< 0.05	
Santa Cruz	37.9	43	ns	< 0.05	ns

the variance in hatchling size of land iguanas is attributable to the water relations during incubation (Snell & Tracy, in prep.). The natural nests on Plaza Sur were always drier than the conditions for incubation of eggs from the other populations. In addition, land iguana hatchlings in natural nests remain below the surface for several days before emerging, and decline in mass during this time (Snell & Tracy, in prep.). Both of these factors could contribute to making the hatchlings from Plaza Sur appear smaller than hatchlings from other populations. While the effects of remaining in the nest would also be true for the Fernandina and Santa Fé samples, we know nothing of the hydric conditions on those islands. We suggest, however, that hatchling size varies among populations so little (if at all) that the variation has little biological significance.

Data pertaining to the seasonality of reproduction were gathered for seven populations (Fig. 11). The periods for Santa Fé and Fernandina have been reported previously (Christian & Tracy, 1982; Werner, 1982, 1983; Tui DeRoy-Moore, pers. comm.). There are generally two different seasonal patterns. One concentrates oviposition January to March, resulting in hatching in April to June. The other concentrates oviposition in September to November with subsequent hatching in January to March (approximately the same time as the oviposition period of the other populations). Reproduction among iguanas from Fernandina is slightly earlier than that for other populations following the second pattern, with oviposition on Fernandina concentrating in July and resultant hatching from October to December. Whether or not Fernandina is interpreted as a third pattern, or simply as a variant of the second pattern, is not important for our discussion.

Comparisons of patterns of ecological adaptation to phylogenetic relationships

Patterns of morphological similarity do not follow the phylogenetic dendrogram established for Galapagos land iguanas. Santa Fé is phylogenetically the most divergent population (Fig. 4), but females from Santa Fé are morphologically indistinguishable (based upon the measurements made in this study) from those of Cerro Azul, Cartago Bay, Santa Cruz, and Los Letreros (Fig. 6). Males from Santa Fé are morphologically indistinguishable from those of Cartago Bay, and most similar to males from Santa Cruz, and then to males from Fernandina/Cerro Azul. Females from Fernandina and Cerro Azul are analytically the most phylogentically similar of the distinct

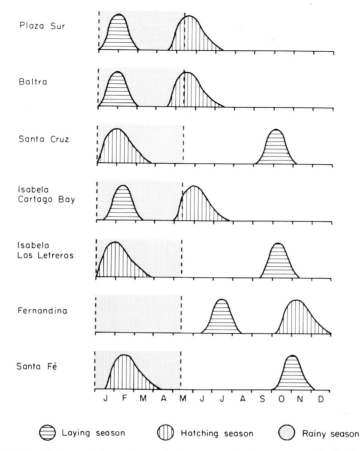

Figure 11. Graphical comparison of reproductive seasons. The curves are idealized in that the amplitudes and the temporal ranges are all equal. The peaks are based on various sources of data (see text). There is no evidence for multiple broods in any *Conolophus* population.

populations, but they are morphologically quite dissimilar. Individuals from Santa Cruz and Baltra are also apparently closely related, but morphologically divergent (both males and females). Comparison of Figs 6 & 7 with Fig. 4 provide many such examples.

Patterns of similarity in clutch sizes are also at variance with the derived phylogenetic relationships. The three populations from the eastern islands (Plaza Sur, Santa Cruz, and Baltra) form a closely-related phylogenetic group. However, their respective clutch sizes are all statistically distinct from one another, but Baltra's is not significantly different from Santa Fé's and Santa Cruz's is not significantly different from Fernandina's (Table 5). Differences between Baltra and Santa Fé clutch sizes may not be discernible due to small sample sizes, but even if they were significantly distinguishable, clutch sizes from Baltra would still be more similar to those from Santa Fé and Fernandina than to those of Plaza (although possibly not different from those of Santa Cruz). The same lack of agreement between phylogeny and clutch size is apparent among populations from Fernandina and Cartago Bay. Of the populations sampled for clutch size, the population from Cartago Bay is phylogenetically most closely-related to the population from Fernandina. However, clutch sizes from Fernandina are more similar to those from Santa Cruz and from Baltra than to those from Cartago Bay.

Hatchling size shows little, if any, variation among populations, so it cannot be related to the phylogenetic scheme.

Variation among populations in the seasonality of reproduction is not great. There are only two (possibly three) seasonal patterns used by land iguanas. However, variations in seasonal patterns of reproduction are not what would be predicted by phylogeny. Early oviposition (January to March) is found in Plaza Sur, Baltra and Cartago Bay, and late oviposition (July on Fernandina; September to November for the others) occurs on Fernandina, Los Letreros, and Santa Fé. Two of these populations are phylogenetically indistinguishable (Los Letreros and Cartago Bay), yet they have different reproductive seasons. The population on Santa Cruz reproduces at the same time as those from Santa Fé, Los Letreros, and Fernandina, to which it is only distantly related; but at a different time from populations on Plaza Sur and Baltra, to which it is closely related.

Comparisons of both morphology (size and shape) and reproductive characteristics (clutch size, hatchling size, and reproductive seasonality) with phylogeny show little agreement. While a few closely-related populations are also similar in morphology or reproduction, the majority are more similar to populations which show little phylogenetic relatedness.

DISCUSSION

Variation in the scale characteristics we examined seemed to result from genetic drift rather than natural selection. Such is not always the case. Variation in similar characters among populations of Andean lizards *(Stenocercus)* has been correlated with variation in ecological parameters of the habitat (Fritt, 1974), and variation in subdigital lamallae of at least one genus of gecko has been proposed as adaptive (Hecht, 1952). In the case of *Stenocercus* at least some of the ecological variables measured may have also been correlated with geographical isolation (river-drainage, latitude, and altitude). In addition, the range of habitats occupied by *Stenocercus* is much greater than that occupied by *Conolophus* and consequently the range of selection pressures may be great enough to act differentially on scale characteristics. The adaptive value of subdigital lamallae is obvious given the locomotory habits of geckos. Hence, it is essential to determine the adaptive potential of a character before assuming that it is selectively neutral and suitable to serve as a phylogenetic indicator.

The phylogenetic scheme derived here makes biogeographical sense. The most closely-related populations are the most geographically proximate (with the single exception of Fernandina and Cerro Azul). During periods of glacial maxima in the Pleistocene, the degree of isolation of these populations was considerably different. While estimates of the magnitude of change in the sea level in the Galapagos region vary, even the most conservative estimates agree that the central islands (Plaza Sur, Santa Cruz, and Baltra) formed a single island (Flint, 1971; Muller, 1973; Simpson, 1974, 1975). With a decrease in sea level of 100 m Fernandina and Isabela would also be joined, and with a decrease between 100 m and 150 m the central islands could be joined to the western islands, only Santa Fé remaining isolated. A sea level approximately 200 m lower than present would be necessary to join Santa Fé with the others (Muller, 1973).

Galapagos land iguanas make weak attempts to swim when placed in the water (Todd Gleason, Carl Angermeyer, pers. comm.), but they could probably float passively between islands (they must have colonized Galapagos originally by floating from a source population). Land iguanas are rarely found in the immediate vicinity of the coast (except on the smallest of islands like Plaza Sur) and such dispersal must be extremely infrequent. Evidence from natural situations supports this view. Land iguanas did not naturally occur on Seymour Norte which is approximately 1 km north of Baltra (Banning, 1933), nor do they naturally occur on Plaza Norte which is 200 m north of Plaza Sur. If land iguanas were adept at marine dispersal we would expect them to have colonized islands such as Seymour Norte and Plaza Norte. While there may be ecological reasons that *Conolophus* propagules could not become successfully established on these islands, the short distance separating Plaza Norte and Plaza Sur should allow adults to reach Plaza Norte if 200 m of water is not a barrier. Nevertheless, the only land iguanas ever seen on Plaza Norte were introduced there by local fishermen (Bernardo Gutierrez, pers. comm.).

The probable mode of dispersal among populations of the central islands and among populations of the western islands was overland during periods of glacial maxima in the Pleistocene. Dispersal between the central and western islands may have been terrestrial at the same time, or if sea levels were not low enough to provide a land connection between islands, dispersal may have occurred *via* passive floating across narrower stretches of water than currently exist. Dispersal between Santa Fé and all other islands populated by land iguanas must certainly have been by passive floating because there was probably always a long stretch of ocean separating this island from the others.

The close association of the population on Fernandina with that on Cerro Azul does not fit the biogeographical scheme outlined above. We would expect all Isabela populations to cluster together, or if one population was to cluster with Fernandina, we would have expected it to be a population from Volcáns Wolf or Darwin, areas much closer to the Pleistocene junction between Fernandina and Isabela (Simpson, 1974). Rather than attempt to explain the association between populations on Cerro Azul and Fernandina, we would like to see a comparison of iguanas from Cerro Azul and the other populations of the western islands based on larger sample sizes than we were able to gather.

Estimates of the age of the Galapagos islands vary considerably. Some studies propose that the archipelago is only 1.5 million years old, while others show definitively that some of the islands have been subaerial for at least 3 million years (Hall, 1983; Cox, 1983, cited in Wright, 1983). Age estimates for individual islands are available (Cox, 1983, cited in Wright, 1983). Among islands with land iguana populations, Santa Cruz is the oldest at approximately 3 millions years (My) (Plaza Sur and Baltra are probably similar in age, but this is not certain), followed by Santa Fé at approximately 2.7 My, Isabela at approximately 1 My and Fernandina at approximately 0.7 My. Thus, the land iguana populations of the western islands are the youngest. The initial colonization of the Galapagos by land iguanas most likely occurred on Santa

Cruz or Santa Fé. Soon after, either Santa Fé was colonized from Santa Cruz or the converse (Fig. 4). Later Isabela was colonized from Santa Cruz, and the smaller radiations then took place. There is no reason to suspect multiple invasions of ancestral *Conolophus* to Galapagos, although such may have occurred with other Galapagos lizards (Wright, 1983).

The relationship between land and marine iguanas is currently in question. Many investigators have concluded that land and marine iguanas diverged after the Galapagos islands were colonized by a common ancestor (e.g. Muller, 1973; Higgins, 1978), but recent electrophoretic data implies that divergence between those two genera occurred at a time earlier than the oldest estimates of the age of the archipelago (Patton, this volume). That would obviously require both forms to invade Galapagos separately. However, during our fieldwork in the Galapagos, we encountered a hybrid land/marine iguana, the description of which will appear in a later paper (Snell *et al.*, in prep.).

The ecotypic variation in characteristics such as body shape, body size, and reproductive parameters is well documented (see review by Ballinger, 1983). In this paper, we make no attempt to identify the selection pressures producing ecotypic variation, but simply compare similarity in ecology with relatedness in phylogeny. The overall lack of concordance between ecology and phylogeny in Galapagos land iguanas is indicative of the greater importance of local adaptations over phylogenetic constraint as explanations for observed variance in ecologically influenced characteristics.

Taxonomically closely-related genera of carabid beetles have been shown to be ecologically more similar than less closely related genera (Den Boer, 1980), in apparent support of the assumption that taxonomic affinity can predict ecological similarity. However, some of the characteristics used in taxonomic treatments of this group of beetles are features of external morphology that are apparently highly selected for specialized environments (Thiele, 1977; Nagel, 1979). If the characteristics used to establish phylogenetic relationships are influenced by ecotypic selective pressures, then the supposed phylogenetic classification is actually an ecological classification.

In *Conolophus* there appear to be few parallels between ecological similarity and taxonomic affinity, thus we conclude that adaptation to local conditions by land iguana populations is relatively free of phylogenetic constraints within the genus. The assumption of ecological similarity reflecting phylogenetic similarity may be valid at higher taxonomic levels where phylogenetic constraints may be greater due to greater genetic separation between the groups being compared. However, competition between groups of organisms as phylogenetically distant as rodents and ants has been well documented (Brown & Davidson, 1977), indicating that ecological similarities can be expected to cross considerable phylogenetic boundaries.

Systematists using electrophoretic characteristics have obtained results different from systematists using gross morphological characteristics when investigating the phylogenies of the same groups of organisms, although at other times the same phylogenies emerge (Van Denburgh, 1912; Van Denburgh & Slevin, 1913; Wright, 1983). Electrophoretic characteristics have been shown to be variable depending upon the immediate environment of the organism sampled (McGovern & Tracy, 1981) and variation in gross morphological characteristics has been related to the environment during embyronic development (Fox, 1948; Osgood, 1978). Hence, both biochemical and morphological characteristics can be affected by the environment. Neither of these examples indicates that all morphological or all electrophoretic characteristics cannot yield useful information about genetic relationships, but simply that interpretation must include an awareness of potential environmental influences.

The divergence between phylogenies based on morphological and biochemical characters is not solely due to differences in the degree of genetic versus environmental determination of the characteristics, but is more likely due to the selective neutrality of the characteristics. The use of ecologically-influenced characters is often encouraged in taxonomic study (Sneath & Sokal, 1973; Platnick, 1976). When such characters are used, convergence will often be a confusing factor in establishing phylogenies. For example, if we had used clutch size as a character suitable for establishing phylogenetic relationships, we would have Santa Fé and Plaza Sur as apparently closely-related populations when actually they appear more likely to be simply convergent in this character. On the other hand, ecologically-influenced characters can provide useful information when the goal is not simply to derive a lineage, but to provide an ecological framework to explain a particular radiation (Fritts, 1974). As long as the derived phylogenetic lineage is based on neutral characters, then further examination of adaptive characteristics can be enlightening in a biological interpretation of the evolution of a group of organisms. Researchers need to be aware, however, of the types of characters used to answer different types of questions to avoid being misled by character convergence and circularity.

ACKNOWLEDGMENTS

We thank Scott Lacour, Randy and Sue Jennings, Scott Steckbauer, Barbara Best, Kati Belt, and Betina von Hagen who provided valuable assistance in fieldwork on Plaza Sur. Miguel Cifuentes and Fausto Cepeda, successive Intendentes of the Servico Parque Nacional Galapagos, granted permission and encouragement for this research, for which we thank them. The Charles Darwin Research Station provided logistic and financial support though past and present directors Craig McFarland, Hendrik Hoeck, and Friedemann Köster whose spirited help we gratefully acknowledge. Alan and Tui Moore provided hospitality, volleyball and humour during some difficult times in Galapagos. Warwick Reed and Bob and Donna Reynolds aided in fieldwork, administrative chores, and all came through in the hard spots. The senior authors wish to express our greatest appreciation to Tom and Patt Fritts whose friendship, guidance, and perseverance gave us the start we needed; and to Sylvia Harcourt whose truely unbelievable patience was often tested by the senior author. Additional financial support for this study was provided by World Wildlife Grant 1544, the Ecuadorian government, the Smithsonian Institution, and the U.S. Peace Corps.

REFERENCES

ATCHLEY, W. R., 1978. Ratios, regression intercepts, and the scaling of data. *Systematic Zoology, 27:* 78–83.
ATCHELY, W. R., GASKINS, C. T. & ANDERSON, D., 1976. Statistical properties of ratios. I. Empirical results. *Systematic Zoology, 25:* 137–148.

ATCHELY, W. R. & ANDERSON, D., 1978. Ratios and the statistical analysis of biological data. *Systematic Zoology, 27:* 71–78.

BALLINGER, R. E., 1983. Life history variations. In R. B. Huey, E. R. Pianka & T. W. Schoener (Eds), *Lizard Ecology: Studies of a Model Organism:* 241–260. Massachusetts: Harvard University Press.

BANNING, G. H., 1933. Hancock expedition to the Galápagos Islands, 1933. General report. *Bulletin of the Zoological Society of San Diego, 10:* 1–30.

BELL, T., 1843. *The Zoology of the Voyage of the H.M.S. Beagle, Part V. Reptiles.* London: Smith, Elder, and Company.

BOWMAN, R. I., 1963. Evolutionary patterns in Galápagos finches. *Occasional Papers California Academy of Sciences, 44:* 107–140.

BROWN, J. H. & DAVIDSON, D. W., 1977. Competition between seed-eating rodents and ants in desert ecosystems. *Science, 196:* 880–882.

CARPENTER, C. C., 1966a. Comparative behaviour of the Galápagos lava lizards *(Tropidurus)*. In R. I. Bowman (Ed.), *Proceedings of the Symposium on the Galápagos International Scientific Project:* 269–273. Berkeley: University of California Press.

CARPENTER, C. C., 1966b. The marine iguana of the Galápagos Islands, its behavior and ecology. *Proceedings of the California Academy of Sciences, Fourth Series, 34:* 329–376.

CARPENTER, C. C., 1969. Behavioral and ecological notes on the Galápagos land iguana. *Herpetologica, 25:* 155–164.

CARPENTER, C. C., 1970. Miscellaneous notes on Galápagos lava lizards *(Tropidurus*—Iguanidae). *Herpetologica, 26:* 377–386.

CHRISTIAN, K. A., 1980. Endangered iguanas. *Bio Science, 30:* 76.

CHRISTIAN, K. A. & TRACY, C. R., 1981. The effect of the thermal environment on the ability of hatchling Galápagos land iguanas to avoid predation during dispersal. *Oecologia, 49:* 218–223.

CHRISTIAN, K. A. & TRACY, C. R., 1982. Reproductive behavior of Galápagos land iguanas *(Conolophus pallidus)* on Isla Santa Fé, Galapagos. In G. M. Burghardt & A. S. Rand (Eds), *Iguanas of the World: Their Behavior, Ecology, and Conservation:* 366–379. New Jersey: Noyes Publications.

CHRISTIAN, K. A., TRACY, C. R. & PORTER, W. P., 1983. Seasonal shifts in body temperature and use of microhabitats by Galápagos land iguanas *(Conolophus pallidus)*. *Ecology, 64:* 463–468.

COLLIER, B. D., COX, G. W., JOHNSON, A. W. & MILLER, P. C., 1973. *Dynamic Ecology.* New Jersey: Prentice-Hall, Inc.

CONGDON, J. D., GIBBONS, J. W. & GREENE, J. L., 1983. Parental investment in the chicken turtle *(Deirochelys reticularia)*. *Ecology, 64:* 419–425.

CORRUCCI, R. S., 1977. Correlation properties of morphometric ratios. *Systematic Zoology, 26:* 211–214.

COX, A., 1983. Ages of the Galápagos Islands. In R. I. Bowman, M. Berson & A. E. Leviton (Eds), *Patterns of Evolution in Galápagos Organisms:* 11–23. San Francisco: Pacific Division, American Association for the Advancement of Science.

DARWIN, C., 1839. *The Voyage of the Beagle.* London: Colburn.

DAWSON, E. Y., 1966. Cacti in the Galápagos Islands with special reference to their relations with tortoises. In R. I. Bowman (Ed.), *Proceedings of the Symposium on the Galápagos International Scientific Project:* 204–214. Berkeley: University of California Press.

DEN BOER, P. J., 1980. Exclusion or coexistence and the taxonomic or ecological relationship between species. *Netherlands Journal of Zoology, 30:* 278–306.

DIXON, W. J., BROWN, M. B., ENGLEMAN, L., FRANE, J. W., HILL, M. A., JENNRICH, R. I. & TOPOREK, J. D., 1981. *BMDP Statistical Software 1981.* Berkeley: University of California Press.

DODSON, P., 1978. On the use of ratios in growth studies. *Systematic Zoology, 27:* 62–67.

EIBL-EIBESFELDT, I., 1962. Neue unterarten der Meerechse, *Amblyrhynchus cristatus*, nebst weiteren Angaben zur Biologie der Art. *Senckenbergiana Biologica, 43:* 177–199.

ETHERIDGE, R. E., 1982. Checklist of the iguanine and Malagasy iguanid lizards. In G. M. Burghardt & A. S. Rand (Eds), *Iguanas of the World: Their Behavior, Ecology, and Conservation:* 5–45. New Jersey: Noyes Publications.

FITCH, H. S., 1970. Reproductive cycles of lizards and snakes. *University of Kansas Museum of Natural History Miscellaneous Publication 52:* 1–247.

FLINT, R. F., 1971. *Glacial and Quarternary Geology.* New York: Wiley and Sons.

FOX, W. W., 1948. Effect of temperature on development of scutellation in the garter snake, *Thamnophis elegans astratus*. *Copeia, 1948:* 252–262.

FRITTS, T. H., 1974. A multivariate evolutionary analysis of the Andean iguanid lizards of the genus *Stenocercus*. *San Diego Society of Natural History, Memoir 7:* 1–89.

FRITTS, T. H., 1983. Morphometrics of Galápagos tortoises: evolutionary implications. In R. I. Bowman, M. Berson & A. E. Leviton (Eds), *Patterns of Evolution in Galapagos Organisms:* 107–122. San Francisco: Pacific Division, American Association for the Advancement of Science.

FRITTS, T. H. & FRITTS, P. R., 1982. Race with extinction. Herpetological fieldnotes of J. R. Slevin's journey to the Galápagos 1905–1906. *Herpetologists' League Herpetological Monograph, 1:* 1–98.

GRAY, J. E., 1831. *Zoological Miscellany.* London.

HALL, M. L., 1983. Origin of Española island and the age of terrestrial life on the Galápagos islands. *Science, 221:* 545–547.

HAMANN, O., 1978. Recovery of vegetation on Pinta and Santa Fé islands. *Noticias de Galápagos, 27:* 19–20.

HECHT, M. K., 1952. Natural selection in the lizard genus *Aristelliger. Evolution, 6:* 112–124.

HELLER, E., 1903. Papers from the Hopkins Stanford Galápagos expedition, 1898–1899. XIV. Reptiles. *Proceedings of the Washington Academy of Sciences, 5:* 39–98.

HIGGINS, P. J., 1978. The Galápagos iguanas: models of reptilian differentiation. *BioScience, 28:* 512–515.

HILLS, M., 1978. On ratios—a response to Atchely, Gaskins, and Anderson. *Systematic Zoology, 27:* 61–62.

HONEGGER, R. E., 1979. *Red Data Book. Volume 3: Amphibia and Reptilia.* Switzerland: International Union for Conservation of Nature and Natural Resources.

LANZA, B., 1973. *On some Phyllodactylus from the Galápagos Islands (Reptilia, Gekkonidae).* Firenze: Museum of Zoology, University of Firenze.

McGOVERN, M. & TRACY, C. R., 1981. Phenotypic variation in electromorphs previously considered to be genetic markers in *Microtus ochrogaster. Oecologia, 51:* 276–280.

MULLER, P., 1973. *The Dispersal Centres of Terrestrial Vertebrates in the Neotropical Realm.* The Hague: Dr W. Junk B. V., Publishers.

NAGEL, P., 1979. The classification of Carabidae. In P. J. Den Boer, H. U. Thiele & F. Weber (Eds), *On the Evolution of Behavior in Carabid Beetles:* 7–14. *Miscellaneous Papers L. H. Wageningen, 18:* 1–222.

OSGOOD, D. M., 1978. Effects of temperature on the development of meristic characters in *Natrix fasciata. Copeia, 1978:* 33–47.

PAULL, D., WILLIAMS, E. E. & HALL, W. P., 1976. Lizard karyotypes from the Galápagos Islands: chromosomes in phylogeny and evolution. *Breviora, 441:* 1–31.

PIANKA, E. R., 1974. *Evolutionary Ecology.* New York: Harper and Row, Publishers.

PLATNICK, N. I., 1976. Are monotypic genera possible? *Systematic Zoology, 25:* 198–199.

RICK, C. M., 1963. Biosystematic studies on Galapagos tomatoes. *Occasional Paper of the California Academy of Science, 44:* 59–77.

RICKLEFS, R. E., 1979. *Ecology.* New York: Chiron Press.

ROTHSCHILD, W. & HARTERT, E., 1899. A review of the ornithology of the Galápagos Islands. With notes on the Webster-Harris expedition. *Novites Zoologicae, 6:* 85–205.

SIMPSON, B. B., 1974. Glacial migrations of plants: biogeographical evidence. *Science, 185:* 698–700.

SIMPSON, B. B., 1975. Glacial climates in eastern tropical south Pacific. *Nature, 253:* 34–36.

SLEVIN, J. R., 1935. An account of the reptiles inhabiting the Galápagos Islands. *Bulletin of the New York Zoological Society, 38:* 2–24.

SNEATH, P. H. A. & SOKAL, R. R., 1973. *Numerical Taxonomy: the Principles and Practice of Numerical Classification.* San Francisco: W. H. Freeman and Company.

SNELL, H. & SNELL, H., 1981. Rare land iguana breeds in captivity. *World Wildlife Fund Monthly Report, August 1981.*

STEBBINS, R. C., LOWENSTEIN, J. M. & COHEN, N. W., 1967. A field study of the lava lizard *(Tropidurus albernarlensis)* in the Galápagos Islands, *Ecology, 48:* 839–851.

STEBBINS, R. C. & WILHOFT, W. C., 1966. Influence of the parietal eye on activity in the lizards. In R. I. Bowman (Ed.), *Proceedings of the Symposium on the Galápagos International Scientific Project:* 269–273. Berkeley: University of California Press.

THIELE, H. U., 1977. *Carabid Beetles in Their Environments: A Study on Habitat Selection by Adaptations in Physiology and Behavior.* Berlin: Springer–Verlag.

VAGVOLGYI, J. & VAGVOLGYI, A. E., 1978. Feeding habits of the Galápagos land iguana. *Copeia, 1978:* 162–163.

VAN DENBURGH, J., 1912. The geckos of the Galápagos archipelago. *Proceedings of the California Academy of Sciences, Fourth Series, 1:* 405–430.

VAN DENBURGH, J., 1914. The gigantic land tortoises of the Galápagos archipelago. *Proceedings of the California Academy of Sciences, Fourth Series, 2:* 203–374.

VAN DENBURGH, J. & SLEVIN, J. R., 1913. The Galapagoan lizards of the genus *Tropidurus;* with the notes on the iguanas of the genera *Conolophus* and *Amblyrhynchus. Proceedings of the California Academy of Sciences, Fourth Series, 2:* 133–202.

WERNER, D. I., 1978. On the biology of *Tropidurus delanonis* Baur (Iguanidae). *Zeitschrift fur Tierpsychologie, 47:* 337–395.

WERNER, D. I., 1982. Social organization and ecology of land iguanas, *Conolophus subcristatus,* on Isla Fernandina, Galápagos. In G. M. Burghardt & A. S. Rand (Eds), *Iguanas of the World: Their Behavior, Ecology, and Conservation:* 342–365. New Jersey: Noyes Publications.

WERNER, D. I., 1983. Reproduction in the iguana *Conolophus subcristatus* on Fernandina island, Galápagos: clutch size and migration costs. *American Naturalist, 121:* 757–775.

WRIGHT, J., 1983. Evolution and biogeography of the lizards of the Galápagos archipelago: evolutionary genetics of *Phyllodactylus* and *Tropidurus* populations. In R. I. Bowman, M. Berson & A. E. Leviton (Eds), *Patterns of Evolution in Galápagos Organisms:* 123–155. San Francisco: Pacific Division, American Association for the Advancement of Science.

Biological Journal of the Linnean Society (1984) *21:* 209–216.

The mating systems of pinnipeds and marine iguanas: convergent evolution of polygyny

FRITZ AND KRISZTINA G. K. TRILLMICH

Max-Planck-Institut für Verhaltensphysiologie, Abt. Wickler, D-8131 Seewiesen, W. Germany

The convergent polygynous mating systems of marine iguanas and otariid pinnipeds depend on the existence of large female aggregations. These can build up where abundant marine food resources occur around oceanic islands which harbour fewer predators than continental areas. For marine iguanas distribution of food resources appears to determine the location of colonies, while for pinnipeds habitat choice is more decisive. In marine iguanas females benefit from gregariousness through reduced predation risk and social thermoregulation. In pinnipeds, sea lions may derive thermoregulatory benefits from gregariousness, while fur seals appear to be largely non-gregarious. In both groups males defend territories in areas of high female density. Large sexual size dimorphism presumably evolved in response to strong selection for high fighting potential of males. The capability to fast for prolonged periods of territory tenure is considered a secondary benefit of large male size, but not the driving force behind its evolution. We hypothesize that marginal males, through continuous sexual harassment of females that stay outside territories, have exerted pressure towards the evolution of female gregariousness.

KEY WORDS:—Sexual selection – evolution of polygyny – female gregariousness – Marine iguana – Pinnipeds.

CONTENTS

INTRODUCTION

The main characteristics of the mating systems of marine iguanas and polygynous pinnipeds are surprisingly similar. In comparing these systems we place emphasis on comparison of marine iguanas with the two pinniped species found on Galapagos, the Galapagos fur seal (*Arctocephalus galapagoensis*) and the Galapagos sea lion (*Zalophus californianus wollebaeki*), although we are aware that these two are probably less polygynous and territorial than more

0024–4066/84/010209+08 $03.00/0

temperate pinniped species. By studying similar social systems of widely different taxonomic groups, the selective forces shaping these systems often become more obvious and problem areas pinpointed (Wickler, 1973; Bradbury & Vehrencamp, 1977). Furthermore, understanding important factors in one system may lead to the formulation of useful hypothesis for the other.

As female aggregation or gregariousness is basic to the evolution of polygyny in both groups (Bartholomew, 1970), factors moulding these traits will be considered first. In this article we distinguish between aggregation as a passive accumulation due to circumstances external to the animals, and gregariousness as an active tendency to search the proximity of others. While within aggregations the proximity of others need not be beneficial to the animals, any sufficient explanation of female gregariousness has to be based on the benefits females derive from actively approaching each other (Wittenberger, 1980).

FACTORS RESPONSIBLE FOR FEMALE AGGREGATION

Aggregations, if they are to be sustained over long periods of time, must have access to abundant food resources. Marine feeding provides pinnipeds, marine iguanas and seabirds alike with such a food resource which can be used by many animals.

Otariid pinnipeds (eared seals) with their extensive adaptations to an aquatic mode of life could spend most of their time in the water; however, they must come ashore for reproduction because the young are unable to survive in the water for some time after birth. For pinnipeds, reproduction is therefore one of the main reasons for coming ashore. This is clear for arctic and antarctic species, but is much less pronounced in the Galapagos species where even females without young haul out all year round in habitual colonies. In contrast to pinnipeds, marine iguanas are primarily terrestrial and colony sites, used by the animals throughout the year, are separated from egg-laying beaches where females aggregate only once a year. Thus, in marine iguanas reproduction cannot be the primary cause for the build-up of colonies (K. Trillmich, 1983); rather colonies seem to build up close to rich foraging grounds in the intertidal zone.

Given the necessity to come or to stay ashore, aggregation in pinnipeds and marine iguanas is furthered by the limited mobility of the animals on land, making it costly to disperse inland where no great benefits can be obtained. Furthermore, tropical pinnipeds need to stay close to the water to be able to cool down during the heat of the day (White & Odell, 1971; Gentry, 1973; F. Trillmich, in press). This leads to a linear spread of colonies along the coasts wherever suitable habitat is available.

As dense colonies of animals of limited mobility are likely to attract predators, only areas with low or no predation pressure can harbour female aggregations over extended periods. Such conditions are provided by oceanic islands devoid of mammalian predators. Due to their high mobility in water, pinnipeds can exploit a large area (which can support great numbers of animals) from one point on land; but the amount of coastline available on predation-free oceanic islands is small and this combination increases density on land.

These considerations make food and availability of suitable habitat the factors most likely to limit and therefore responsible for female aggregation. In marine iguanas, every colony has its own intertidal feeding area with very little or no overlap between colonies. Colonies on undisturbed islands are found almost everywhere where reefs and intertidal flats provide a rich feeding area. The distribution and abundance of food resources therefore seems to determine the location and size of marine iguana aggregations. Animals within an aggregation choose an area close to the foraging ground that meets their thermoregulatory requirements and this further increases density. There is no evidence that good foraging grounds would not be used because of lack of adequate land habitat nearby.

In pinnipeds feeding areas of adjacent colonies may largely overlap due to the high mobility of the animals in the water. Thus distribution of food resources is less likely to determine the location of female aggregations in pinnipeds. Feeding conditions may, however, determine the total number of animals that can successfully raise young by foraging in the vicinity of any one island. Pinnipeds show clear preference for the wind exposed side of oceanic islands and this factor alone already limits the area they can settle on land. In Galapagos, the sea lion and fur seal show very different habitat preferences. Sea lions prefer sandy beaches and appear to be using most of the ones suitably exposed to wind while fur seals need shade of caves or large boulders and are thus restricted to very rugged coasts. In the Galapagos fur seal female competition for preferred resting sites in the shade appears to limit population density on land (F. Trillmich, in press).

This brief comparison shows that food resources and habitat choice are of differing importance in the determination of the size and distribution of aggregations of marine iguanas and pinnipeds.

FACTORS RESPONSIBLE FOR FEMALE GREGARIOUSNESS

Marine iguana and sea lion females are attracted to others of their species. When in a choice situation they prefer to approach other females rather than resting on an unoccupied, but apparently equally suitable site. While feeding conditions and habitat choice provide the necessary basis for build-up of female aggregations, these ecological factors do not explain why females are actually attracted to each other, i.e. are truly gregarious.

K. Trillmich (1983) suggested two reasons for true gregariousness in marine iguanas. (Firstly, marine iguana females reduce the risk of being preyed upon by hawks (*Buteo galapagoensis*) by living in large colonies. Such clumping together reduces the risk per individual of falling victim to a predator (Hamilton, 1971). Secondly, living in dense colonies helps marine iguanas thermoregulate. It has been shown that piling reduces overnight heat loss in marine iguanas (White, 1973; Boersma, 1982). Furthermore, during the night groups of females enter holes and cracks in the lava. In these poorly circulated micro-environments temperature and humidity may be increased, further minimizing the gradient for heat and water loss (White & Lasiewski, 1971). The functional significance of keeping warm overnight presumably lies in more efficient food assimilation. These two advantages are true benefits available to marine iguanas only through gregariousness. As they also apply to marine iguana males it is not surprising to observe that males, outside the breeding season, are just as gregarious as females. We believe that these benefits were not the primary causes for build-up of colonies in marine iguanas, but led to the evolution of gregariousness only once aggregations were established for the reason discussed above.

In contrast to marine iguanas, no adequate explanation for gregariousness in pinnipeds has been suggested. Bartholomew (1970) only refers to an "inherent tendency to aggregate" in pinnipeds but offers no explanation for the phenomenon. Gentry (1975) has pointed out that sea lions and fur seals show very different patterns of gregariousness. Galapagos sea lion females returning from foraging trips push into the middle of clumps of resting sea lions. This behaviour contrasts sharply with that of fur seal females which aggressively space out as much as possible (Mattlin, 1978; Francis & Gentry, 1981; F. Trillmich, in press).

One explanation for this difference between sea lions and fur seals may lie in different thermoregulatory requirements. While at sea, constriction of peripheral blood vessels allows the blubber layer to act as a strong heat retaining insulator. On land full circulation of blood to the skin is restored. This is necessary for epidermal cell production since these cells undergo mitosis only in warm skin (Feltz & Fay, 1966). Because of their sparse fur, sea lions can dissipate large amounts of heat to cooler surroundings while resting on land. Heat loss is reduced by resting in contact with other sea lions. Since these animals spread out and lie singly when air temperatures or solar radiation increase (Gentry, 1973; own obs.) the primary function of huddling appears to be reduction of heat loss. In fur seals, however, the skin is always kept warm by the insulation provided by air trapped in the dense underfur. This makes huddling unnecessary and may even make it dangerous by preventing fur seals from unloading metabolic heat via their (exposed) flippers; this might explain their avoidance of body contact. Thus, as with marine iguanas, gregariousness in sea lions may have a thermoregulatory basis. This hypothesis cannot explain why the South African fur seal (*Arctocephalus pusillus*), which in many ways resembles sea lions, huddles although it has underfur.

Sea lion females with young pups, which are susceptible to attacks by other animals, become less gregarious and exhibit territoriality until the pups are strong enough to join the crowd (Peterson & Bartholomew, 1967). Fur seal females are also most aggressive shortly after birth of pups (Francis & Gentry, 1981) and the same has been observed for elephant seal females (Bartholomew, 1952). Thus reproduction brings pinnipeds on land, but inhibits gregarious behaviour.

Another benefit of gregariousness for sea lion, fur seal and perhaps marine iguana females alike has presumably developed as the polygynous system evolved and is referred to as the 'ecological marginal male effect'. This is discussed below.

MALE COMPETITION FOR ACCESS TO FEMALES AND THE EVOLUTION OF SEXUAL
DIMORPHISM

In pinnipeds males cannot monopolize access to females in an economical manner by defending feeding territories, since females forage singly and their high mobility in the water allows them to forage over a wide area, making the density of foraging females extremely low. In marine iguanas, ectothermy prevents males from monopolizing females by means of feeding territoriality in the intertidal or in shallow water close to the coast since they cool down rapidly when in the water (Bartholomew, 1966) and a cold marine iguana is too sluggish to fight and perhaps even to copulate.

In pinnipeds and marine iguanas the dense aggregations of more or less gregarious females therefore form the basis for male–male competition for access to females. Another necessary precondition for the evolution of a high degree of male polygyny is the complete absence of male parental care (Trivers, 1972; Emlen & Oring, 1977). The example of sea birds, which also live in densely aggregated colonies, shows that freedom from paternal care really is a necessary condition. No polygyny has evolved in seabirds which cannot raise young successfully without substantial brood care by the male. Bartholomew (1970) mentioned the absence of paternal behaviour in pinnipeds, but unfortunately did not include it in the explicit formulation of his model. The absence of paternal care has been questioned by Barlow (1972, 1974) for the Galapagos sea lion. Eibl-Eibesfeldt (1955) and Barlow (1972) each observed one instance where territorial bulls appeared to be herding pups ashore and attacked a shark. During more than seven months of observations in sea lion colonies we did not observe a single instance of this behaviour, suggesting that it is very rare. During this time, however, resting and feeding non-territorial sea lions and fur seals, some of them immatures, were observed to mob sharks. As even adult sea lions are often seen with shark-bite injuries, sharks can apparently be very dangerous even to adult animals. But a detected shark can be attacked and chased away without great risk, due to the superior manoeuvering powers of seals. We therefore prefer to interpret the observations by Barlow and Eibl-Eibesfeldt as self-protective mobbing behaviour. Consequently we accept the absence of paternal care as a general and important pre-adaptation for the evolution of polygyny in pinnipeds, just as in marine iguanas where males do not even come close to egg-laying sites and have little or no contact with hatchlings.

Given that female aggregations exist and males have no responsibilities for their offspring, selection will favour males which gain access to and copulate with the largest number of females in the aggregation. In response to this selective pressure marine iguanas and pinnipeds have evolved strikingly similar patterns of male territoriality. During the mating season males, through often violent fights, establish territories in areas of high female density excluding a large section of the male population from access to sexually receptive females. This leads to very skewed sex ratios in the breeding colonies and potentially high copulatory success of territorial males. Females are essentially free to move between territories and do so quite frequently. Therefore the term 'harem' often applied to the females found on a male's territory is inappropriate (Peterson, 1968). Rather, males defend suitable habitat which is a valuable resource for females (K. Trillmich, 1983; F. Trillmich, in press). Only the best fighting males succeed in the competition for access to females. Non-territorial males may occasionally intercept a receptive female, but their probability of copulation is extremely low in comparison to that of territorial males. Reproductive success of a male thus becomes dependent on his territorial success.

One of the most important factors promoting male territorial success is fighting potential, which largely depends upon size. As a consequence only the largest and most successful fighting males sire offspring. This situation creates a positive feedback of selection for ever larger size of males (Bartholomew, 1970). In marine iguanas territorial males are on average about 2.5 times as heavy as females (K. Trillmich, 1983) and the same may be true for Galapagos sea lions, although measurements of male weights do not exist. In the Galapagos fur seal, size dimorphism is least pronounced, with territorial males averaging about twice as heavy as females. This size dimorphism is small in comparison to the northern fur seal (*Callorhinus ursinus*) where males are about six times as heavy as females.

As size dimorphism in pinnipeds has remained relatively constant since the Miocene (Bartholomew, 1970) the loop of positive feedback of selection for increased male size must have been opened. The same probably applies to marine iguanas. The maximal size which a male marine iguana can reach may be constrained by competition for limiting food resources during the less productive warm season (K. Trillmich, 1983). In Galapagos fur seals, large territory size and the accompanying high cost of locomotion, as well as thermoregulatory constraints on highly active bulls are assumed to have decreased the sexual size dimorphism to less than that observed in the South American fur seal (*Arctocephalus australis*), from which the Galapagos species evolved (F. Trillmich, in press). The factors which have opened the positive feedback loop in Galapagos sea lions are unknown. Galapagos sea lions largely circumvented the problems of thermal stress and high locomotion cost by establishing partly aquatic territories which are patrolled by swimming back and forth in front of the beach.

Large size, with its accompanying lower weight-relative metabolism, leads to an increased capacity for prolonged fasting. By defending their territories for prolonged periods territorial males exclude even more males from the females. In fasting pinnipeds an enormous blubber layer provides a large store of energy. Marine iguanas have much smaller fat stores, but their lower metabolic rate (compared to mammals) also allows them to fast for long periods. One exceptional territorial marine iguana was observed to fast for at least 49 days. Their physiology thus provides marine iguanas with a different solution to the same problem which territorial pinnipeds face. In Bartholomew's model, fasting capability and fighting potential of males are of about equal importance in the evolution of size dimorphism and polygyny. In the light of observations on marine iguanas and Galapagos pinnipeds, this seems questionable. Most successfully copulating territorial marine iguana males fasted for only relatively short bouts and left their territories several times during the mating season to forage. The same was observed in Galapagos sea lions in low density colonies. As the breeding season of this species lasts for about 6–10 months in any given colony no male can possibly fast for such a long period. Nevertheless, the polygynous breeding system and male territoriality have not disintegrated. Rather, in low-density colonies, males leave their territories during the day, when most females are away from the rookery foraging, and return in the late afternoon shortly before the main arrival of females. In this way they manage to hold territories for up to three months. In high-density colonies territorial male Galapagos sea lions conform more to the usual pattern by staying for a shorter, continuous period. They may, however, return several times during a season to claim the same territory. Although Galapagos fur seals have a shorter breeding season, of about three months, males holding territories early in the season may come back to reclaim their territories after having fattened up at sea for 4–6 weeks.

These observations show that in marine iguanas and Galapagos sea lions continuous attendance on territory is not as important for male reproductive success as fighting potential. As K. Trillmich (1983) postulated for marine iguanas, increased fasting ability does not appear to be the driving force behind the selection for large size in males, but rather a consequence of the selection for high fighting potential.

THE GENETICAL AND ECOLOGICAL MARGINAL MALE EFFECT

Bartholomew (1970) suggested that females will choose territorial males and avoid copulation with excluded (marginal) males. In so doing they presumably increase the probability of bearing young with 'good genes' as territorial bulls are of proven (phenotypical) quality. This factor, called the (genetical) marginal male effect, should increase female gregariousness as females would benefit from gathering around territorial males. There are some theoretical problems with this idea (Williams, 1966; Maynard-Smith, 1978), but Cox & LeBoeuf (1977) have described a mechanism of female choice in northern elephant seals (*Mirounga angustirostris*) and females of marine iguanas and Galapagos fur seals were observed to try to avoid copulations with marginal males. There is, however, no indication of female choice between territorial males in marine iguanas (K. Trillmich, 1983) or Galapagos fur seals (pers. obs.).

Another marginal male effect was observed in Galapagos sea lions and fur seals which selects for increased female gregariousness via ecological effects. We therefore call it the ecological marginal male effect. In most polygynous pinniped species for which data exist, females away from territories, whether in oestrus or not, are constantly harassed by marginal males. This sexual harassment can lead to the death of a female, but even under usual circumstances frequent disturbances by marginal males create high costs for females. These costs consist of expenditure of time and energy in fending off these over-zealous males, and risk of injury and separation from pups. This contrasts with the small benefit females may derive from decreased competition with other females. Choosing an area defended by a strong territorial male protects pinniped females from copulation attempts by marginal males. This benefit will act to concentrate females into the areas claimed by territorial males. It results in selection for gregariousness in species where females otherwise do not derive direct benefits from association with each other, as suggested above for fur seals. This effect could also explain why, of two apparently equally suitable beaches, only the beach occupied by territorial males is used by females. A similar effect may be operative in marine iguanas but has not been very clearly observed.

ACKNOWLEDGEMENTS

The work on which this paper is based was done in Galapagos between 1976 and 1981 with the permission of the Galapagos National Park and the support of the Charles Darwin Research Station. We would like to thank the Park Intendentes Miguel Cifuentes and Fausto Cepeda and the directors of the Charles Darwin Station Drs. Craig MacFarland, Hendrik N. Hoeck and Friedemann Koester for their continuous friendship and help. We greatly appreciate the critique and suggestions that George A. Bartholomew, Jack Bradbury, Fred N. White and Terry Williams made to various versions of this manuscript. This study was supported by the Max-Planck Gesellschaft and is contribution no. 363 of the Charles Darwin Foundation for the Galapagos.

REFERENCES

BARLOW, G. W., 1972. A paternal role for bulls of the Galapagos islands sea lion. *Evolution, 26:* 307–308.

BARLOW, G. W., 1974. Galapagos sea lions are paternal. *Evolution, 28:* 476–478.

BARTHOLOMEW, G. A., 1952. Reproductive and social behaviour of the northern elephant seal. University of *California Publications in Zoology, 47:* 369–472.

BARTHOLOMEW, G. A., 1966. A field study of temperature relations in the Galapagos marine iguana. *Copeia, 2:* 241–250.

BARTHOLOMEW, G. A., 1970. A model for the evolution of pinniped polygyny. *Evolution, 24:* 546–559.

BOERSMA, P. D., 1982. The benefits of sleeping aggregation in marine iguanas. *Amblyrhynchus cristatus.* In G. M. Burghardt & A. S. Rand, (Eds), *Iguanas of the World:* 292–299. Park Ridge, New Jersey, U.S.A.: Noyes Publications.

BRADBURY, J. W. & VEHRENCAMP, S. L., 1977. Social organization and foraging in Emballuronid bats. III. Mating systems. *Behavioral Ecology and Sociobiology, 2:* 1–17.

COX, C. R. & LEBOEUF, B. J., 1977. Female incitation of male competition: a mechanism of mate selection. *American Naturalist 111:* 317–335.

EIBL-EIBESFELDT, I., 1955. Ethologische Studien am Galapagos-Seelowen, *Zalophus wollebaecki* Sivertsen. *Zeitschrift für Tierpsychologie, 12:* 286–303.

EMLEN, S. T. & ORING, L. W., 1977. Ecology, sexual selection, and the evolution of mating systems. *Science, 197:* 215–223.

FELTZ, E. T. & FAY, F. H., 1966. Thermal requirements in vitro of epidermal cells from seals. *Cryobiology, 3:* 261–264.

FRANCIS, J. M. & GENTRY, R. L., 1981. Interfemale aggression in the northern fur seal, *Callorhinus ursinus.* Abstract, Fourth Biennial Conference Biology of Marine Mammals, San Francisco, 1981.

GENTRY, R. L., 1973. Thermoregulatory behaviour of eared seals. *Behaviour, 46:* 73–93.

GENTRY, R. L., 1975. Comparative social behaviour of eared seals. *Rapports et Procès-Verbaux des Réunions. (Conseil permanent International pour l'Exploration de la Mer.)*, 169: 188–194.

HAMILTON, W. D., 1971. Geometry for the selfish herd. *Journal of Theoretical Biology, 31:* 295–311.

MATTLIN, R. H., 1978. *Population biology, thermoregulation and site preference of the New Zealand fur seal,* Arctocephalus forsteri *(Lesson, 1828) on the Open Bay Islands, New Zealand.* Ph.D.Thesis, University of Canterbury, U.K.

MAYNARD-SMITH, J., 1978. *The Evolution of Sex.* Cambridge: Cambridge University Press.

PETERSON, R. S., 1968. Social behaviour in pinnipeds with particular reference to the northern fur seal. In R. J. Harrison, R. C. Hubbard, R. S. Peterson, C. E. Rice & R. J. Schusterman (Eds), *The Behaviour and Physiology of Pinnipeds:* 3–53. New York: Appleton-Century-Crofts.

PETERSON, R. S. & BARTHOLOMEW, G. A., 1967. The natural history and behaviour of the California sea lion. *American Society of Mammalogist,* Spec. Publ. No. 1.

TRILLMICH, F., in press. The natural history of the Galapagos fur seal (*Arctocephalus galapagoensis,* Heller 1904). In R. Perry (Ed.), *Key Environment Series: Galapagos.* Oxford: Pergamon Press.

TRILLMICH, K. G. K., 1983. The mating system of the marine iguana (*Amblyrhynchus cristatus*). *Zeitschrift für Tierpsychologie, 63:* 141–172.

TRIVERS, R. L., 1972. Parental investment and sexual selection. In B. Campbell (Ed.), *Sexual Selection and the Descent of Man:* 139–179. Chicago: Aldine.

WHITE, F. N., 1973. Temperature and the Galapagos marine iguana. Insights into reptilian thermoregulation. *Comparative Biochemistry and Physiology, 45:* 503–513.

WHITE, F. N. & LASIEWSKI, R. C., 1971. Rattlesnake denning: theoretical considerations in winter temperatures. *Journal of Theoretical Biology, 30:* 553–557.

WHITE, F. N. & ODELL, D. K., 1971. Thermoregulatory behaviour of the northern elephant seal, *Mirounga angustirostris. Journal of Mammalogy, 52:* 758–774.

WICKLER, W., 1973. Ethological analysis of convergent adaptation. *Annales of the New York Academy of Sciences, 223:* 65–69.

WILLIAMS, G. C., 1966. *Adaptation and Natural Selection.* Princeton, N. J.: Princeton University Press.

WITTENBERGER, J. F., 1980. Group size and polygamy in social mammals. *American Naturalist, 115:* 192–222.

Biological Journal of the Linnean Society (1984) *21:* 217–227. With 4 figures

Distribution of bulimulid land snails on the northern slope of Santa Cruz Island, Galapagos

G. COPPOIS

Laboratoire de Zoologie Systématique et d'Ecologie Animale,
Université Libre de Bruxelles, 50. av. F.D. Rosevelt, 1050-Brussels, Belgium

The observations made along south–north transects on Santa Cruz Island's northern slope show that the distribution of bulimulid land snails follows a distinct zonation. Their zonation pattern is related to the well-marked plant zonation, the substrate and the important climatic gradient characterizing this area.

Live snails and shells belonging to 14 bulimulid taxa were collected along a transect (11 km long) starting from Cerro Puntudo, one of the island's highest points. Their distribution, the vegetal zonation and the rapid change in the climate observed along the transect are described.

KEY WORDS:—Mollusca – Pulmonata – Bulimidae – *Bulimulus* – *Naesiotus* – Galapagos Islands – distribution – zonation – climatic gradient.

CONTENTS

INTRODUCTION

The group of Galapagos Bulimulidae includes about 65 species of land snails, all endemic and distributed on most of the islands of the archipelago. They all belong to the same genera *Bulimulus (Naesiotus)*. Eighty-five taxa have been described since 1832 but the taxonomy of the group still needs investigation. In many cases, the extent of intraspecific variation has been ignored, making identification hazardous (revision under study).

On Santa Cruz alone, 29 taxa were identified which account for at least 24 species. A comparison of these taxa (biometry) has been presented in a previous paper and shows the wide range of morphological variation encountered in the

217

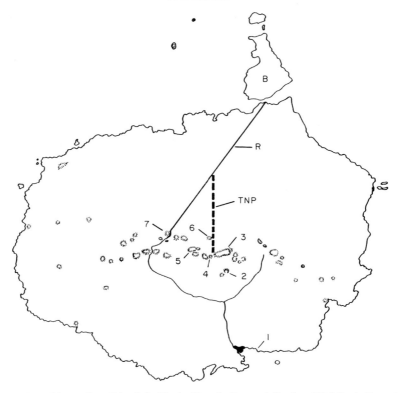

Figure 1. Map of Santa Cruz Island: 1, Charles Darwin Research Station (CDRS); 2, Cerro Media Luna; 3, Cerro Crocker, top of the island, altitude 864 m; 4, Cerro Puntudo; 5, Cerro Coralon; 6, Cerro Colorado II; 7, Cerro Maternidad; B, Baltra; R, Road; TNP, Transect line north of Cerro Puntudo.

shape of the shell within this group (Coppois & Glowacki, 1982b). These variations of shell morphology correspond to the adaptation of species to particular conditions of life. The distribution of bulimulid land snails on Santa Cruz Island is related both to the important climatic gradient (which is clearly indicated by the plant zonation) and to the nature of the substratum.

The study of south–north transects on the island shows that the distribution of Bulimulids also follows a distinct zonation, most obvious on the northern slope of the island in the vicinity of the 'summit zone'. In the south of the island distribution areas of neighbouring species are more imbricated.

MATERIAL AND METHODS

The transect displayed (Figs 1, 2) starts at the bottom of the Cerro Puntudo (altitude 730 m) and heads North, crossing well-marked vegetal zones. Its total length is 11 km; samples were taken regularly at 200 m ($\pm$ 10 m) intervals (April–July 1974). Distances were measured with a Topofil—CHAIX (lost cotton-string meter). The altitude of each station was read on a THOMMEN—Everest altimeter (precision: 10 m). At each station, samples of present bulimulid species, (most of the time empty shells) were collected (three people, 5–10 min, qualitative). Collections were also made in the upper part of the transect, using 1 m² quadrats at 100 m intervals, between 1400 m and 2600 m from Cerro Puntudo, and additional 1 m² quadrats at 500 m (9 quadrats), 850 m (5 quadrats), 1700 m (6 quadrats), 2200 m (12 quadrats), and 2475 m (7 quadrats) (July–September 1974). The densities of shells given on Fig. 4 and in the text include live snails and empty shells (average/m², see discussion).

As no weather stations were run by the Darwin Station (C.D.R.S.) on the northern slope of Santa Cruz, we installed maximum–minimum thermometers and simple rain gauges at two places along the transect; in the upper part (SN: distance 850 m, altitude 685 m); at the lower limit (Cerro Colorado II, CC: distance 2200 m. altitude 640 m, also known as Mt Cavagnaro, Smith, 1972) of the Scalesia forest, and at Media Luna (ML: altitude 620 m, southern slope, Fig. 1) (see records on Fig. 3). The rain gauges were made of 9.5 cm diameter polyethylen funnels attached to 5 or 1 litre bottles covered with aluminium paper. With such rain gauges, evaporation cannot be controlled and precipitation levels presented are probably slightly underestimated. Comparisons were made with the rain gauge of the Darwin Station at Media Luna and the results were satisfactory. One major difference was noted for February 1974—the Station's record was only 20.6 mm, and ours 185 mm. Alan Franklin (pers. comm. and 1979) recorded 112 mm at the same place from 1 to 15 February 1974, therefore our record seems to be the more accurate.

DESCRIPTION OF THE TRANSECT

Vegetal zonation

The profile of the transect and the main characteristics of the plant zonation are shown on Fig. 2. The transect follows a gentle slope crossing five well-marked vegetal zones. In the upper part of the transect the transitions from one zone of vegetation to another occur over very short distances. The most striking changes are at the upper limit of the 'Scalesia forest' and at the narrow stretch of the 'upper transition zone'. Santa Cruz Island's vegetal zonation has already been described by botanists, and although classical descriptions by Bowman (1961) and Wiggins & Porter (1971) concern vegetal zones on the southern slope of the island (see also Reeder & Riechert, 1975), they also apply to the northern slope. Nevertheless there is no 'Miconia zone' or 'Brown zone' between the 'Fern-Sedge zone' and the 'Scalesia forest'. Our observations fit quite well with the phytosociological descriptions proposed by van der Werff (1978) and we refer to them in the following brief and tentative description.

Dr Adsersen (plant ecology) established permanent quadrats of vegetation along this transect: PQ1187 (distance 2000 m), PQ1186 (distance 2600 m), PG1185 (distance 3350 m), PQ1184 (distance 6000 m), PG1183 (distance 4650 m). The descriptions of these quadrats are available (Adsersen, in prep. and C.D.R.S. reports) and, as both studies were simultaneous, give an accurate idea of the vegetal zonation at the time this transect was made. (See also Adsersen, 1976.)

The 'Fern-Sedge zone' or 'Evergreen broad leafed weedy vegetation zone' (van der Werff, 1978) is found at altitude 730–685 m, distance 0–700 m. In this 'pampa' region, *Pteridium aquilinum* and *Trisetum howellii* are the common species. The soil is acid (measured pH 4.4–6.2) and never deep, greyish to brown in the

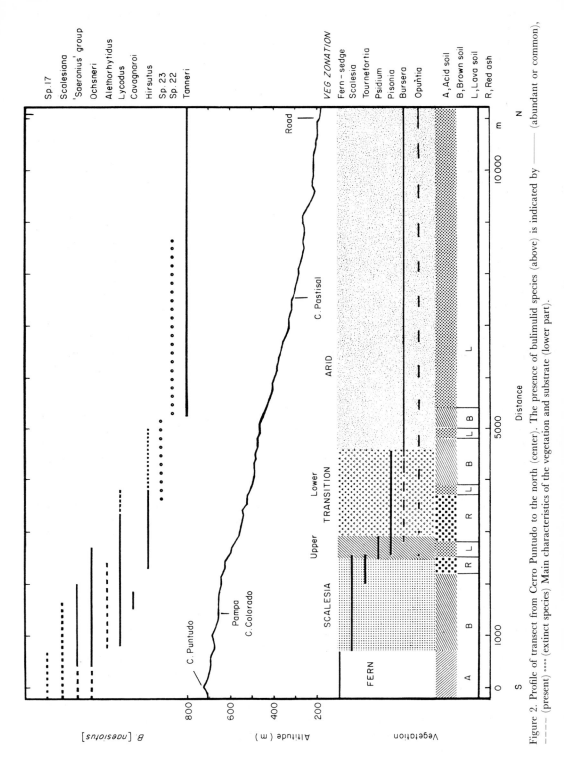

Figure 2. Profile of transect from Cerro Puntudo to the north (center). The presence of bulimulid species (above) is indicated by ——— (abundant or common), ––– (present) •••• (extinct species) Main characteristics of the vegetation and substrate (lower part).

lower part of this zone, developing on weathered pyroclastic deposits or superficial lava beds.

The 'Scalesia forest' or 'Evergreen soft wood orthophyll forest' (van der Werff, 1978) is found at altitude 685–615 m, distance 700–2500 m. The most common tree is *Scalesia pedunculata* which forms an almost continuous canopy 5–10 m above ground level. At a distance of 1400–1500 m the forest opens in a natural glade (or 'pampa' on Fig. 2, altitude 660 m). The shrub and herb layer is dense and dominated by *Psychotria rufipes*, *Tournefortia rufo-sericea*. *Alternanthera halimifolia*, *Justicia galapagana*, and numerous ferns. Epiphytic ferns, bryophytes and liverworts are abundant. The brown forest soil is deep and pH varies between 6 and 7 in the upper part of the forest, then between pH 7 and 8.5 for the lower part where soil turns to reddish brown or red ash in the area of Cerro Colorado II.

The 'Transition zone' or 'Dry season semi-deciduous forest' (van der Werff, 1978) can be divided in two distinct sub-zones. The upper part (altitude 615–580 m, distance 2500–2800 m) is mainly characterized by abundant *Psidium galapageium* trees whose branches are covered by liverworts and by *Pisonia floribunda*. The shrub layer is represented mainly by *Tournefortia pubescens* and *Ciococca alba* and herb layer by *Alternanthera halimifolia*. There are frequent basaltic outcrops and the pH of the soil is about 7.5. The first small and scarce *Opuntia echios* var. *echios* are already found at ±620 m altitude. In the lower part of this zone (altitude 580–490 m, distance 2800–4500 m) the forest is more open, the prevailing tree is *Pisonia floribunda* mixed with an increasing amount of *Bursera graveolens* downslope. At the same time one passes from red ash to lava outcrops and then to brown soil, the pH remaining, as in the arid zone, above 7.5.

The 'Arid zone' or 'Dry season deciduous forest' (van der Werff, 1978) starts at about 490–500 m altitude (distance 4500 m) and goes down to the seashore. The upper limit is not easy to define, the most abundant trees are *Bursera graveolens* and *Piscidia carthagenensis*, and the giant *Opuntia echios* var. *echios* are common but not abundant. Basalt boulders are abundant, often representing 50% or more of the soil surface. The transect is limited to a distance of 11 000 m (altitude 195 m), where the transect reaches the road to Baltra. The remaining part of the arid zone (from 195 m altitude to the seashore) was not considered, as previous collections in this area have shown that only one bulimulid species was uniformly present.

Climate

The vegetal sequence briefly described above is obviously a consequence of the climatic gradient along the transect. The upper ferns-edge zone is the wettest part, but the climate becomes rapidly drier with decreasing altitude. During the cool season (April–December) clouds usually accumulate on the southern slope of Santa Cruz, producing drizzle ('garua') and precipitation, both being much more important in the highlands than along the coast. Blown by the southern winds, clouds make their way over the island's highlands (Cerro Crocker, the top of the island is only at 864 m altitude) and saturate the atmosphere of the upper part of the northern slope with garua. Rainfall occurs eventually, but is much less than that recorded at Media Luna (620 m) on the southern slope (Fig. 3).

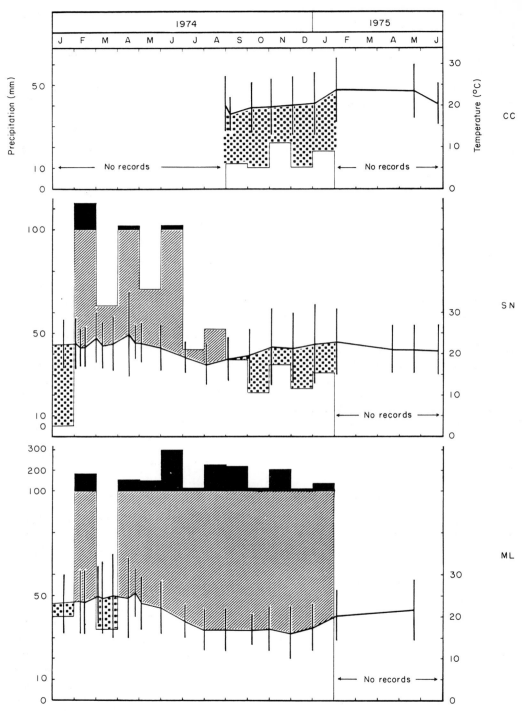

Figure 3. Climate diagrams, Santa Cruz highlands (dotted areas: dry periods; hachured areas: wet periods). CC. Cerro Colorado II, northern slope, altitude 640 m. SN. upper limit of the Scalesia forest, northern slope, altitude 685 m. ML. Media Luna, southern slope, altitude 620 m.

Only during the warm season (January–April) are there sometimes heavy showers from the north. This explains why more precipitation was recorded in February and March 1974 in the northern Scalesia forest (SN: 227 mm, and 63 mm) than in Media Luna (ML: 187 mm and 34 mm). One can see in Fig. 3 that precipitation is much more significant at Media Luna (ML) than on the northern slope (SN and CC). Apart for two short drier periods (January and March 1974), precipitation measured at Media Luna remains high throughout the year. Although records are limited to only five months at Cerro Colorado II weather station (CC) the precipitation is obviously less there than at the upper limit of the Scalesia forest (SN). For the month of September 1974 (Table 1) only 12 mm was recorded by CC rain gauge, and 37 mm for SN. During this same month, D.A. & D.B. Clark (pers. comm.) recorded 29 mm at their camp on the 'pampa' which has an intermediate position, (at 600 m from SN and 750 m from CC.) For the same period 218 mm rainfall was recorded at Media Luna.

No major differences appear on the month's average, maximum and minimum temperature curves (Table 1 and Fig. 3). During the garua season, the temperature is slightly higher on the northern slope than in Media Luna (which is usually lost in fog) (Table 1). If daily curves of temperatures were available, differences would have been noticed, e.g. between simultaneous readings at the different stations. But such microclimatic records (thermography, evaporimetry) could not be collected at the time we made the field work because of lack of equipment.

Although limited, the records presented prove the existence of an important climatic gradient, at least in a short distance of the upper area of the northern slope of Santa Cruz island. The upper zone is cooler and wetter, and when walking down the transect it is obvious that temperature and aridity increase. The lower part of the transect has a typical arid climate. No meteorological records are available from the northern arid zone of Santa Cruz, but good estimates can be made from the records taken on the nearby Baltra Island. (Alpert, 1946, 1961; Walter, Harnickell & Mueller-Dombois, 1975.)

DISTRIBUTION OF BULIMULID LAND SNAILS

The bulimulid land snails collected along the transect belong to fourteen taxa. Their distribution is shown on Fig. 2. Six of these taxa are not yet identified, and we will refer to them by code numbers as in Coppois & Glowacki (1982b,

Table 1. Meterological records for September 1974 at Media Luna and at three places along the transect north of Cerro Puntudo (Abbreviations as for Fig. 3.)

September 1974	Southern slope		Northern slope	
	ML	SN	Pampa	CC
Altitude (m)	620	685	660	640
Rainfall (mm)	218	37	29	12
Max. temperature (°C)	20.5	26.0	28.3	25.5
Month's average temperature °C)	17.0	19.5	20.25	19.5
Min. temperature (°C)	13.5	13.0	12.5	13.5

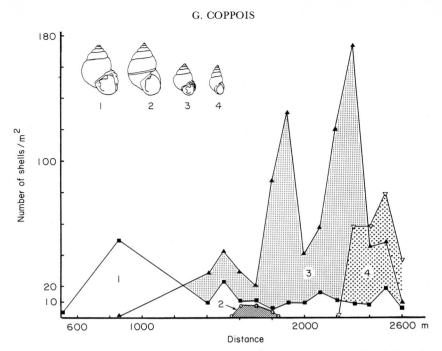

Figure 4. Abundance of shells along the upper part of the transect (number of shells by m², see text); 1, *B. (N.) ochsneri;* 2, *B. (N.) cavagnaroi;* 3, *B. (N.) lycodus;* 4, *B. (N.) hirsutus.*

taxa 14, 15, 17, 21, 22, 23). Under the name 'saeronius group' (Fig. 2) individuals of *B. (N.) saeronius* and of three other taxa (14, 15, 21) close to *B. (N.) saeronius* by their general morphology and ground dwelling habits, are gathered. Although taxon 21 is almost certainly a different species and taxa 14 and 15 could be variations of *B. (N.) saeronius*, it is wiser to wait for the result of a taxonomic study (in prep.) in order to be precise about the extent of variations for each of these taxa and to be able to give them proper names.

Bulimulus (N.) saeronius is mainly found (less than 10 shells/m^{-2}, live snails and empty shells) in the fern-sedge zone, along with *B. (N.) ochsneri, B. (N.) scalesiana*, taxon 15, and taxon 17 which seems to be restricted to that zone. In the fern-sedge zone, living specimens of *B. (N.) ochsneri* are rare but are sometimes found with a highly decalcified shell easily deformed or broken by pressure between the fingers (never observed elsewhere).

Bulimulus (N.) scalesiana, found only occasionally in the fern-sedge zone, has a wide distribution in all the humid zones on Santa Cruz island (like *B. (N.) ochsneri*). This species is regularly found in the upper part of the Scalesia forest but prefers bushes or a 'pampa' type of vegetation and is often seen on *Sida rhombifolia* bushes and *Paspalum conjugatum.*

Bulimulus (N.) saeronius is also found, like taxa 14, 15 and 21, in the upper part of the Scalesia forest, but it is much less abundant there than the other members of the 'saeronius group'. Taxon 14 is only found occasionally and taxon 15 is most abundant at a distance of 1600–1700 m on the transect. Progressively, and down to the lower limit of the Scalesia forest, taxon 15 is replaced by another litter snail, *B. (N.) alethorhytidus* which is found mainly between 2200–2400 m (upper red ash areas, Fig. 2). This species is never abundant (5.5 shells/m², average for 12 quadrats) and is occasionally found higher in the Scalesia forest.

Figure 4 shows the abundance of four of the most conspicuous species collected along the upper section of the transect: *B. (N.) ochsneri, B. (N.) cavagnaroi, B. (N.) lycodus, B. (N.) hirsutus. B. (N.) ochsneri*, already reported from the fern-sedge zone, is very abundant (51 shells/m², average for 20 quadrats at 850 m distance) in the upper part of the Scalesia forest and between the distances 700–1400 m along the transect. It lives in the forest litter and often hides in small cavities under fallen logs, roots or rhizomes. It is also seen on the vegetation of the herb and shrub layer. *Bulimulus (N.) ochsneri* is less abundant in the lower part of the Scalesia forest; only a few were found alive between the 1700 and 1800 m distance marks. Lower down, only empty shells were found: these old shells, often deprived of their periostracum, are evidence of previously wetter climates at lower altitudes on the northern slope of Santa Cruz.

Bulimulus (N.) cavagnaroi, although never abundant (7.3 shells/m²), is by its distribution probably the most surprising species found along the transect. It is restricted to a very narrow band, 300 m wide (from 1550 to 1850 m distance on the transect), extending from the north of Cerro Crocker (eastern limit, altitude 600–650 m) to Cerro Maternidad (western limit, altitude 600–640 m. Fig. 1). Three main colour variations of *B. (N.) cavagnaroi* have been described (Smith, 1972), but only the 'banded brown form' was noticed along the transect. A 'white unbanded form' is encountered (5%) in the population of Cerro Maternidad (and nowhere else); and a population of a 'brown-banded yellow form' (Van Mol, 1972) is restricted to the tiny valley of Cerro Coralon (altitude 650 m, Fig. 1, paper in prep.) at a distance of 850 m south of the narrow band of *B. (N.) cavagnaroi* crossed by the transect. *B. (N.) cavagnaroi* is a litter snail often found at the same places as *B. (N.) ochsneri*. The present observations do not allow to define their respective niches or to determine if competition occurs between these two species.

Bulimulus (N.) lycodus is abundant between 1400 m and 2500 m: up to 123 shells/m² were counted at the 2200 m sample station (average for 12 quadrats). It is rare in the upper part of the Scalesia forest. This species is arboreal, light coloured (ivory to whitish-grey) and lives exposed on the vegetation. It is often seen aestivating on the bark of *Scalesia* and *Psidium* trees. Its abundance decreases rapidly in the upper part of the transition zone.

Bulimulus (N.) hirsutus, easily identified by its shape and the minute hairs on the shell's surface, is found at the lower limit of the Scalesia forest (from the distance 2200 m and down) and replaces *B. (N.) alethorhytidus* in the forest litter of the transition zone. Old shells of *B. (N.) hirsutus* are found in the upper part of the arid zone, but recent empty shells and live snails are mostly abundant between 2300–2600 m (Fig. 4). At the 2500 m sample station, up to 80 shells were collected on 1 m² (average for seven quadrats, dead and alive).

In the arid zone (Fig. 2) only three species were observed. Two of them are extinct (taxa 22, 23, not identified). Shells of taxon 23 were found under blocks of lava, a usual hiding place for the arid zone land snail. Shells of taxon 22, white-coloured and elongated, were found dispersed on the substratum and under the lava blocks, which seems to indicate they were mainly living exposed on the vegetation (like *B. (N.) reibishi* in the southern arid and transition zones of Santa Cruz).

Bulimulus (N.) tanneri is the only species found alive and in abundance along the transect in the arid zone. This species lives hidden in cavities under the blocks of lava, often aestivating in the small alveoles of the lower surface of the basaltic boulders. It is distributed over all the arid zone on the northern slope of Santa Cruz.

The intraspecific variation is important in *B. (N.) tanneri* and modifications in the size and shape of the shell occur within short distances along the transect near the upper limit of its distribution. This confirms similar observations of variations in cline already presented in previous papers while analyzing the distribution of *B. (N.) tanneri* along a transect parallel to the present one, extending from Cerro Maternidad to the north (Coppois & Glowacki, 1982 a, c).

DISCUSSION AND CONCLUSIONS

The zonation pattern in the distribution of the bulimulid land snail species observed along the transect can be seen clearly in Figs. 2 and 4. Equivalent zonation was observed for the same species along two other transects, parallel to the one presented in this paper, when exploring other areas of Santa Cruz island's northern slope. One started from Cerro Maternidad (Coppois & Glowacki, 1982a, c), the other from Cerro Coralon.

This zonation can be described in the field as a succession of bands extending East-West and delimited in altitude, in each of which one bulimulid species predominates. This zonation is most obvious along the transect for *B. (N.) ochsneri, B. (N.) cavagnaroi, B. (N.) lycodus, B. (N.) hirsutus, B. (N.) tanneri* and sp. 22.

The conditions of life of each species are determined by the altitude and by the local climate, vegetation and substrate which change within very short distances when going down the northern slope. Most of the species present along the transect are also found on the southern slope of the island (with the exception of *B. (N.) cavagnaroi, B. (N.) tanneri*, taxa 22 and 23), but on this slope the vegetal zones are more extensive, and changes in the climatic gradient occur more regularly and are spread over longer distances. The distribution areas of bulimulid species are more patchy and their limits more diffuse and overlapping. Therefore, the zonation in the distribution of bulimulid species along the transect provides a unique opportunity to study the limiting factors of the habitat corresponding to each species. This study started in January 1974 but had to be abandoned after two months. Bulimulid land snails seemed to have suffered greatly from the very dry year 1973, and after March 1974 live snails became rare in the Scalesia forest north of Cerro Puntudo. As we have not had other opportunities to visit this area since then, we do not know if the bulimulid land snail populations recovered or if further investigations concerning them will be prevented by local extinction.

ACKNOWLEDGEMENTS

We wish to extend our thanks to: the Charles Darwin Foundation for the Galapagos Islands and especially to Drs Peter Kramer and Craig MacFarland for their constant support during our stay at the Darwin station; the authorities of the Parque Nacional Galapagos for granting us the authorizations to work in the Park and for their help and cooperation; Professors Jean Bouillon and Jean-Jacques Van Mol of the Université Libre de Bruxelles who supplied the necessary guidance; and Drs Anne and Henning Adsersen for their patient friendship and their invaluable collaboration during our many common field trips, and for providing much of the basis of the botanical information. Financial support for this research was provided by a grant from the Belgian Ministère de l'Education Nationale.

REFERENCES

ADSERSEN, H., 1976. *Ombrophytum peruvianum* (Balanophoraceae) found in the Galapagos Islands. *Botaniska Notiser, 129:* 113–117.

ALPERT, L. A., 1946. Notes on the weather and climate of Seymour Island, Galapagos Archipelago. *Bulletin of the American Meterological Society, 27:* 200–209.

ALPERT, L. A., 1961. Climate of the Galapagos Islands. *Occasional Papers of the Californian Academy of Sciences 44:* 21–44.

BOWMAN, R. I., 1961. Morphological differentiation in the Galapagos finches. *University of California Publications in Zoology 58:* 1–302.

COPPOIS, G. & GLOWACKI, C., 1982a. Factor analysis of intraspecific biometrical variations of *Bulimulus (Naesiotus) tanneri* (Pulmonata, Bulimulidae) in the Galapagos Islands. Proceedings of the Seventh International Malacological Congress *Malacologia, 22* (1–2): 495–497.

COPPOIS, G. & GLOWACKI, C., 1982b. Bulimulid land snails from the Galapagos: 1. Factor analysis of Santa Cruz Island species. *Malacologia, 23* (2): 209–219.

COPPOIS, G. & GLOWACKI, C., 1982c. Etude systématique et écologique des Bulimulidae (Gastéropodes, Pulmonés) de l'Archipel des Galapagos. Contribution n°2: Analyse factorielle des variations biométriques intraspécifiques chez *Bulimulus (Naesiotus) tanneri. Annales de la société r. Zoologique de Belgique, 112* (2): 175–195.

FRANKLIN, A. B., CLARK, D. A. & CLARK, D. B., 1979. Ecology and behaviour of the Galapagos rail *Laterallus spilonotus. Wilson Bulletin, 91* (2): 202–221.

REEDER, W. C. & RIECHERT, S. E., 1975. Vegetation change along an altitudinal gradient. Santa Cruz. Galapagos Islands. *Biotropica, 7:* 162–175.

SMITH, A. G., 1972. Three new land snails from Isla Santa Cruz (Indefatigable Island), Galapagos. *Proceedings of the Californian Academy of Science, (4) 39:* 7–24.

VAN DER WERFF, H. H., 1978. *The vegetation of the Galapagos Islands.* Doctoral dissertation. Rijksuniversiteit te Utrecht (Nederland): 1–102.

VAN MOL, J. J., 1972. Au sujet d'une nouvelle et remarquable espèce de Bulimulidae des îles Galapagos. (Mollusca, Gasteropoda, Pulmonata). *Bulletin de l'Institute des Sciences Naturelles Belgique, 48* (11): 1–7.

WALTER, H., HARNICKELL, E. & MUELLER-DOMBOIS, D., 1975. *Climate-diagram Maps of the Individual Continents and the Ecological Climatic Regions of the Earth.* Berlin. Heidelberg. New York.: Springer-Verlag. 1–36, 9 maps.

WIGGINS, I. L., & PORTER, D. M., 1971. *Flora of the Galapagos Islands.* Stanford: Stanford Univ. Press.: 1–998.

Biological Journal of the Linnean Society (1984) *21:* 229–242. With 5 figures

Changes in the native fauna of the Galápagos Islands following invasion by the little red fire ant, *Wasmannia auropunctata*

YAEL D. LUBIN

Charles Darwin Research Station, Galápagos, Ecuador

CONTENTS

INTRODUCTION

Island biotas may undergo rapid and irreversible changes in species composition and diversity as a consequence of accidental colonizations by agressive and rapidly-spreading species. Such accidental colonization incidents are by no means due solely to man's activities. Nonetheless, some of the most dramatic introductions known have been caused by man, either deliberately (rabbits to Australia; goats, pigs and cattle to Galápagos) or accidentally (rats to the Hawaiian Islands and Galápagos).

Ants are among the most ubiquitous and destructive of invading species. The invertebrate fauna of the lowland forests of the Hawaiian Islands have undergone a drastic reduction in species diversity, with concomitant extinction of many endemic species as a result of the introduction of an ant of African origin, *Pheidole megacephala* (Fabricius) (Myrmicinae) (Illingworth, 1917; Zimmerman, 1970). In Bermuda, there has been a succession of invasions of cosmopolitan 'tramp' ants (i.e., species which are readily transported by human commerce), most recently *P. megacephala* and the Argentine ant, *Iridomyrmex humilis* (Mayr) (Myrmicinae), with each new introduction changing the distributions and abundances of previous invaders and of native species (Haskins & Haskins, 1965; Crowell, 1968). Likewise, a number of widely distributed 'tramp' ants have made their way to the Galápagos Islands since the advent of permanent human settlements in the islands (and perhaps some even earlier, with the buccaneers and whalers). One of the most recent introductions

229

0024–4066/84/010229 + 14 $03.00/0

8

is that of *Wasmannia auropunctata* (Roger) (Myrmicinae), the 'little fire ant', which was brought to Santa Cruz Island (Indefatigable) sometime in the early part of this century (Silberglied, 1972; Clark *et al.*, 1982) and has since spread to other inhabited islands (see below). Thought to have originated on continental South America (Kusnezov, 1951), *Wasmannia* now occurs widely throughout the Neotropics and in parts of the Old World tropics and Pacific basin (see Clark *et al.*, 1982 and Fabrés & Brown, 1978 for reviews of the current distribution). *Wasmannia* is considered a pest species in many areas due to its painful sting (disproportionately so, considering that workers are a mere 1.2 mm long) and to its occurrence in high population densities, particularly in agricultural crops, to the exclusion of other ants (Spencer, 1941).

Clark *et al.* (1982) determined the distribution of *W. auropunctata* on Santa Cruz Island in 1975–76 and showed that in areas of high population densities, it has displaced many other species of ants. They suggested that these displacements may have come about primarily through interference competition (see also Meier, in press). I have expanded the scope of these previous studies by examining the distribution of *Wasmannia* on other islands in the archipelago, the distribution of other species of ants, and some of the faunal changes that can be attributed to the presence of *Wasmannia*.

METHODS

The Galápagos Islands need no introduction. Descriptions of the topography, climate, and vegetation zones of these primarily desert equatorial islands may be found in Darwin (1860) and Wiggins & Porter (1971). Collections of ants were

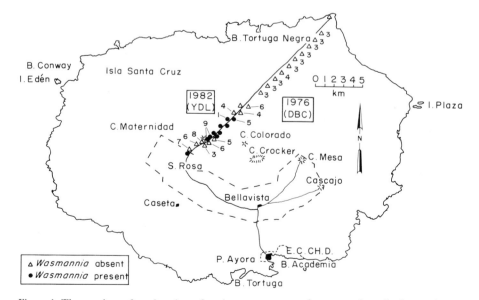

Figure 1. The numbers of species of ants found on two censuses of transects along the Puerto Ayora–Baltra Road, Santa Cruz Island, one in 1976 (Clark *et al.*, 1982) (symbols and numbers on the E side of the road), and another in 1982 (present study) (symbols and numbers on the W side of the road). Numbers beside the symbols are the numbers of other species of ants (excluding *Wasmannia*) observed at a site.

made in as many different habitats as possible, ranging from mangroves, beaches and lava on the coast, low-elevation arid zone vegetation (cacti and scrub forest), to transition and humid zone forests, agricultural areas and high elevation grasslands. Studies of the impact of *Wasmannia* on other invertebrates were carried out on two islands: Santa Cruz Island, in transition zone scrub forest and lower humid zone *Scalesia* forest (from 375–600 m elevation) along the Puerto Ayora-Baltra Road (Fig. 1), and San Salvador Island, in lower transition zone scrub forest (*Psidium-Bursera*) at Guayabillos (230 m) and denser *Psidium-Psychotria* forest at Trágica (330 m) (Fig. 2).

During 1981–82 ants were collected on the following islands (for a list of synonymies of island names, see Linsley & Usinger 1966): Pinta, Marchena, Isabela (Volcán Sierra Negra), San Salvador, San Cristóbal, Santa Cruz, Plaza Sur, Eden, Pinzón, Rábida and Chámpion. D. J. Anderson collected ants on Fernandina Island and M. Alvarez and M. Coulter on Floreana Island; these collections were kindly made available. Preliminary identifications have been made by R. R. Snelling and specimens are deposited at the Los Angeles County Museum of Natural History. A more detailed study of the Galápagos ant fauna is currently in progress.

During March–April 1982, I searched for and collected ants along nine transects at 0.5 to 1 km intervals along the Puerto Ayora–Baltra Road on the northern side of Santa Cruz Island, repeating part of a similar census conducted in 1976 (Clark *et al.*, 1982). Each transect was 100 m long by 2 m wide and divided into ten 10 m sections, and was oriented perpendicular to the north–south axis of the road. Soil samples taken from the transects sometimes yielded species of ants that were not found in the visual search.

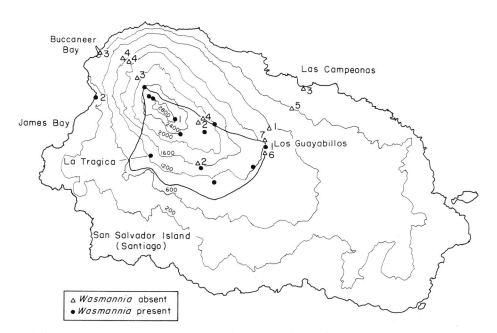

Figure 2. Map of San Salvador Island showing the approximate distribution of *W. auropunctata* (heavy line) and collection sites. Numbers beside the symbols indicate the number of other ant species collected at each site.

Five similar transects were searched along the north-western boundary of *Wasmannia* at Guayabillos on San Salvador Island, (see Fig. 2). In addition, 43 plots each 10 m × 2 m, were searched (under rocks, logs and debris) for scorpions, *Hadruroides maculata galapagoensis* Maury (Scorpionida, Iuridae), and for nests or groups (clusters of workers with brood) of *Wasmannia*. At the same site, a random sample of 50 guayabillo (*Psidium galapageium*) trees was examined for webs of two species of spiders, *Tidarren sisyphoides* (Walckenaer) and *Theridion calcynatum* Holmberg (Araneae, Theridiidae).

Two types of sticky traps were used to sample insects at two localities (Guayabillos and Trágica, see Fig. 2) on San Salvador Island: tree traps were sheets of plastic 19 cm × 10 cm tacked to tree trunks 1–2 m above ground; aerial traps were metal frames 25 cm × 25 cm, each strung with 25 threads (3 lb weight nylon fishing line), and hung from branches about 2 m above ground. The plastic sheets and nylon threads were coated with 'Tack-Trap' (Osticon Co.) and left out for 48 h. Insects were removed from the traps, immersed in kerosene to dissolve the adhesive and transferred to alcohol for counting and identification.

RESULTS

The distribution of Wasmannia and of other ant species

Twenty-nine ant taxa are currently known from the Galápagos Islands (Table 1, Fig. 3), four of which may be endemic species (Wheeler, 1919, 1924, 1933). Of the remaining, nearly half are common 'tramp' species, distributed by commerce throughout the tropics. While one can roughly date the entry of *Wasmannia* into the islands, it is more difficult to do so with the other less

Figure 3. The distribution of *Wasmannia auropunctata* in the Galápagos Islands and the total number of species of ants known from each island based on the combined data from early and recent collections (Wheeler, 1919, 1924, 1933; Clark *et al.*, 1982; present study).

Table 1. A list of species of ants known from the Galápagos Islands, based on Wheeler (1919, 1924, 1933) and on recent collections (Clark *et al.*, 1982, present study). A, Endemic species[1]; B, native species[2]; C, possible recent introductions[3]; D, species of unknown origin. Species which have spread recently (in the last 60 years) among the islands are marked (X). * = species that are affected or eliminated altogether by the presence of *Wasmannia*

Subfamily	A	B	C	D
Ponerinae	*Hypoponera beebei*	*Hypoponera opaciceps* *Odontomachus bauri** *Cylindromyrmex striatus*		*Hypoponera* sp. *Leptogenys* sp.*
Myrmicinae	*Pheidole williamsi**	*Solenopsis pacifica*[4]* *S. geminata* *Crematogaster chathamensis*[4] *Tetramorium simillimum** *T. bicarinatum* (X)* *Monomorium floricola* (X)* *M. pharaonis*	*Wasmannia auropunctata* *Cardiocondyla nuda** *C. emeryi** *Strumigenys louisianae*	*Solenopsis* sp. A *Solenopsis* sp. B *Pheidole* sp. A* *Pheidole* sp. B* *Cyphomyrmex* sp.*
Formicinae	*Camponotus planus** *C. macilentus*	*Paratrechina nesiotis*[4]* *P. longicornis* (X)		
Dolichoderinae		*Conomyrma albemarlensis*[4]* *Tapinoma melanocephalum* (X)*		
Totals (%)	4 (13.8)	14 (48.3)	4 (13.8)	7 (24.1) 29

[1]Endemics are those species that have not been found outside the Galápagos, and may, therefore, be derived from now-extinct mainland populations.

[2]Native species include those found on early expeditions and may, therefore, have arrived on the islands by natural means of dispersal.

[3]Possible recent introductions include those species which were not collected on early expeditions.

[4]These species may be synonomous with mainland forms (*Solenopsis globularia, Crematogaster brevispinosa, Paratrechina fulva* and *Conomyrma pyramica*) of which they have been considered subspecies.

conspicuous species. Early expeditions of 1905, 1923, 1925, and 1932 encountered neither species of *Cardiocondyla* (Wheeler, 1919, 1924, 1933; Stitz, 1932), both currently conspicuous elements of transition and arid-zone faunas. Common, diurnal ants such as *Monomorium floricola* (Jerdon) and *Tapinoma melanocephalum* (Fabricius) have apparently increased their distribution in the islands since these early expeditions: the former was collected only on Floreana and Genovesa Islands and is now known from 11 islands; the latter was found on Santa Cruz, Española and Genovesa, and now occurs on six other islands as well.

Two species/area (S/A) curves can be drawn for Galápagos ants, one based on early collections up until 1932 (log S = 0.204 + 0.166 log A, $\mathcal{N}$ = 15 islands; r = 0.232, ns) and another for recent collections from 1981/82 (log S = 0.709 + 0.114 log A, $\mathcal{N}$ = 12 islands; r = 0.722, $P < 0.01$). There has been a 41% increase in the number of species collected since 1933.

Wasmannia auropunctata is currently found on the inhabited islands of Santa Cruz, Floreana, San Cristóbal and Volcán Sierra Negra on Isabela, as well as on San Salvador Island (previously inhabited) and at two isolated sites, one at Point Albermarle on the northern tip of Isabela and another at James Bay on San Salvador Island (Fig. 3).

The impact of Wasmannia on other species of ants

The boundaries of *Wasmannia* on Santa Cruz Island have not yet stabilized. When the 1976 census of ant species along the Puerto Ayora–Baltra Road (Clark *et al.* 1982) was repeated six years later, we found that although *Wasmannia* had not advanced northward (into the arid zone), it had spread south along the road into humid *Scalesia* forest at a rate of about 170 m per year (Fig. 1).

As was found to be the case in the 1976 census, *Wasmannia* overlapped with few or no other species of ants, except on those transects near the southern edge of its distribution, in areas that had been invaded since 1976 (Figs 1, 4). *Wasmannia* occurred on 36 of the 90 transect sections (9 transects, each containing ten 10 m sections) and on 81% of these sections it occurred alone, not overlapping with any other species of ant. Only three other species occurred on more than one-third of the sections: *Pheidole* sp. A on 37 sections, *Solenopsis* sp. A on 43 and *Paratrechina nesiotis* (Wheeler) on 52 sections, each co-occurring with 1–5 other species of ants (with a median of 2.0–2.8 species). *Paratrechina nesiotis* occurred alone on only one transect section, while the other two species were never found alone.

Only four other ant taxa were found in those areas where *Wasmannia* occurred in high densities: *Solenopsis* spp. A and B, *Hypoponera* sp. and *Strumigenys* sp., all hypogeic ants which were collected in soil samples. In six soil samples taken in the Giant Tortoise Reserve (Caseta, Fig. 1) on Santa Cruz, at 150 m elevation in lower transition zone vegetation, the average relative abundances of the different species were (*Wasmannia/Solenopsis/Hypoponera*) 1/0.008/0.0006. Average relative abundances in five soil samples from the agricultural and lower *Miconia*-guava zone (180–430 m elevation) were (*Wasmannia/Solenopsis/Hypoponera/-Strumigenys*) 1/0.08/0.01/0.001.

The distribution of *Wasmannia* on San Salvador was patchier than on Santa Cruz. Areas of highest densities occurred in the SW portion of its distribution,

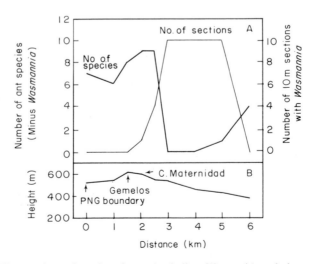

Figure 4. A, The numbers of species of ants (excluding *Wasmannia*) and the numbers of 10 m sections per transect on which *Wasmannia* was found on 9 transects along the Puerto Ayora–Baltra Road. B, Elevations of the transects and distance, in km, from the boundary of the Galápagos National Park on the northern side of Santa Cruz I. (see Fig. 1).

near its original site of introduction at La Trágica around 1967 (J. Villa, pers. comm.) and the ants appeared to be expanding their range to the NNE (Fig. 2). Within this area, there were pockets of primarily denser vegetation from which *Wasmannia* was absent and where other species of ants could be found.

In a census of ants along the NW boundary of *Wasmannia* at Guayabillos, *Wasmannia* occurred on six of the eight transects and on all ten sections of each of these six transects. On 80% of these sections (out of 60), *Wasmannia* occurred alone, not overlapping with any other species of ants. By comparison, of three other relatively common species, two species (*Cardiocondyla emeryi* Forel and *Solenopsis pacifica* Wheeler) never occurred alone, and another (*Conomyrma albemarlensis* (Wheeler)) occurred alone on only one section.

Wasmannia overlapped with as many as six other species of ants along the boundary at Guayabillos: with *Cardiocondyla emeryi* on 10 transect sections, *Tetramorium simillimum* (F. Smith) and *Conomyrma albemarlensis* on two sections each, and with *Solenopsis pacifica*, *Pheidole williamsi* Wheeler and *Paratrechina nesiotis* on one section each. These and other ant species had significantly wider distributions on the transects in the absence of *Wasmannia*, where they were found on 4.9 ± 3.2 (mean ± standard deviation) of the sections (all species combined, $N = 11$ species-occurrences), than in the presence of *Wasmannia*, where they occurred on only 1.7 ± 0.7 sections ($N = 10$ species occurrences; $t = 3.124$, $P < 0.01$).

A different distribution pattern was found in the agricultural zone of Santo Tomás and the nearby slopes of Volcán Sierra Negra and Cerro Grande on Isabela Island (Fig. 5). *Wasmannia* was restricted to a small area encompassing three farms at about 180–250 m elevation, where it had been introduced between 1966 and 1967 in a shipment of clumps of elephant grass (*Pennisetum purpureum*). *Wasmannia* did not overlap in distribution with one of the 'true' fire ants, *Solenopsis geminata* (Fabricius), and in fact, there appeared to be a 'no-man's-land' of several meters between adjacent areas of these two species along their common boundary. Although widespread in the islands, *Solenopsis geminata* may have also been introduced to Volcán Sierra Negra by settlers, as it was not recorded from Isabela by earlier expeditions.

Unlike *Wasmannia*, *S. geminata* overlapped with many other species of ants (up to nine other species in Santo Tomás), including many of the same species that were displaced by *Wasmannia* (e.g. *Pheidole* spp., *Paratrechina nesiotis* and *Tetramorium bicarinatum* (Nylander)). It remains unclear if the boundary between the two species is stable or if one is spreading at the expense of the other.

The impact of Wasmannia on other arthropods

Wasmannia influences the distribution not only of other species of ants, but of numerous other arthropods as well. A native iurid scorpion, *Hadruroides maculata galapagoensis*, and two web-building threridiid spiders, *Theridion calcynatum* and *Tidarren sisyphoides*, were displaced by *Wasmannia* (Table 2). At Guayabillos, on San Salvador I., the population density of scorpions was significantly lower in *Wasmannia*-infested areas, with 0.06 ± 0.14 scorpions per m² (mean ± standard deviation; $N = 25$ plots, each 20 m²), than in adjacent non-infested areas, with 0.41 ± 0.23 scorpions/m² ($N = 18$; $t = 6.324$, $P < 0.001$). On a single 50 m long transect across the boundary of *Wasmannia* at the same site, there was a

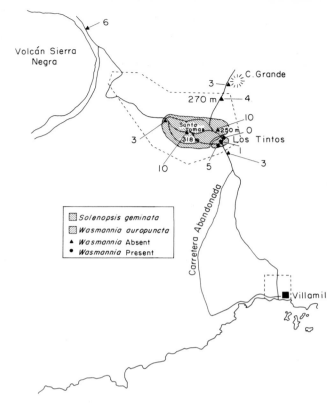

Figure 5. Map of the agricultural sector of Santo Tomás on Volcán Sierra Negra, Isabela Island, showing the distributions of *W. auropunctata* and of *S. geminata*. Numbers beside the symbols indicate the numbers of other species of ants collected at each site.

significant negative correlation between scorpion density and numbers of groups (i.e., queen-right nests and groups of workers with brood) of *Wasmannia* ($r = -0.84, P < 0.05$). Along this transect, the density of scorpions decreased from $0.55/m^2$ (no *Wasmannia* groups) to $0.05/m^2$ (with 0.85 groups of *Wasmannia* per m^2).

At the Guayabillos and Trágica sites on San Salvador Island, both the numbers of species of insects caught in sticky traps and their overall abundances were lower in *Wasmannia*-infested areas than in non-infested areas (Table 3). The increase in numbers of individuals captured in traps (especially in tree-traps) at Trágica was due entirely to the emergence of large numbers of flies (Diptera) from a nearby temporary pond. At Trágica, 63.6% (54.2–72.9%) of the total captures were small flies, in comparison with 19.2% (9.5–29.2%) at Guayabillos.

Certain species may actually benefit from the spread of *Wasmannia*, including nest inquilines such as the spider *Ischnothyreus* sp. (Oonopidae) and some plant-feeding species of Homoptera that produce honeydew and are tended by *Wasmannia* (Silberglied, 1972). In the transition zone of Santa Cruz Island, we found a significant positive correlation ($r = 0.48, P < 0.05$) between the numbers of *Wasmannia* and the numbers of coccids (Homoptera, Coccoidea) on 10 branches

Table 2. Displacement of a scorpion, *Hadruroides maculata galapagoensis,* and two species of spiders, *Theridion calcynatum* and *Tidarren sisyphoides,* by *Wasmannia* at Guayabillos on San Salvador Island A, Numbers of 20 m² plots that contined scorpions in adjacent areas with and without *Wasmannia.* B, numbers of *P. galapageium* trees with spiders (both species combined) in adjacent areas with and without *Wasmannia.* (Fisher's exact probability test was used in both instances)

A	Scorpions			
Wasmannia	Present	Absent	Total	*P*
Present	7	18	25	0.004
Absent	18	0	18	
Total	25	18	43	

B	Spiders			
Wasmannia	Present	Absent	Total	
Present	0	25	25	0.009
Absent	17	8	25	
Total	17	33	50	

each of five trees each of *Croton scouleri, Scalesia pedunculata, Psychotria rufipes* and *Psidium galapageium.*

DISCUSSION

For island faunas in general, the 'faunal coefficient', or the slope of the logS/logA curve, falls within the range of $z = 0.3$–0.4 (Gorman, 1979). Possible reasons for the low faunal coefficient observed for Galápagos ants based on 1981–82 collections ($z = 0.114$) are: (1) the islands are still poorly sampled and more species of ants remain to be discovered, particularly on the larger islands; (2) the island fauna is not yet fully saturated and larger islands in particular have fewer species than expected, and (3) most of the species are 'tramps' which have been able to colonize most of the islands. Perhaps a similar case is that of the Hawaiian Islands, where $z = 0.16$ (calculated from Wilson & Taylor, 1967) for an entirely introduced ant fauna of 36 species. The high proportion, in recent collections, of 'tramp' species that are newly recorded for the Galápagos suggests that the islands have the potential to sustain a higher diversity of species of ants than is currently present.

Wasmannia has successfully invaded a number of different habitats in the Galápagos Islands, displacing other species of ants and at least some other arthropods as well (see also Clark *et al.,* 1982; Meier, in press). Clearly, *Wasmannia* has not had the same impact in all other parts of its range; in lowland monsoon rainforest in Panamá, it occurred together with numerous other species of terrestrial ants at densities of 0.05–0.13 nests per m² (Levings & Franks 1982). Likewise, on Cocos Island (Costa Rica), *Wasmannia* co-occurs

Table 3. Comparison of insects caught in sticky traps at two sites on James Island. Shown are total numbers of species and individuals of all taxa captured in aerial traps and in tree traps at each location. χ^2 tests with equal expected values were used to compare captures at sites in *Wasmannia*-infested and non infested areas

	Aerial traps	
Location	No. species	No. individuals
Guayabillos: WA	28 ⎱ $p<0.02$	66 ⎱ $p<0.001$
Guayabillos: WP	13 ⎰ ⎱ $p<0.01$	21 ⎰ ⎱ $p<0.001$
Trágica: WP	9 ⎰	32 ⎰
	Tree traps	
	No. species*	No. individuals*
Guayabillos: WA	34 ⎱ ns	107 ⎱ $p<0.001$
Guayabillos: WP	24 ⎰ ⎱ $p<0.05$	46 ⎰ ⎱ ns
Trágica: WP	21 ⎰	133 ⎰

*All ants excluded from totals of species and individuals. WA, *Wasmannia* absent; WP, *Wasmannia* present.

with 22 other species, although the spatial relationship between *Wasmannia* and these other species has not been described (Hogue & Miller, 1981). The rapid expansion of this ant in the Galápagos may be due, on the one hand, to a suite of behavioural and ecological characteristics which give *Wasmannia* certain advantages over other species of ants in dispersal, colonization and competition, and on the other hand, to the nature of the native Galápagos invertebrate fauna and particularly that of the relatively species-poor native ant fauna.

Dispersal

Wasmannia has efficient short-range terrestrial dispersal. The nests contain numerous queens (more than 100 queens were found in a nest of *c.* 24 000 workers; L. Endara, pers. comm.) and new nests are started by 'budding', i.e. one or more queens move out on foot with a complement of workers and establish a new nest (Hölldobler & Wilson 1977; pers. obs.). *Wasmannia* nests are shallow, and workers and queens will relocate readily. Long-range dispersal is effected by human commerce and perhaps by rafting; small nests may be transported in the soil around plant roots and in produce (pers. obs.). Apparently, however, *Wasmannia* alates do not fly large distances and thus have limited self-propelled, long-range dispersal. Unlike alates of many other species of ants, those of *Wasmannia* were never trapped in aerial sticky traps, nor were they attracted to lights.

Colonization

Habitat requirements (temperature and humidity, nest sites) and food requirements determine successful colonization. *Wasmannia* is capable of nesting in a wide array of substrates and in habitats ranging from semi-arid to humid (Spencer, 1941; Kusnezov, 1951). In the Galápagos, *Wasmannia* occurs in most habitats, but is most abundant in the moist transition and lower humid zones (Clark *et al.*, 1982; pers. obs.) and in habitats disturbed by man (pastures, fruit crops, villages). Its distribution does, however, appear to be restricted both geographically and altitudinally by extreme conditions of either high temperature and low humidity or low temperature and high humidity. Numerous dessicated groups of *Wasmannia* were found under rocks in open, treeless areas on San Salvador, particularly as the dry season progressed. On Santa Cruz Island, *Wasmannia* invaded the cool, wet *Miconia* zone during the hot season, but disappeared during the cold, wet season (Clark *et al.*, 1982; pers. obs.).

Wasmannia workers feed primarily on honeydew and invertebrates (scavenged or killed) (Clark *et al.*, 1982). Although honeydew appeared to be the major food item by weight (ibid.), when offered a choice of four baits (tuna, marmalade, sugar-water and milk), recruitment was consistently highest to the oil-rich tuna (I. de la Vega, pers. comm.). How dependent is *Wasmannia* on honeydew and nectar? The correlation between numbers of coccids and *Wasmannia* ants on four native tree species was noted earlier; one other tree species, *Hippomane mancinella*, had few or no coccids, and virtually no ants. In the arid zone on Santa Cruz I., *Wasmannia* fed on honeydew produced by coccids on *Opuntia echios* (Meier, in press). At Guayabillos, on San Salvador Island, *Wasmannia* foraged during the hot season almost exclusively on two species of trees, *Psidium galapageium* and *Castela galapageia*, both of which have abundant extra-floral nectaries. The presence of coccids, however, was not noted. While it is possible that *Wasmannia* would be less successful in areas with poor nectar sources and few coccids, the range of foods actually taken by these ants clearly establishes *Wasmannia* as a generalist or opportunistic feeder.

Competition

Wasmannia might gain advantage over other species of ants by interference competition, predation, exploiting some resources more efficiently, or by all of these means (Clark *et al.*, 1982). There is evidence that *Wasmannia* displaces some species (e.g. *Pheidole williamsi*) by utilizing available nesting sites, while in other cases (e.g. *Conomyrma albemarlensis*, which has nests deep in the soil) competition may be for nectar and prey. While few encounters were observed at baits, *Wasmannia* recruited much faster and in larger numbers to food than did most other species and effectively prevented other ants from approaching baits (I. de la Vega, pers. comm.). *Wasmannia* workers were also observed to attack any foreign arthropod in the vicinity of the nest—other ant species (Clark *et al.*, 1982) as well as large scorpions and centipedes (*Scolopendra galapageia*).

Behavioral characteristics which may favour successful competition include mass recruitment to food, aggressiveness towards other ants (and possibly repelling other ants by means of chemical defenses, Howard *et al.*, 1982), and activity during both day and night (Clark *et al.*, 1982: I. de la Vega pers. comm.). The unicolonial social organization (Hölldobler & Wilson, 1977), with numerous small workers and interconnected nests, may ensure rapid transmission of information about sources of food throughout an extended colony consisting of one or more main nests plus satellite groups of workers and brood.

Few species of ants overlap with *Wasmannia* in areas of high population density. Those that do so are all small ants which are either entirely hypogeic, nesting and foraging in the soil (*Solenopsis* spp. A and B), or partially hypogeic nesting inside rotting wood and foraging in the soil and leaf litter (*Strumigenys louisianae* (Roger) and *Hypoponera* spp.). In either case, there is little overlap in foraging and nesting habits between these species and *Wasmannia*.

Only one species of ant, *Solenopsis geminata*, may be a successful competitor of *Wasmannia*, at least in agricultural habitats. *Solenopsis geminata*, like *Wasmannia*, is a mass-recruiting species with large colonies and aggressive workers. Given these behaviours, and the fact that *S. geminata* soldiers and workers are considerably larger than the workers of *Wasmannia*, the latter may be unable to invade areas where *S. geminata* is already firmly established.

Some general predictions can be made concerning the outcome of interactions between *W. auropunctata* and certain elements of the native Galápagos fauna. The taxa most likely to be affected are transition and lower humid zone species (in areas of high *Wasmannia* densities); terrestrial or arboricolous species (broadly overlapping *Wasmannia* in habitat); species of small–medium body size and soft exoskeleton (susceptible to *Wasmannia* attacks); species lacking active defence mechanisms, such as chemical defences, or that rely on crypsis for defence (particularly those that rest under bark or stones); species lacking extended brood care; and insectivorous or nectar and honeydew feeders (overlapping in food requirements). Most spiders, for example, are insectivorous and have soft exoskeletons. Despite their large size, relatively hard exoskeletons and extended brood care, scorpions live under stones and debris and are rather sedentary. Furthermore, in view of the reduced diversity and abundance of insects in areas heavily infested by *Wasmannia*, it is likely that predatory arachnids such as scorpions and spiders will face a shortage of prey. The elimination of prey species by *Wasmannia* is the likely explanation for the observed absence from *Wasmannia*-infested areas on San Salvador island of *Tmarus stoltzmanni* Keyserling (Thomisidae), a crab spider that specializes on formicine and dolichoderine ants (*Camponotus* spp. and *Conomyrma albemarlensis*) (Lubin, 1983).

Species which should, to some extent, escape the deleterious influences of *Wasmannia* include strictly arid-zone adapted species on the one hand and high elevation, *Miconia* and fern-sedge zone species on the other (lower population densities of *Wasmannia*); hypogeic, aquatic, cavernicolous and wood or plant-boring species (non-overlapping micro-habitats); large species with hard exoskeletons or chemical defences (resistant to *Wasmannia* attacks); and primarily herbivorous species (non-overlapping food requirements).

SUMMARY

The little fire ant, *Wasmannia auropunctata*, occurs on five islands in the Galápagos archipelago. It is still in the process of expanding its range on at least two of these islands (Santa Cruz and San Salvador). At least 17 of the remaining 28 ant taxa currently known from the Galápagos are affected by the presence of *Wasmannia*. On Santa Cruz and San Salvador few other species of ants co-occurred with *Wasmannia*, except at the edges of its distribution or in areas which it had only recently invaded. *Wasmannia* was also found to reduce population densities, or eliminate altogether, three species of arachnids (a scorpion and two theridiid spiders) as well as reducing the overall abundance and species diversity of flying and arboricolous insects at two sites on San Salvador. The mechanisms by which these species are displaced are currently being investigated.

Certain arthropods may escape the detrimental influences of *Wasmannia* through non-overlap of habitat and food requirements (as documented in the case of certain hypogeic ants), while others may actually benefit from the presence of *Wasmannia*, as appears to be the case for some coccids.

ACKNOWLEDGEMENTS

I wish to thank M. Alvarez and M. J. Campos for field assistance; F. Köster for unfailing enthusiasm and support for the project; R. J. Berry for the invitation to present these results at the Symposium; D. J. Anderson and M. Coulter for collections of ants; R. R. Snelling for identification of ants and H. W. Levi and W. A. Shear for identification of arachnids; and last, but not least, J. Gordillo, A. Tupiza, A. Kastdalen and many other islanders who helped in numerous ways to trace the history and distribution of *Wasmannia* in the islands. I thank R. R. Snelling and P. Grant for reviewing the manuscript.

REFERENCES

CLARK, D. B., GUAYASAMIN, C., PAZMIÑO, O., DONOSO, C. & PÁEZ DE VILLACÍS, Y., 1982. The tramp ant *Wasmannia auropunctata:* Autecology and effects on ant diversity and distribution on Santa Cruz Island, Galapagos. *Biotropica, 14:* 196–207.

CROWELL, K. L., 1968. Rates of competitive exclusion by the Argentine ants in Bermuda. *Ecology,* 49: 551–555.

DARWIN, C., 1860. *The Voyage of the Beagle.* Natural History Library Edition, 1962. New York: Doubleday and Co.

FABRÉS, G. & BROWN, W. L., 1978. The recent introduction of the pest ant *Wasmannia auropunctata* into New Caledonia. *Journal of the Australian Entomological Society, 17:* 139–142.

GORMAN, M., 1979. *Island Ecology.* Outline Studies in Ecology. London: Chapman and Hall. 79 pp.

HASKINS, C. P. & HASKINS, E. F., 1965. *Pheidole megacephala* and *Iridomyrmex humilis* in Bermuda—equilibrium or slow replacement? *Ecology, 46:* 736–740.

HOGUE, C. L. & MILLER, S. E., 1981. Entomofauna of Cocos Island, Costa Rica. *Atoll Research Bulletin No. 250, Smithsonian Institute:* 1–29.

HÖLLDOBLER, B. & WILSON, E. O., 1977. The number of queens: an important trait in ant evolution. *Naturwissenschaften, 64:* 8–15.

HOWARD, D. F., BLUM, M. S., JONES, T. H. & TOMALSKI, M. D., 1982. Behavioral responses to an alkylpyrazine from the mandibular gland of the ant *Wasmannia auropunctata. Insectes Sociaux, 29:* 369–374.

ILLINGWORTH, J. F., 1917. Economic aspects of our predaceous ant (*Pheidole megacephala*). *Proceedings of the Hawaiian Entomological Society 4:* 349–368.

KUSNEZOV, N., 1951. El género *Wasmannia* en la Argentina. Acta Zool. *Lilloana, 10:* 173–182.

LEVINGS, S. C. & FRANKS, N. R., 1982. Patterns of nest dispersion in a tropical ground ant community. *Ecology 63:* 338–344.

LINSLEY, E. G. & USINGER, R. L., 1966. Insects of the Galapagos islands. *Proceedings of the Californian Academy of Science, ser. 4, 33 (7):* 113–196.

LUBIN, Y. D., 1983. An ant-eating crab spider from the Galapagos. *Noticias de Galapagos, 37:* 18–19.

MEIER, R. E., In press. Ecological and behavioral aspects of ants foraging on giant cacti at the Darwin Station and at Tortuga Bay (Island of Santa Cruz). Charles Darwin Research Station, Annual Report, 1982.

SILBERGLIED, R., 1972. The "little fire ant," *Wasmannia auropunctata*, a serious pest in the Galapagos Islands. *Noticias de Galapagos, 19/20:* 13–15.

SPENCER, H., 1941. The small fire ant *Wasmannia* in citrus groves—a preliminary report. *Florida Entomology,* 24: 6–14.

STITZ, H., 1932. Formicidae. *Nyt magazin for naturvidenskaberne, 71:* 367–372.

WHEELER, W. M., 1919. The ants of the Galapagos Islands. *Proceedings of the Californian Academy of Science, ser. 4, 2:* 259–310.

WHEELER, W. M. 1924. The Formicidae of the Harrison Williams Galapagos Expedition. *Zoologica, 5:* 101–122.

WHEELER, W. M., 1933. The Templeton Crocker Expedition of the California Academy of Sciences, 1932, no. 6. Formicidae of the Templeton Crocker Expedition. *Proceedings of the Californian Academy of Science ser. 4, 21:* 57–64.

WIGGINS, I. & PORTER, D., 1971. Flora of the Galapagos Islands. Stanford, CA: Stanford Univ. Press.

WILSON, E. O. & TAYLOR, R. W., 1967. The ants of Polynesia (Hymenoptera: Formicidae). *Pacific Insects, 14:* 1–109.

ZIMMERMAN, E. C., 1970. Adaptive radiation in Hawaii with special reference to insects. *Biotropica, 2:* 32–38.

Biological Journal of the Linnean Society (1984) *21:* 243–251. With 2 figures

Relationships of the Galapagos flora

DUNCAN M. PORTER

*Department of Biology, Virginia Polytechnic Institute &
State University, Blacksburg, Virginia 24061, U.S.A.*

Joseph Dalton Hooker's pioneer 1847 paper on Galapagos plants and their relationships is a classic in the field of phytogeography. It was the first study of its kind to be published, comparing the islands' flora with island and continental floras elsewhere, hypothesizing on the dispersal mechanisms of the plants, and pointing out anomalies in the inter-island distributions of the native species. These are still three of the primary concerns of contemporary phytogeographers, and the present paper contrasts Hooker's findings with those of today. Despite the accumulation of a large amount of data since his time, many of Hooker's conclusions regarding Galapagos phytogeography remain valid.

KEY WORDS:—Island biogeography – Galapagos Islands – Joseph Dalton Hooker – Charles Darwin.

CONTENTS

INTRODUCTION

Our knowledge of the vascular plants of the Galapagos Islands and their relationships has, not surprisingly, markedly increased since Charles Darwin's six-week visit on H.M.S. *Beagle* in 1835. Darwin was not the first to collect plants in the archipelago, having been preceded by the British naturalists Archibald Menzies in 1795, David Douglas, Dr John Scouler and James McRae in 1825, and Hugh Cuming in 1829. The first flora of the islands (Hooker, 1847a) was based on their collections, except for those of Menzies which were labelled "Sandwich Islds.", and a few others. Of these collections, Darwin's was by far the largest, not being surpassed until the California Academy of Science's year-and-a-day expedition in 1905–1906. Darwin's plant collections also are the most important for typification of Galapagos endemics (Porter, 1980b).

Although he played down his prowess as a plant collector while on the *Beagle*, Darwin did admit to his mentor the Rev. John Stevens Henslow, Professor of Botany in the University of Cambridge, that he had collected everything that he

0024-4066/84/010243 + 09 $03.00/0

saw in flower in the islands. The 210 collections of 173 taxa that Darwin made in 1835 represent a sample of about 24% of the presently known flora.

PHYTOGEOGRAPHIC RELATIONSHIPS

Not only did Joseph Dalton Hooker write the first floristic study of the Galapagos Islands, he soon produced the first phytogeographic study as well (Hooker, 1847b). As with the former, the latter was also primarily based on Darwin's collections. After making the point that about half of the species are endemic; Hooker (1847b: 235–236) continued as follows.

> The results of my examination have been, that the relationship of the Flora to that of the adjacent continent is a double one, the peculiar or new species being for the most part allied to plants of the cooler parts of America, or the uplands of the tropical latitudes, whilst the non-peculiar are the same as abound chiefly in the hot and damper regions, as the West Indian islands and the shores of the Gulf of Mexico; also that, as is the case with the Fauna, many of the species, and these the most remarkable, are confined to one islet of the group, and often represented in others by similar, but specifically very distinct congeners.

A few pages later in the paper, Hooker (1847b: 239) slightly modified his remarks on the relationships of the flora.

> Here, as in other countries, the vegetation is formed of two classes of plants; the one peculiar to the group, the other identical with what are found elsewhere. In this there are even indications of the presence of two nearly equal Floras, an indigenous and introduced, and these of a somewhat different stamp; for the introduced species are for the most part the plants of the West Indian islands and of the lower hot parts of the South American coast; whilst the peculiar Flora is chiefly made up of species not allied to the introduced, but to the vegetation which occurs in the Cordillera or the extra-tropical parts of South America.

Here, Hooker is using "indigenous" or "peculiar" for endemic, a term not widely used in a geographical context until a few years later (DeCandolle, 1855). He uses "introduced" to mean non-endemic, rather than those species introduced by human activity.

A further clarification is made in his introduction to a more detailed discussion of the archipelago's phytogeographical relationships (Hooker, 1847b: 250).

> In this second part of the essay I propose to treat of the Flora of the Galapagos as divisible into two types: these are the West Indian (including Panama), to which the plants common to other countries and the dubious species almost universally belong; and the Mexican and temperate American type, or that under which the great majority of the peculiar species will rank.

These relationships, recognized by Hooker on the basis of about 25% of today's known vascular plant flora, still hold. The endemic plants have their closest relatives primarily (56%) in the Andean region (Table 1). In addition, 52% of the original introductions that have given rise to the present endemic flora presumably have come from the Andean region. The Neotropical region, which overlaps with much of the Andean region, accounts for 27% of endemic taxa and of introductions. The third-largest element is the endemics with Pantropical affinities. Hooker did not think that many species with Pantropical relationships were represented in the islands, clearly because of a lack of knowledge of the archipelago's entire flora and of these species' distributions

Table 1. Geographical relationships of the endemic vascular plants of the Galapagos Islands

	Neotropical	Pantropical	Andean	Mexico & C. America	South America	Caribbean	North America	Total
Pteridophytes	3	2	1	2				8
Monocotyledons	4	1	9		4	1		19
Dicotyledons	54	15	118	4		8	3	202
Total	61 (27%)	18 (8%)	128 (56%)	6 (3%)	4 (2%)	9 (4%)	3 (1%)	229
Number of single original introductions from each area	31 (27%)	10 (9%)	60 (52%)	5 (4%)	1 (1%)	6 (5%)	3 (3%)	116

Geographical areas are defined as follows: *Neotropical* (distributed generally in the American tropics); *Pantropical* (distributed in both the Old and New World tropics); *Andean* (occurring only in western South America from Venezuela to Chile, generally or in part); *Mexico and Central America* (occurring only in Mexico and/or Central America, and in one case also in northern Colombia and Venezuela); *South America* (occurring only in extra-Andean South America); *Caribbean* (occurring in the West Indies and often also on the edges of the surrounding continents); *North America* (occurring in the south-western United States and ajacent northern Mexico).

elsewhere. Those that were present he thought to be derived from South America, not from the westward. Pantropical relatives apparently provided 9% of the original introductions, which evolved into 8% of the presently known endemics.

The indigenous non-endemics have their greatest numbers occurring elsewhere in the Neotropics (Table 2). Here are to be found 48% of the indigenes; 25% are Pantropical, and 21% Andean. The Neotropical, Pantropical, and Andean groups all overlap in adjacent South America, so it is possible that 88% of the introductions that have given rise to the endemic flora, and 94% of introductions of indigenes, have come from this nearby land mass.

The importance of the Pantropical element in the Galapagos flora was not recognized until the publication of the second floristic study by N. J. Andersson (1855). Andersson's work was based on his own collections, made during a 10-day visit to the islands in 1852. Interestingly, the first to indicate the flora's relationship to adjacent South America was Charles Darwin (1839). This was done before Hooker had seen Darwin's collections, although Henslow had examined them. However, this insight came about through Darwin's own field observations (see Barlow, 1933), and not from information supplied by Henslow. Henslow did a superb job in handling Darwin's collections, shipped to him from South America, but he failed in his promise to identify the plants (Porter, 1980a).

Subsequent studies on Galapagos phytogeography (Robinson & Greenman, 1895; Robinson, 1902; Stewart, 1911, 1915; Svenson, 1935, 1942, 1946; van Balgooy, 1960; Porter, 1976, 1979, 1983, in press) have continued to provide evidence for Hooker's hypotheses of relationship. The close relationship of the vascular flora with that of adjacent South America was first documented by Svenson (1935, 1942, 1946), and is supported by my own research (Porter, 1976, 1979, 1983, in press).

Hooker's estimate of about 50% endemism still holds if all subgeneric taxa are considered. Given 541 indigenous species, subspecies, and varieties, 229 endemics and 312 indigenes, endemism would be 42%. However, at the species level, this drops to 34%, 170 endemic species out of a total native flora of 497 species. If flowering plants only are considered, species endemism is 41%, while that for species, subspecies, and varieties rises to 51%. The relationships of each endemic taxon are discussed in detail elsewhere (Porter, 1979, 1983), as are those of the entire flora (Porter, 1983).

The first section of Hooker's paper concludes with a family-by-family discussion of geographical relationships. In it he makes the observation that the

Table 2. Geographical relationships of the indigenous, non-endemic vascular plants of the Galapagos Islands

	Neotropical	Pantropical	Andean	Mexico & C. America	Caribbean	Total
Pteridophytes	52	18	22	1	6	99
Monocotyledons	35	24	5		4	68
Dicotyledons	63	35	39	2	6	145
Total	150	77	66	3	16	312
Percentage	48%	25%	21%	1%	5%	

flora is basically a disharmonic one. That is, one comprised of a skewed sample of easily dispersed taxa and not of a random sample of all taxa in the flora of the adjacent continental area:

> the more an island is indebted to a neighbouring continent for its vegetation, the more fragmentary does its flora appear, migration being effected by the transport of isolated individuals, generally in no wise related, while an independent flora is generally made up of groups, the lowest order of which we call genera. (Hooker, 1847b: 247).

The Galapagos flora is indeed disharmonic, as is the fauna, providing further biological evidence that it has been derived by long-distance dispersal.

DISPERSAL

Hooker's paper was not only a pioneering study in the geographical relationships of island floras, it also was the first to speculate on how such a flora might be derived. Several pages were devoted to a family-by-family discussion of adaptations for dispersal. Hooker (1847b: 253) concluded that "The means of transport which may have introduced these plants are, oceanic and aërial currents, the passage of birds, and man". Subsequent botanists (Andersson, 1855; Robinson, 1902; Stewart, 1911; Svenson, 1942) added comments on dispersal mechanisms, but none provided as detailed observations as those of Hooker.

More recently, in a study based on Stewart's (1911) flora, Carlquist (1967) determined that the flowering plants were derived as follows: 73% by bird dispersal, 23% by oceanic drift, and 4% by wind currents. With the publication of a modern flora for the islands (Wiggins & Porter, 1971), it became possible to more accurately portray the relationships of the plants and their methods of dispersal. For the vascular plants, this revealed that 40% had been derived through birds, 32% by human carriage, 21% by wind, and 6% through drift (Porter, 1976). Natural means of dispersal were 60% by birds, 31% wind, and 9% drift.

Since the publication of the *Flora of the Galapagos Islands,* a series of papers have appeared, adding to and subtracting from the flora (see Schofield, 1973, 1980, for references). Addition of this information does not significantly change my previously calculated figures (Table 3).

If only the flowering plants are calculated, then the overwhelming importance of bird dispersal for this group is revealed. Of the total successful introductions that have given rise to the presently known angiosperms, 48% have been by birds, 39% by humans, 7% by drift, and 6% by wind. However, if

Table 3. Original introductions that have resulted in the present vascular plant flora of the Galapagos Islands

	Birds	Humans	Wind	Oceanic Drift	Total
Pteridophytes	1		106		107
Monocotyledons	64	39	14	3	120
Dicotyledons	178	156	14	33	381
Total	243 (40%)	195 (32%)	134 (22%)	36 (6%)	608
Total for natural introductions	243 (59%)		134 (32%)	36 (9%)	413

only natural agencies are considered, this changes to 79% of the introductions having arrived on or in birds, 12% floating in on oceanic currents, and 9% wafted on the wind. Taxon by taxon discussion of dispersal mechanisms is published elsewhere (Porter, 1979, 1983).

Hooker was the first botanist, and the only one until recently, to recognize the importance of human impact on the islands' flora. He wrote that currents, winds, and birds played their various roles in the derivation of the plants, then added (Hooker, 1847b: 254).

> Man is the last agent to which I alluded: that he has been already active is very perceptible from the fact, that Charles Island, the only colonized island, contains the smallest proportion of peculiar plants, and numerically far the most of these common to and probably introduced from the coast with cultivation.

So far as is known at present, humans have introduced 195 weeds and escapes from cultivation that reproduce themselves in the islands. Indeed, humans are now the most important dispersers of plants, both to and within the archipelago. For example, the Pantropical (originally tropical Asian) weed *Cleome viscosa* was first collected on the island of Baltra in 1963. Personal observation in 1977 showed it to be common in the vicinity of the Baltra airport, and in 1978 it formed large populations across the island. In 1981 I found it along the trail above Tagus Cove on Isla Isabela (Fig. 1). Tagus Cove is usually visited by

Figure 1. *Cleome viscosa* (Capparidaceae), a pantropical weed introduced into the Galapagos Islands through human activity. The fruits are thickly covered with sticky glandular hairs and readily stick to clothing. It was first collected on Isla Baltra in 1963 and has since spread to Isla Isabela.

tourists soon after their arrival in the islands at the airport on Baltra. Obviously, *Cleome viscosa*, easily dispersed because of the many sticky hairs on the fruits, is now a prime candidate for distribution throughout the islands through the inadvertent intervention of humans.

A more alarming example is that of the endemic grass, *Cenchrus platyacanthus*. Like other members of its genus, the fruits are enclosed in a burr-like involucre that is provided with numerous retrorsely-barbed spines, which make them easily dispersed. Presumably, they are naturally dispersed by birds. However, I know by personal experience that these disseminules are as likely to be found attached to one's trousers or socks as to the feathers of a bird (Fig. 2).

In 1978 I found a specimen of *Cenchrus platyacanthus* growing along a path on Isla Plaza Sur, a small, well-collected island much visited by tourists, from which it had not hitherto been reported. Thus, humans are not only introducing new elements into the flora of the islands from elsewhere, they are interfering with the distributions of the endemic plants as well. Their effects on the vegetation, and those of their introduced animals, also are profound (see Hamann, 1975; van der Werff, 1979).

Figure 2. *Cenchrus platyacanthus* (Poaceae), an endemic grass that occurs on a number of islands in the archipelago. The fruit is surrounded by a burr-like structure that aids in dispersal. Natural dispersal is presumably by birds, but dispersal by humans is also possible. In 1978 it was found along a trail frequented by tourists on Isla Plaza Sur, a small, well-known island on which it had not hitherto been seen.

INTER-ISLAND RELATIONSHIPS

Another subject which fascinated Hooker was the apparently restricted ranges of various species from island to island within the archipelago. He stated (Hooker, 1847b : 239),

> In the third place, I shall allude to the most singular feature in the botany of the group, the unequal dispersion of the species, the restriction of most of them to one islet, and the representation of others by allied species in two or more of the other islets.

Hooker reckoned that 112 of 128 (88%) of his "peculiar" (that is, endemic) taxa occurred on only a single island. Subsequent investigators, particularly Robinson (1902) and Kroeber (1916), also have commented on the number of endemics in the islands with restricted distributions.

This line of investigation begun by Hooker has been extended by several recent studies, which attempted to discover correlations between species numbers on the various islands and such parameters as island area, elevation, distance to the nearest island, distance to the center of the archipelago, etc. Island elevation, the area of the adjacent island, and the log of the island area were found to be significant variables in predicting species numbers (Hamilton *et al.*, 1963; Johnson & Raven, 1973; Simpson, 1974). However, Connor & Simberloff (1978) repeated the calculations of the former investigators and, adding more precise information on island areas, elevations, and species numbers, found that only area contributed significantly to explaining variance in species numbers from island to island, a point made by Kroeber in 1916. Connor & Simberloff (1978 : 219) also showed that, "The number of botanical collecting trips to each of the Galapagos islands is a better predictor of species numbers than are area, elevation, or isolation".

As knowledge of species' distributions has advanced through further collecting and observation, many of the single-island taxa found by Hooker have disappeared as well. According to my latest calculations, no pteridophyte, two monocotyledon, and 65 dicotyledon taxa (30% of the endemics) occur on only a single island. A majority of these taxa of restricted distribution are subspecies or varieties, and some are questionably distinct entities. These presumably are recently evolved taxa, and they provide biological evidence for the hypothesis that the islands themselves are geologically young (Porter, in press), which is confirmed by geological evidence discussed in another paper at this symposium.

ACKNOWLEDGEMENTS

Field trips to the Galapagos Islands in 1977, 1978, and 1981 were made courtesy of Harvard University's Friends of the Museum of Comparative Zoology. I was able to present the gist of this paper at the Linnean Society symposium "Evolution in the Galapagos Islands" through a travel grant from the Society. Both organizations are gratefully acknowledged.

REFERENCES

ANDERSSON, N. J., 1855. Om Galapagos öarnes Vegetation. *Kongelige Vetenskaps-academiens Handlingar, 1853:* 61–120.
BALGOOY, M. M. J. VAN, 1960. Preliminary plant–geographical analysis of the Pacific as based on the distribution of Phanerogam genera. *Blumea, 10:* 385–430.

BARLOW, N. (Ed.), 1933. *Charles Darwin's Diary of the Voyage of H.M.S. Beagle*. Cambridge: University Press.

CARLQUIST, S., 1967. The biota of long-distance dispersal. V. Plant dispersal to Pacific islands. *Bulletin of the Torrey Botanical Club, 94:* 129–162.

CONNOR, E. F. & SIMBERLOFF, D., 1978. Species number and compositional similarity of the Galápagos flora and avifauna. *Ecological Monographs, 48:* 219–248.

DARWIN, C., 1839. *Narrative of the Surveying Voyages of His Majesty's ships* Adventure *and* Beagle, *between the Years 1826 and 1836, describing their Examination of the Southern Shores of South America, and the Beagle's Circumnavigation of the Globe.* III. *Journal and Remarks. 1832–1836.* London: Colburn.

DeCANDOLLE, A., 1855. *Géographie Botanique Raisonnée.* Paris.

HAMANN, O., 1975. Vegetational changes in the Galápagos Islands during the period 1966–73. *Biological Conservation, 7:* 37–59.

HAMILTON, T. H., RUBINOFF, I., BARTH R. H. & BUSH, G. L., 1963. Species abundance: Natural regulation of insular variation. *Science, N.Y., 142:* 1575–1577.

HOOKER, J. D., 1847a. An enumeration of the plants of the Galapagos Archipelago; with descriptions of those which are new. *Transactions of the Linnean Society of London, 20:* 163–233.

HOOKER, J. D., 1847b. On the vegetation of the Galapagos Archipelago as compared with that of some other tropical islands and of the continent of America. *Transactions of the Linnean Society of London, 20:* 235–262.

JOHNSON, M. P. & RAVEN, P. H., 1973. Species number and endemism: The Galápagos Archipelago revisited. *Science, N.Y., 179:* 893–895.

KROEBER, A. L., 1916. Floral relations among the Galapagos Islands. *University of California Publications in Botany, 6:* 199–220.

PORTER, D. M., 1976. Geography and dispersal of Galapagos Islands vascular plants. *Nature, Lond., 264:* 745–746.

PORTER, D. M., 1979. Endemism and evolution in Galápagos Islands vascular plants. In D. Bramwell (Ed.), *Plants and Islands:* 225–256. London: Academic Press.

PORTER, D. M., 1980a. Charles Darwin's plant collections from the voyage of the *Beagle. Journal of the Society for the Bibliography of Natural History, 9:* 515–525.

PORTER, D. M., 1980b. The vascular plants of Joseph Dalton Hooker's *An enumeration of the plants of the Galapagos Archipelago; with descriptions of those which are new. Botanical Journal of the Linnean Society, 81:* 79–134.

PORTER, D. M., 1983. Vascular plants of the Galapagos: Origins and distribution. In R. I. Bowman, M. Berson & A. Levitan (Eds), *Patterns of Evolution in Galapagos Organisms:* 33–96. San Francisco: American Association for the Advancement of Science Pacific Division.

PORTER, D. M., in press. Endemism and evolution in terrestrial plants. In R. Perry (Ed.), *Key Environments: Galapagos Islands.* Oxford: Pergamon Press.

ROBINSON, B. L., 1902. Flora of the Galapagos Islands. *Proceedings of the American Academy of Arts and Sciences, 38:* 77–269.

ROBINSON, B. L. & GREENMAN, J. M., 1895. On the flora of the Galapagos Islands as shown by the collection of Dr. Baur. *American Journal of Science, 150:* 135–149.

SCHOFIELD, E. K., 1973. Annotated bibliography of Galapagos botany. *Annals of the Missouri Botanical Garden, 60:* 461–477.

SCHOFIELD, E. K., 1980. Annotated bibliography of Galapagos botany. Supplement I. *Brittonia, 32:* 537–547.

SIMPSON, B. B., 1974. Glacial migrations of plants: Island biogeographical evidence. *Science, N.Y., 185:* 698–700.

STEWART, A., 1911. A botanical survey of the Galapagos Islands. *Proceedings of the California Academy of Sciences, Series 4, 1:* 7–288.

STEWART, A., 1915. Further observations on the origin of the Galapagos Islands. *Plant World, 18:* 192–200.

SVENSON, H. K., 1935. Plants of the Astor Expedition, 1930 (Galapagos and Cocos Islands). *American Journal of Botany, 22:* 208–277.

SVENSON, H. K., 1942. Origin of plants on the Galapagos Islands. *Proceedings of the 8th American Science Congress, 3:* 285–286.

SVENSON, H. K., 1946. Vegetation of the coast of Ecuador and Peru and its relation to the Galapagos Islands. I. Geographical relations of the flora. *American Journal of Botany, 33:* 394–426.

VAN DER WERFF, H., 1979. Conservation and vegetation of the Galápagos Islands. In D. Bramwell (Ed.), *Plants and Islands:* 391–404. London: Academic Press.

WIGGINS, I. L. & PORTER, D. M., 1971. *Flora of the Galápagos Islands.* Stanford: Stanford University Press.

Biological Journal of the Linnean Society (1984) *21:* 253–258.

Man and other introduced organisms

P. KRAMER

Department of Biology, FB9, Universität Essen, Postfach 103764, 4300 ESSEN 1, Federal Republic of Germany

---

Man's tremendously increased migratory potention, coupled with his ability to transport any material, causes ecological revolutions on most islands of this world—Fernandina and most smaller islands of the Galapagos being fortunate exceptions to that rule. It is proposed to make distinctions between species colonizing the Galapagos. We can distinguish between those immigrant species which do not profit from man as a transport medium (independent immigrants) and those who do depend on him (man-dependent immigrants). These immigrants, in turn, may or may not be able to settle and these settlers may either gain a footing with or without depending on man's direct or indirect influence on the habitat (non-settlers, primary resource-using settlers, and secondary resource-using settlers). Introduced species represent a terrible attack on the biotic uniqueness of the Galapagos ecosystems. However, it is proposed to make better use of the scientific value of these introductions. For example, a case of selective impact of an introduced on an indigenous species was investigated: lava lizards of the genus *Tropidurus* seem to be more wary of moving objects on islands where cats have been introduced by man than on islands free of cats.

Under the peculiar Galapagos conditions it may turn out that science's most difficult and important task is to investigate and interpret man's role in such places.

KEY WORDS:—Galapagos – island management – behavioural evolution.

One of the more entertaining pastimes of biologists on Galapagos is to speculate how organisms managed to get there; and why so many of them never made it or never got established. The question of how those which populate the islands today reached them, excites dispersion biologists; and the question of whether those which are absent really never got there, or sometimes do, but fail to get established, is of relevance to the theory of island biogeography. In the terms of MacArthur & Wilson (1963, 1967) the number of species which reach an island is expressed by the rate of immigration, while the number of those who make it but do not get established, contributes to the rate of extinction.

It is of some interest to include in such speculations the question of whether man belongs to those species who are able to become truly resident on the Galapagos or whether he is a straggler unable to become established in a biological sense. My judgment is that, in contrast to the case of Hawaii, man was and remains a straggler on the Galapagos. That was particularly obvious in the days of Tomás de Berlanga. The reason why man is now found permanently in the archipelago is not that he forms a stable population, but because he immigrates so frequently. Of course you will tell me that under such criteria the city of London and probably all cities and industrialized countries are settled only by stragglers. I would not disagree. It is characteristic of industrial man that he makes himself independent of any specific ecosystem. He is not 'at home' in any particular biocoenosis, but depends only on the biosphere as a whole, and therefore tends to alter and exhaust one ecosystem after the other. Ray Dasmann's (1976) distinction between 'Ecosystem People' and 'Biosphere People' highlights this distinction.

It is precisely industrial man's tremendously increased migratory potential, coupled with his ability to transport any material, that causes ecological revolutions on most islands of this world—Fernandina and most smaller islands of the Galapagos being fortunate exceptions to that rule.

The number of species on a given island is, as we have seen, the result of the rates of immigration and extinction on that island. The rate of immigration is in turn influenced by the degree of isolation—in itself a complex entity including the distance of an island from its sources of immigration and the specific qualities of the transporting media, whether these be air, water, or other migrating animals. This of course is the critical point where man primarily exerts his influence on the Galapagos ecology: he lowers the degree of isolation, and thus causes an increase of the number of species, and a reduction in the speciation rate. In the long run the exterminations brought about by man and other colonists are more than counterbalanced by the number of introduced species, which either would not have reached the islands at the former level of isolation or are able to gain a firm footing only since ecological conditions have been changed by those colonists which owe their establishment on the islands to the lowering of the degree of isolation, (i.e. an increased immigration rate).

As a consequence and in view of management needs, I would propose to make the following distinctions between species colonizing the Galapagos. Firstly, we can distinguish between those immigrant species which do not profit from man as a transport medium, and those who depend on him. The former would have made it (and presumably continue to make it) under so-called 'natural' conditions; the latter would not. We might call these the independent immigrants and the man-dependent immigrants respectively. Secondly, these immigrants, be they independent or man-dependent, may or may not be able to settle, i.e. make use of the Galapagos resources and survive for more than a generation. These settlers again may either gain a footing with or without depending on man's direct or indirect influence on the habitat. In other words, among independent, as well as among man-dependent immigrants, we may distinguish between non-settlers, primary resource-using settlers, and secondary resource-using settlers.

Such distinctions may seem an irrelevant academic exercise. However, I think it helps formulate questions that must be answered in order to devise adequate strategies to manage problems caused by introduced species and to prevent or at least slow down the influx of man-dependent immigrants. A few examples may explain what I mean.

All introduced feral mammals are man-dependent immigrants, otherwise they would have reached the archipelago long ago and Galapagos would have become quite a different place. Furthermore, goats and black rats, at least on Pinta and Pinzón, must be primary resource-using settlers. The same is probably true for donkeys, cattle, and pigs, wherever they have been introduced. In the case of cats, I would say, it is uncertain (they may depend on introduced rodents); and in dogs, the extent to which they depend on man-induced habitat changes (the infrastructure and resources of human settlements and populations of feral herbivores) is only now being examined. If they really are secondarily resource-dependent, then the dog eradication programme, which has been carried out with so much success, must be followed by management of dogs in human settlements or it may have unwanted consequences on feral goat, pig, and cattle populations.

What is the situation of exotic birds on the Galapagos? Cattle egrets and anis seem to have become or are becoming established in the islands. My guess is that they are independent immigrants and secondary resource-settlers (in spite of the recent observation of five cattle egrets on the Island of Wolf; F. Koester, pers. comm.). What effect do they have on the environment? Is their presence undesirable? If so, can they be controlled? If they are frequent independent immigrants, any forms of direct control are senseless.

The problem of introduced and potentially introduced invertebrates and plants, many (although not all) of which may be man-dependent immigrants, has not really been tackled yet. Not even the indigenous invertebrate fauna has been thoroughly investigated. Probably most of those continental plants and invertebrates, which have flying or drifting propagules, do not constitute a serious threat to the completely unaltered Galapagos habitat, unless they meet with close relatives with which they may interbreed. This is also a problem with respect of inter-island transport of indigenous invertebrate with different island forms. Silberglied (1978) documented such transport on tourist vessels. Are there ways to predict which species constitute a danger, as the fire ant and the earth worms do? The investigation of the distribution and resource utilization of introduced (or presumably introduced) species on islands, parts of which have altered habitats, could be important. Such results in turn would help to devise at least some effective barriers against plant and invertebrate introductions, and hence improve the protection of those islands which are not settled by man. In many cases such protection must simply be the prevention of the introduction of propagules. In other cases I believe it is a matter of carefully avoiding habitat changes. For example, the establishment of cockroaches in camping sites on islands like Santiago may be caused primarily by transport by campers or it may be the result of environmental modifications brought about by campers or by man-introduced species. In the first case one must avoid the introduction of cockroaches; in the second case one may have to avoid camping, or remove the introduced species.

Of course introduced species represent a terrible attack on the biotic uniqueness of the Galapagos ecosystems. They contribute to ecological uniformity and their presence undoubtedly reduces the aesthetic and scientific interest of the islands. However I am surprised that the introduced species' significance as a scientific resource has been recognized in so few cases and therefore in closing I would like to point out some possibilities to use this resource.

The first type of research is directed toward the investigation of populations of introduced organisms. Recently discussion on the concept of bimodality of evolutionary rates and the importance of the founder effect has intensified among biologists (e.g. Williamson, 1981). And palaeontologists (e.g. Stanley,

1979) are constructing hypotheses concerning 'quantum speciation' in order to account for the longevity of so-called 'chronospecies' and the sudden appearance of new lineages without prior transitional stages in the palaeontological record. Berry (1964, 1975), in his classical investigation of British mouse populations demonstrated that those inhabiting islands diverge markedly from their mainland ancestors. He attributed that divergence to the founder effect. Others believe that a reduction of the genetic spectrum and the subsequent establishment of a new genetic stability in a founder population is not as significant as the strong selection pressure affecting a small group of immigrants. I am sure that Berry would be delighted if the Vikings would have preserved for him a record and a sample of those mice which according to him, they brought to the Hebrides. Now, in the case of the Galapagos, we are the Vikings. All introductions are relatively recent and we know that they continue to occur. Should we not take a very precise record of those introductions? And if we argue that we need at least 100 generations in order to conclude whether the founder effect is of great relevance, should we not take such a record, particularly in invertebrates, some of which may produce 100 generations in a matter of decades? My point is that recording and monitoring of as many introduced invertebrates as possible may be less glamorous than studies on the various endemic giants, but I believe it would contribute at least as much to our knowledge of evolutionary processes.

There is a second chance for evolutionary research, which is due to man's influence on the biogeography of the Galapagos. Since the beginning of regular tourism, cruise ships have produced a constant flow of individuals of moths and beetles between the Islands (Silberglied, 1978). This transport obviously results in a certain amount of gene flow between closely related taxa. Of course this process is another attack on the 'naturalness' and the diversity of Galapagos biota and I hope that methods will soon be devised to stop this flow of propagules. But meanwhile, why are we not using it to test concepts of evolutionary theory under natural conditions, for example, the relative significance of allopatry in speciation or in disruptive selection? If a moth species is occuring in distinct forms on different islands, what is the measurable gene transport via tourist vessels doing to that distinctiveness?

And finally, it is highly desirable to investigate the selective impact of introduced on indigenous species. Simple cases, such as predator–prey relationships may be most rewarding. Nine years ago I conducted a pilot study into the general assumption that introduced cats must exert a strong selective force on their indigenous Galapagos prey. Wherever feral cats are found on Galapagos they feed on lava lizards of the genus *Tropidurus*. I believe that in many places these lizards are their primary food source. I once shot a cat close to the Darwin Station which had nothing but eleven lizards in its stomach.

Everybody knows the tameness of Galapagos lava lizards, against which cat predation presumably selects. I devised a simple measure of this tameness in island populations. I used 13×24 cm plywood board, which was placed in a horizontal position one metre in front of a resting lizard. The board was attached to a spring which pushed it to a vertical position when a trigger was released. This little machine was mounted on a 3 m stick and the trigger could be released by pulling a string. So I was able to quietly place the horizontal board 1 m in front of the lizard and then try to scare it by letting this black plate appear. Afterwards I measured the exact distance between the flip board and the resting place of the lizard (there was no difference of error between samples) and I took the temperature of the rock or wood on which the lizard sat. This I did with samples on Santa Cruz, San Cristóbal, Floreana, Española, and Santa Fe. All lizards were adult, with the exception of four animals on Española which were subadults. In all cases about half of the sample were females, and about half males. Temperatures were between 22 and 29°C and experiments were made between 9 and 17 hours. My results are recorded in Table 1.

Table 1. Distance moved by Galapagos lava lizards on being startled

	Floreana	San Cristóbal	Santa Cruz	Santa Fe	Española
Moved four feet	14	23	22	8	7
Moved less than four feet	5	10	18	32	17

The difference between lizard populations suffering from cat predation (Floreana, San Cristóbal, Santa Cruz) and those which do not (Santa Fe and Española) is striking, and although I have no absolute proof, I would predict that it is the cats' presence which causes it. The question, of course, is whether lizards on islands with cats tend to be more wary because of their *individual* experiences or whether they are the product of a selection process. Comparative examination of the variability of the flight distance (particularly Kaspar–Hauser experiments) would help to establish this. If it can be proved that selection pressure exerted by cats has influenced the inborn mechanisms releasing flight behaviour of Galapagos lizards, then it would be of considerable interest to relate this change to the period and strength of that pressure. I have almost inevitably speculated that the apparent difference in flight distance between Santa Cruz lizards on the one hand and San Cristóbal and Floreana lizards on the other may be associated with the length of time since cats were introduced on the different islands. My guess is that cats have been present on San Cristóbal and Floreana for 40–60 lizard generations and on Santa Cruz for 20–30 lizard generations. Further data must be collected before continuing such speculation.

Finally, let me return to man's role in the game: man, as in most places on this earth (despite and maybe because of his position as a straggler), is the dominant species on the Galapagos. I do not deplore that. There is no reason why he should not dominate and manage all the resources of this earth for his purposes. My impression is that the danger is not caused by man's domination, but by his inability to fully assume his role as a responsible dominant—in spite of his unique ability to act consciously. In a way he continues to act as though there are still ecological constraints, as though Mother Nature is still there to put him in his place for his own good. So he continues to consume and destroy the resources he depends on. Sometimes he takes too high risks in his ignorance, sometimes he actively destroys, because he can load the consequences on conspecifics living elsewhere or in future times.

Galapagos and similar wildlife conservation areas of worldwide significance are peculiar places. Local interests have to be reconciled with the interests of a country and of all humanity against the background of the historic conflict between dominant industrialized countries and so-called developing nations.

Under these peculiar conditions it may turn out that science's most difficult and important task is to investigate and interpret man's role in such a place. But scientists must avoid being pushed into the role of magicians who are expected to have all the solutions, or the role of court fools, who know and tell the truth, but are never taken seriously. We should rather assume a role of enlighteners and critics, declaring what we know and even more particularly what we do not know. In this way we can claim a full place when decisions about the management of nature are taken.

REFERENCES

BERRY, R. J., 1964. The evolution of an island population of the house mouse. *Evolution 18:* 468–483.
BERRY, R. J., 1975. On the nature of genetical distance and island races of *Apodemus sylvaticus*. *Journal of Zoology, 176:* 292–295.
DASMANN, R., 1976. National parks, nature conservation, and "future primitive". *Ecologist, 6:* 164–167.
MACARTHUR, R. H. & WILSON, E. O., 1963. An equilibrium theory of insular zoogeography. *Evolution, 17:* 373–387.
MACARTHUR, R. H. & WILSON, E. O., 1967. *The Theory of Island Biogeography*. New York: Princeton, University Press.
SILBERGLIED, R. E., 1978. Inter-island transport of insects aboard ships in the Galápagos Islands. *Biological Conservation, 13:* 273–278.
STANLEY, S. M., 1979 *Macroevolution, Pattern and process*. San Francisco: Freeman.
WILLIAMSON, M., 1981. *Island Populations*. Oxford: University Press.

INDEX

Compiled by A. S. Thorley F.L.S.

*denotes a keyword
Fig.—Figure
n.—note
Tab.—Table

Abington Island   32 Fig. 1, *see also* Pinta 79 Fig. 1, 116 Fig. 1, 126 Fig. 7, 232 Fig. 3
Acrita and *Vermes*, MacLeay's   13–14
affinity and analogy, differences between   13–14
Africa: antelopes, giraffes, pachyderms and zebras, distribution   24; relationship to Madagascar 23
Agassiz, Louis: Centres of creation   19–20, 21, 22; *Essay on Classification* (1859) 7, 13, 17
age assignments of fossil deposits   86, 87
aggregation, female, marine iguanas and pinnipeds   210–211, 215
agonistic behaviour in tortoises   168–170 Figs 2–4 & Tab. 1; fighting potential in male marine iguanas and pinnipeds   213–215
albatrosses, see also *Diomedia:* Laysan 144; waved. *Diomedea irrorata* 138 Tab. 1, 139–141. 144–145, 148, 151–152, 154
Albemarle Island   32 Fig. 1 *see also* Isabela 79 Fig. 1, 100 Fig. 1, 116 Fig. 1, 126 Fig. 7, 187–188 Figs 1–2, 232 Fig. 3, 236 Fig. 5; Darwin and the Galapagos 32 Fig. 1, 35 n.⁵, 36, 51 Fig. 11
Alcedo, central volcano of Isabela 188 Fig. 2; geology 67, 70, 71; land iguanas 188 Fig. 2, 189, 191 Tab. 1, 193 Figs 3–4, 194
Aldabra, Indian Ocean: frigate birds 145, 147; tortoise, *Testudo indicus* 35
Allodesminae, Miocene group of Pinnipedia   182
allopatry, Darwin's finches   125–129 Figs 7–8, 131
Alternanthera halimifolia, Santa Cruz herb layer 221
Amadina included by Sclater in the Fringillidae (1858)   15–16
Amblyrhynchus, marine iguanas 186, 204; genic variability 105 Tab. 1
Andean origin of plants   244, 245 Tab. 1, 246 Tab. 2
Anderson, N. J. on the Galapagos flora (1855)   246
anis, degree of establishment in the islands   255
Anomia, jingle-shells   14–15
Anous stolidus, sea-birds   138 Tab. 1, 139, 145, 148
ants, Galapagos, see also *Companotus, Cardiocondyla, Conomyrma, Crematogaster, Cylindromyrmer, Cyphomyrmex, Hyponera, Leptogenys, Monomorium, Odontomachus, Paratrechina, Pheidole, Solenopsis, Strumigenys, Tapinoma, Tetramorium* and *Wasmannia* 233 Tab. 1; Argentine, *Iridomyrmex humilis* 229; little red fire, *Wasmannia auropunctata* 229–242 Figs 1–5 & Tabs 1–3; 'tramp' 229–230, 232, 237
Aphrodite, C. Collingwood on resemblance to a *Chiton* 18

Aplacophora, malacofauna   82, 84 Tab. 1
Apteryx of New Zealand   23
Archeogastropoda, malacofauna   84 Tab. 1
Archipelago, Galapagos: age   63–64, 203; allopatric model of speciation 125–126 Fig. 7; Darwin and the Galapagos 32 Fig. 1, 55; distribution of mockingbirds 115–117 Fig. 1; flora 51 Fig. 11; maps 32 Fig. 1, 79 Fig. 1, 116 Fig. 1, 126 Fig. 7, 232 Fig. 3; official discoverer of (1535) 2 & n, 253
Archipelagos: Azores   21, 22, 23–24, 64; Canaries 9, 21, 23, 64; Samoa islands 64
Arctocephalus, Galapagos seals   177–184; *A. australis*, South American fur seal 178, 179, 182, 183, 214; *A. australis galapagoensis* 178; *A. galapagoensis*, Galapagos fur seal 178, 179, 182, 183, 209–210; *A. gazella*, Antarctic fur seal 182; *A. pusillus*, South African fur seal 178, 212; *A. townsendi* 179
Ascension Island, Atlantic Ocean, *Fregata aquila* 145
Asio flammeus, short-eared owls   114 & Tab. 1
Australia, evolution and islands   15–16, 19, 23, 24
Australasia, evolution and islands   15, 20
Azores: former Lusitanian land postulated to explain biotic affinities with Madeira 22; high degree of endemism 21, 23–24; still-active ocean archipelago (volcanic) 64

Baker, J. G., numbers of world's ferns (1867)   10
Baltra   69 Fig. 2, 79 Fig. 1, 100 Fig. 1, 116 Fig. 1, 126 Fig. 7, 187 Fig. 1, 218 Fig. 1, 232 Fig. 3 *see also* Seymour Is. 32 Fig. 1; black rats, *Rattus rattus* 100 Fig. 1, 105 Fig. 4, 106; *Cleome viscosa*, pantropical weed 248–249 Fig. 1; climate 223; geology 64, 68, 69 Fig. 2; land iguanas 186–187 & Fig. 1, 189, 191–194 Figs 3–4 & Tabs 1–2, 196–203 Figs 6–9, 11 & Tabs 5–6; late Cenozoic marine invertebrate localities 79 Fig. 1, 86, 87–88
barnacles, stranded, following recent uplift on Fernandina 70
Barrington Island   32 Fig. 1 *see also* Santa Fé 79 Fig. 1, 116 Fig. 1, 126 Fig. 7, 187 Fig. 1, 232 Fig. 3
Bartolomé   100 Fig. 1, 116 Fig. 1, 126 Fig. 7 (Bartholomew 32 Fig. 1); black rats, *Rattus rattus* 100 Fig. 1, 105 Fig. 4; land iguanas 187
Bartholomew, G. A.: evolutionary model for polygynous pinnipeds 182, 183, 214; pinniped female gregariousness 212, 215
Beagle voyage 1, 3, 32–33 Fig. 1, 36, 54;

Biological Journal of the Linnean Society (1984), *21*, 1–4

Darwin was astonished

R. J. BERRY

Department of Zoology, University College London,
Gower Street, London WC1E 6BT

Darwin did not approach the Galapagos with the same enthusiasm and energy as he showed at earlier places visited by the *Beagle*. Notwithstanding, he looked back on the five weeks the *Beagle* spent in the Galapagos as a time when he made observations important for the development of his evolutionary ideas. In retrospect, he was astonished at what he saw there.

KEY WORDS:—Darwin – Galapagos Islands – evolution.

"Considering the small size of these islands, we feel the more astonished at the number of their aboriginal beings, and their confined range." Charles Darwin, writing in the *Origin of Species* (1859) about his visit to the Galapagos Islands.

The *Beagle* was in the Galapagos archipelago from 15 September to 20 October 1835. This period has entered scientific mythology as the time when Darwin was converted to a belief in evolution. In fact, it is more accurately regarded as just one of the more important episodes in Darwin's education, which led over the next decade to his evolutionary ideas. At the time, Darwin failed to grasp the significance of the poverty of species diversity and the differentiation of so many forms which inhabited the islands (Sulloway, 1982). When sorting his ornithological specimens on the way home, he wrote the oft-quoted words (probably in July 1836), "When I see these Islands in sight of each other and possessed of but a scanty stock of animals, tenanted by these birds, but slightly differing in structure and filling the same place in Nature, I must suspect they are varieties... if there is the slightest foundation for these remarks the zoology of archipelagoes will be well worth examining: for such facts would undermine the stability of species". His faith in the immutability of species was only finally broken by the recognition that there were two species of the flightless *Rhea* in South America, and not one as traditionally thought; and then more definitely when in March 1837 John Gould convinced him that the mocking-birds (*Mimus*) on three of the Galapagos Islands were specifically distinct. As a result Darwin wrote in the *Origin*, "when comparing ... the birds

The papers in this number constitute Contribution Number 363 of the Charles Darwin Foundation for the Galapagos Isles.

1

from the separate islands of the Galapagos archipelago, both with one another, and with those from the American mainland, I was much struck how entirely vague and arbitrary is the distinction between species and varieties". As Mayr (1982) comments, "It became clear to Darwin that many populations (as we would now call them) were intermediate between species and variety, and that particular species on islands, when studied geographically, lacked the constancy and clear-cut delimitation insisted on by creationists and essentialists". Although Darwin did not realize it during his visit, these biological phenomena were present for the discerning eye. Darwin did not recognize them at the time but he was a sufficiently good naturalist to record and collect enough information that the Galapagos later became a significant link in the evidence for the *fact* of evolution, and more recently, for the *processes* which bring about evolutionary change (Patton, 1984; Grant, 1984).

Darwin was not first to describe the natural history of the Galapagos. The official discoverer of the archipelago in 1535*, Bishop Tomas de Berlanga of Panama, sent an account of the islands to his emperor, Charles V of Spain, including descriptions of the giant tortoises (or *galapagos*), the iguanas, and the remarkable tameness of the birds. Eleven years later, another Spaniard, Diego de Rivadeneira reported similarly, and became the first person to mention another unique form, the Galapagos hawk (*Buteo galapagoensis*).

The first scientific expedition to the Galapagos was that of the Sicilian Alessandro Malaspina, sent by the Spanish king Carlos IV in 1790, just a generation before the first recorded introduced mammals were released (goats, on Santiago Island in 1813). The last two centuries have been a time of biological scourge on the islands' fauna, by sealers in the 1800s and by feral donkeys, goats, cats, dogs and rats. The first conservation legislation affecting the Galapagos was passed in 1934, and in 1936 the islands were declared a National Park by the Ecuadorian Government.

Then in 1959, at the time of the centenary of the publication of the *Origin of Species*, the Charles Darwin Foundation for the Galapagos Islands was created with international support (largely channelled through UNESCO and IUCN). The Foundation and the Ecuadorian Government signed an agreement permitting the Foundation to set up a Research Station, and this was opened on Santa Cruz Island in 1964. This volume contains the Proceedings of a Symposium organized jointly by the Darwin Foundation and the Linnean Society, held in London on 8th December, 1982.

EVOLUTIONARY RESEARCH AND THE GALAPAGOS

Despite the revolution brought to biological thinking by Darwin, evolutionary research in the fullest sense is relatively new. The early evolutionists regarded evolutionary change to be so slow that it could not be examined experimentally, while the failures of communication between biological disciplines prior to the neo-Darwinian synthesis of the 1930s acted as a deterrent to the experimental study of evolution (Mayr & Provine, 1980). This is not to ignore the patient (if at times, absurd) phylogenetic

*There is a legend that the Inca chieftain Tupac-Yupanqui visited the islands at the beginning of the fifteenth century.

reconstructions of palaeontologists, neontologists, and more recently, biochemists, nor the extensive theoretical investigations of the factors that may bring about genetical change, but to focus on the need to study the interplay of genes and environment in natural situations. These interactions are likely to be stronger on islands than in continental situations, and therefore particularly informative (Mayr, 1954; Miller, 1966). It is somewhat ironical that the recent interest in island biogeography (MacArthur & Wilson, 1963, 1967) has all but ignored questions of divergence and differentiation (Williamson, 1981; Berry, 1983a).

The Galapagos have been put into context by the biogeographers (Johnson & Raven, 1973; Connor & Simberloff, 1978), and they have not been neglected by evolutionary biologists (Lack's, 1947 and Bowman's, 1961, studies of the island finches rapidly achieved classical status; other research is reviewed by Bowman, 1966; Thornton, 1971; Levinton & Bowman, 1983). Notwithstanding, evolutionary biology on the Galapagos has not been as progressive and adventurous as might have been expected from its illustrious beginning and intrinsic possibilities. The Darwin Research Station provides excellent facilities for visiting scientists, but the priorities of the Foundation have had to be directed towards short-term conservation management because of the urgency of the problems facing some of the fauna (particularly the tortoises, land iguanas, and native rodents).

This volume concentrates on evolutionary questions in the Galapagos. The hope is that it will encourage more studies of evolutionary processes and differentiation, for which the islands provide such a marvellous laboratory (Mayr, 1967; Berry, 1979, 1983b). The papers herein review some of the more important results gained so far. There is great potential for more work.

Sulloway (1984) has analysed in detail the impact that the Galapagos had on Darwin. In fact, it was only in retrospect that Darwin came to recognize their importance for his ideas, and the section on the islands in later editions of his *Journal of Researches* was considerably expanded from that in his original *Narrative* (which was largely written after leaving the Galapagos, during the voyage across the Pacific). Although he faithfully recorded the comment of the English resident, Lawson, about the giant tortoises, that he (Lawson) "could at once tell from which island any one was brought", Darwin was, as we have seen, not prepared at the time of his visit to accept that speciation could have taken place on the islands. Indeed, by the time the *Beagle* began her journey across the Pacific, Darwin seemed to want nothing more than to get home. Fitzroy was being particularly tiresome; the last phase in South America (a visit to Peru) was a waste of time because the political situation made it impossible to get out into the country; and the diary entries for Tahiti, New Zealand, Australia and South Africa are little more than perfunctory.

It is a pity that a letter that Darwin wrote from the Galapagos to his sister Caroline seems not to have survived (Barlow, 1945; P. J. Gautrey, *pers. comm.*). That would presumably have given Darwin's first-hand impressions of Galapagos. A note by Darwin in the *Proceedings of the Zoological Society* soon after his return to London shows that he had very vague memories of the islands. All he could say about the finches was that "their general resemblance in character and the circumstance of their indiscriminately associating in large flocks,

4 R. J. BERRY

rendered it almost impossible to study the habits of particular species . . . They appeared to subsist on seeds" (Darwin, 1837).

It smacks of heresy to suggest that Darwin might have been bored and homesick during the five weeks he spent on the Galapagos in 1835. However, the important point was that Darwin was always first and foremost a naturalist. He observed and recorded accurately, whatever his preconceptions (unlike Fitzroy, who noted in his *Narrative* that the variation in the finches "appears to be one of those admirable provisions of Infinite Wisdom by which each created thing is adapted to the place for which it was intended"). The Galapagos did not give Darwin a Damascus Road experience; nor even a "strange warming" such as Wesley experienced at Aldersgate Street. Notwithstanding, they were a major landmark in his pilgrimage. And even today when we know vastly more about island faunas and floras than did Darwin we can still like him, "feel astonished at the number of their aboriginal beings, and their confined range". May we never lose that wonder which is at the root of all scientific curiosity.

REFERENCES

BARLOW, N., 1945. *Charles Darwin and the Voyage of the Beagle.* London: Pilot.
BERRY, R. J., 1979. The Outer Hebrides: where genes and geography meet. *Proceedings of the Royal Society of Edinburgh 77B:* 21–43.
BERRY, R. J., 1983a. Diversity and differentiation: the importance of island biology for general theory, *Oikos, 41:* 523–529.
BERRY, R. J., 1983b. Evolution of animals and plants in the Inner Hebrides. *Proceedings of the Royal Society of Edinburgh, 83B:* 433–447.
BOWMAN, R. I. 1961. Morphological differentiation and adaptation in the Galapagos finches. *University of California Publications in Zoology* no. 58.
BOWMAN, R. I. 1966. *The Galapagos: Proceedings of the Symposia of the Galapagos International Scientific Project.* Berkeley and Los Angeles: University of California Press.
CONNOR, E. F. & SIMBERLOFF, D. S., 1978. Species number and compositional similarity of the Galapagos flora and avifauna. *Ecological Monographs 48:* 219–248.
DARWIN, C., 1837. Remarks upon the habits of the genera *Geospiza, Camarhynchus, Cactornis* and *Certhidea* of Gould. *Proceedings of the Zoological Society of London, 5:* 47.
GRANT, P. R., 1984. Recent research on the evolution of land birds in the Galapagos. *Biological Journal of the Linnean Society, 21:* 113–136.
JOHNSON, M. P. & RAVEN, P. H., 1973. Species number and endemism: the Galapagos revisited. *Science, N.Y., 179:* 893–895.
LACK, D., 1947. *Darwin's Finches.* Cambridge: University Press.
LEVINTON, A. E. & BOWMAN, R. I. (Eds), (1983). *Patterns of Evolution in Galapagos Organisms.* San Francisco: American Association for the Advancement of Science Pacific Division.
MACARTHUR, R. H. & WILSON, E. O., 1963. An equilibrium theory of insular zoogeography. *Evolution, 17:* 373–387.
MACARTHUR, R. H. & WILSON, E. O., 1967. *The Theory of Island Biogeography.* Princeton: University Press.
MAYR, E., 1954. Change of genetic environment and evolution. In J. Huxley, A. C. Hardy & E. B. Ford (Eds), *Evolution as a Process:* 157–180. London: Allen & Unwin.
MAYR, E., 1967. The challenge of island faunas. *Australian Natural History, 15:* 369–374.
MAYR, E., 1982. Epilogue. *Biological Journal of the Linnean Society, 17:* 115–126.
MAYR, E. & PROVINE, W. B., 1980. *The Evolutionary Synthesis.* Cambridge, Mass.: Harvard University Press.
MILLER, A. H., 1966. Animal evolution on islands. In R. I. Bowman (Ed.), *The Galapagos: Proceedings of the Symposia of the Galapagos International Scientific Project:* 10–17. Berkeley and Los Angeles: University of California Press.
PATTON, J. L., 1984. Genetical processes in the Galapagos. *Biological Journal of the Linnean Society, 21:* 97–111.
SULLOWAY, F. J., 1982. Darwin and his finches: the evolution of a legend. *Journal of the History of Biology, 15:* 1–53.
SULLOWAY, F. J., 1984. Darwin and the Galapagos. *Biological Journal of the Linnean Society, 21:* 29–59.
THORNTON, I. 1971. *Darwin's Islands: a Natural History of the Galapagos.* New York: Harper & Row.
WILLIAMSON, M. H., 1981. *Island Populations.* Oxford: University Press.